Harmonic Analysis on Symmetric Spaces and Applications I

Audrey Terras

Harmonic Analysis on Symmetric Spaces and Applications I

With 54 Illustrations

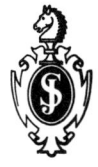

Springer-Verlag
New York Berlin Heidelberg Tokyo

Audrey Terras
Department of Mathematics C-012
University of California at San Diego
La Jolla, California 92093
U.S.A.

AMS Classifications: 42-01, 43-01, 44-01, 10Dxx, 51M10

Library of Congress Cataloging in Publication Data
Terras, Audrey.
  Harmonic analysis on symmetric spaces and applications.
  Bibliography: p.
  Includes index.
  1. Harmonic analysis. 2. Symmetric spaces. I. Title.
QA403.T47   1985   515′.2433   84-23568

© 1985 by Springer-Verlag New York Inc.
All rights reserved. No part of this book may be translated or reproduced in any form without written permission from Springer-Verlag, 175 Fifth Avenue, New York, New York 10010, U.S.A.

Typeset by Asco Trade Typesetting Ltd., Hong Kong.
Printed and bound by Halliday Lithographers, West Hanover, Massachusetts.
Printed in the United States of America.

9 8 7 6 5 4 3 2 1

ISBN 0-387-96159-3 Springer-Verlag New York Berlin Heidelberg Tokyo
ISBN 3-540-96159-3 Springer-Verlag Berlin Heidelberg New York Tokyo

*To all women mathematicians,
to my parents,
to my POSSLQ.*

I have myself always thought of a mathematician as in the first instance an *observer*, a man [sic] who gazes at a distant range of mountains and notes down his observations. His object is simply to distinguish clearly and notify to others as many different peaks as he can.... But when he sees a peak he believes that it is there simply because he sees it. If he wishes someone else to see it, he *points to it*, either directly or through the chain of summits which led him to recognize it himself. When his pupil also sees it, the research, the argument, the *proof* is finished.

The analogy is a rough one, but I am sure that it is not altogether misleading. If we were to push it to its extreme we should be led to a rather paradoxical conclusion; that there is, strictly, no such thing as mathematical proof; that we can, in the last analysis, do nothing but *point*; that proofs are what Littlewood and I call *gas*, rhetorical flourishes designed to affect psychology, pictures on the board in the lecture, devices to stimulate the imagination of pupils. This is plainly not the whole truth, but there is a good deal in it.

From Hardy [2, Vol. 7, p. 598].

Preface

Since its beginnings with Fourier (and as far back as the Babylonian astronomers), harmonic analysis has been developed with the goal of unraveling the mysteries of the physical world of quasars, brain tumors, and so forth, as well as the mysteries of the nonphysical, but no less concrete, world of prime numbers, diophantine equations, and zeta functions. Quoting Courant and Hilbert, in the preface to the first German edition of *Methods of Mathematical Physics*: "Recent trends and fashions have, however, weakened the connection between mathematics and physics." Such trends are still in evidence, harmful though they may be. My main motivation in writing these notes has been a desire to counteract this tendency towards specialization and describe applications of harmonic analysis in such diverse areas as number theory (which happens to be my specialty), statistics, medicine, geophysics, and quantum physics. I remember being quite surprised to learn that the subject is useful. My graduate eduation was that of the 1960s. The standard mathematics graduate course proceeded from Definition 1.1.1 to Corollary 14.5.59, with no room in between for applications, motivation, history, or references to related work. My aim has been to write a set of notes for a very different sort of course.

A second impulse pushing me toward the typewriter was the knowledge that in the past 30 years there have been some really exciting discoveries in the field of harmonic analysis on symmetric spaces and their fundamental domains for discrete isometry groups—the work of Harish-Chandra, Helgason, Langlands, Maass, Selberg, and others. It is time that these ideas received an exposition comprehensible to the average applied mathematician, number theorist, etc. In particular, I believe that many of the results to be described have interesting implications for statistical physics and number theory.

The outline of the book can be sketched as follows. Chapter I concerns Euclidean Fourier analysis and its applications to the solution of the wave and heat equations, the study of potential functions of crystals, as well as zeta functions of algebraic number fields, for example. Chapter II deals with spherical Fourier analysis and its connections with the Euclidean theory. There are applications to CAT scanners, the solar corona, and the Zeeman effect for the hydrogen atom in a magnetic field. Chapter III studies non-Euclidean Fourier analysis on the Poincaré or Lobatchevsky upper half-plane H, with an elementary discussion of the work of Harish-Chandra and Helgason in this special case. The main idea is to use the group invariance under $SL(2, \mathbb{R})$—the special linear group of all 2×2 real matrices of determinant one—to determine the spectral measure in the non-Euclidean Fourier inversion formula via the asymptotics of the special functions involved. Applications include the solution of the Dirichlet problem for cones, wedges, and other domains in Euclidean space, as was discovered by Mehler, Fock, and Lebedev long ago. One can also solve the non-Euclidean heat and wave equations on H itself. These results will be used to obtain a non-Euclidean central limit theorem with applications to the statistics of transmission lines. The results of Chapter III have different interpretations when one realizes that the group $SL(2, \mathbb{R})$ can be identified with many other Lie groups; e.g., $Sp(1, \mathbb{R})$ the group of 2×2 real symplectic matrices g such that ${}^t g J g = J$ if $J = \begin{pmatrix} 0 & -1 \\ 1 & 0 \end{pmatrix}$, or the group $SU(1, 1)$ of 2×2 complex matrices g such that ${}^t \overline{g} K g = K$ if $K = \begin{pmatrix} 1 & 0 \\ 0 & -1 \end{pmatrix}$. And finally, $SL(2, \mathbb{R})$ is locally isomorphic to the Lorentz-type group $SO(2, 1)$ of real 3×3 matrices of determinant one preserving the quadratic form $x_1^2 + x_2^2 - x_3^2$. However, the higher dimensional analogues of these groups are distinct.

Non-Euclidean analogues of Fourier series also make their appearance in Chapter III. This is the Fourier inversion formula for functions on H that are periodic under the group $SL(2, \mathbb{Z})$ of fractional linear transformations with integer entries and determinant one. The methods of the first part of the chapter work in this case to find the spectral measure on the continuous part of the spectrum of the non-Euclidean Laplacian on $H/SL(2, \mathbb{Z})$ just as they did for H itself. However, there is, in addition, a discrete part of the spectrum, which remains as mysterious as the quanta in quantum mechanics. Applications in this section include the solution of the non-Euclidean heat equation on the fundamental domain $H/SL(2, \mathbb{Z})$, the computation of class numbers of imaginary quadratic fields, and Peter Sarnak's use of the Selberg trace formula to say something about the asymptotics of units in real quadratic fields. Recently Hurt [1, 2]* has described applications of the trace formula in quantum statistical mechanics. Chapter V includes another application—Kaori Imai's extension of the converse result in Hecke's theory of the relation between Dirichlet series such as Riemann's zeta function, and modular forms such as the theta function. Imai extends the theory to Siegel modular forms of genus 2. Examples of such modular forms are the theta functions appearing in the study

---

* This reference number style will be used throughout the book. In this instance, reference is made to the first and second Hurt titles in the Bibliography section at the end of the book.

of abelian integrals. These theta functions arise in recent work on the Korteweg-DeVries equation, as well as in Sonya Kovalevsky's solution of the third known case of the problem of the motion of a rigid body about a fixed point.

Chapter IV extends harmonic analysis to the symmetric space $\mathscr{P}_n$ of positive definite real $n \times n$ matrices Y as well as to the Minkowski fundamental domain for $\mathscr{P}_n/GL(n, \mathbb{Z})$, where $GL(n, \mathbb{Z})$ is the unimodular group of $n \times n$ integer matrices of determinant ± 1 (with the action Y goes to ^tAYA, if ^tA = transpose of A). This is the prototype of the general theory for the symmetric space of a noncompact semisimple (or reductive) Lie group. Applications of Chapter IV include the study of the Wishart distribution in statistics, integrals arising in statistical quantum mechanics, lattice packings of spheres and Hilbert's 18th problem, and integral tests for convergence of sums over matrix variables. And the theory that underlies the development of the higher dimensional Selberg trace formula will also be discussed. The most important part concerns various matrix argument analogues of certain useful special functions such as spherical, gamma, Bessel, Whittaker, zeta, and L-functions. There are many thesis topics and intriguing open questions here, some of which are quite basic. Thus Chapter IV has a somewhat unfinished aspect.

Chapter V is much more crude. Perhaps it should be considered merely as a guide to the literature. My goal was to present an introduction to some of the work of Maass and Siegel, because this work had motivated much of the development in Chapter IV (and even some of Chapter I). Another stimulus was the hope that by fitting various examples into the general picture, one would derive a clearer understanding of the examples. There are also quite interesting relations between the various symmetric spaces and fundamental domains, some of which have been of great use in number theory. Here the key word is "base change."

I would like to thank those people who have discussed these notes with me over the years. It is becoming hopeless to name them all. I would also like to note my indebtedness to certain books that influenced me greatly during the time I began to write this manuscript: Courant and Hilbert [1], Dym and McKean [1], Elstrodt [2], Gangolli [1, 2], Hejhal [1, 2], Helgason [1], Kubota [1], Lang [3], Lebedev [1], Maass [1, 2], Mennicke [1], Minkowski [1], Selberg [1], and Siegel [2, 4, 5].

I feel now that I am probably quite far from reaching the goals that motivated my writing. Certainly the average engineer will not want to think about the intricacies of $GL(n, \mathbb{Z})$ presented in Chapter IV, without some more convincing applications. Number theorists are just beginning to see the utility of that chapter. But I have to admit that the chapter is still in quite a preliminary state. The final word cannot yet be said. For this reason, I have decided to split these notes into two parts. The first three chapters constitute the first volume and the last two chapters, the second volume.

I still hope that both volumes will be useful to a beginner in the subject. The book has been used in parts of mathematics graduate courses at U.C.S.D. and

M.I.T. since 1978. These courses had names like Lie groups, harmonic analysis, mathematical physics, and number theory. The last revision was made at the Institute for Advanced Study, Princeton during the first months of the year of Orwell.

These notes do assume some things of the reader. At the beginning advanced calculus (with a little measure theory) suffices, along with the ability to look at a partial differential equation or a Bessel function without flinching. By Volume II, more is required. The reader has to look at a matrix argument Bessel function without flinching.

One final comment: many important details have been consigned to the exercises. This happened principally because the author likes participatory democracy. However, sometimes the author got carried away with the exercises. But I did try to include references to the proofs. Perhaps later an answer book will appear.

I feel impelled to add a P.S. of warning to the reader. This author is incredibly bad at proofreading. So, if something appears weird in a formula, feel free to change a letter, insert a missing Fourier transform, correct my spelling or history, etc., etc. I would be very grateful if you would send me a list of the errors that you have found. And I would like to thank those that have done so in the past.

La Jolla, California   AUDREY TERRAS

Contents for Volume I

CHAPTER I
Flat Space. Fourier Analysis on $\mathbb{R}^m$ — 1

1.1. Distributions or Generalized Functions — 1
1.2. Fourier Integrals — 8
1.3. Fourier Series and the Poisson Summation Formula — 28
1.4. Mellin Transforms, Epstein and Dedekind Zeta Functions — 55

CHAPTER II
A Compact Symmetric Space—The Sphere — 83

2.1. Spherical Harmonics — 83
2.2. $O(3)$ and $\mathbb{R}^3$. The Radon Transform — 107

CHAPTER III
The Poincaré Upper Half-Plane — 120

3.1. Hyperbolic Geometry — 120
3.2. Harmonic Analysis on H — 134
3.3. Fundamental Domains for Discrete Subgroups Γ of $G = SL(2, \mathbb{R})$ — 163
3.4. Automorphic Forms—Classical — 181
3.5. Automorphic Forms—Not So Classical—Maass Waveforms — 204
3.6. Automorphic Forms and Dirichlet Series. Hecke Theory and Generalizations — 229
3.7. Harmonic Analysis on the Fundamental Domain. The Roelcke-Selberg Spectral Resolution of the Laplacian, and the Selberg Trace Formula — 250

Bibliography — 301

Index — 330

Contents for Volume II

CHAPTER IV

The Space $\mathscr{P}_n$ of Positive $n \times n$ Matrices

4.1. Geometry and Analysis on $\mathscr{P}_n$
4.2. Special Functions on $\mathscr{P}_n$
4.3. Harmonic Analysis on $\mathscr{P}_n$ in Polar Coordinates
4.4. Minkowski's Fundamental Domain for $\mathscr{P}_n/GL(n, \mathbb{Z})$
4.5. Automorphic Forms for $GL(n, \mathbb{Z})$ and Harmonic Analysis on $\mathscr{P}_n/GL(n, \mathbb{Z})$

CHAPTER V

The General Noncompact Symmetric Space

5.1. Geometry and Analysis on G/K
5.2. Geometry and Analysis on $\Gamma \backslash G/K$

Bibliography

Index

CHAPTER I

Flat Space. Fourier Analysis on $\mathbb{R}^m$

There was no time to waste. It's possible to grasp alef-null-sized collections once you're in your aethereal body... but you need some to look at. My job right now was to generate infinities.

...

I slipped out of my physical body and began running around the room, I did alef-null laps... and then the whole rest of the class joined in....

"What happened?" a kid with glasses and dandruff asked. "Did you hypnotize us?"

...

"The main thing is that you saw infinity," I said ....

From White Light, by Rudy Rucker, Ace Books; NY, 1980, pp. 248–249. Reprinted by permission.

1.1. Distributions or Generalized Functions

Formalism denies the status of mathematics to most of what has been commonly understood to be mathematics, and can say nothing about its growth. None of the "creative" periods and hardly any of the "critical" periods of mathematical theories would be admitted into the formalist heaven, where mathematical theories dwell like the seraphim, purged of all the impurities of earthly uncertainty. Formalists, though, usually leave open a small back door for fallen angels... On those terms Newton had to wait four centuries... Dirac is more fortunate: Schwartz saved his soul during his lifetime.

From Lakatos [1, p. 2].

In 1927 P.A.M. Dirac [1] introduced a "function" $\delta(x)$ which he postulated to have the property

$$\int_{-\infty}^{+\infty} f(x)\delta(a-x)\,dx = f(a) \tag{1.1}$$

if $f(x)$ is continuous near $x = a$. It is easy to see that no such function exists. However, delta has been legalized by Laurent Schwartz and others. And we shall sometimes find it very convenient to have access to this theory of generalized functions or distributions. Thus we begin by summarizing these results. The reader who wants more details should consult any of the following references: Choquet-Bruhat et al. [1], DeJager [1, 2], Dieudonné [1], F.G. Friedlander [1], Gelfand and Shilov [1, Vol. 2], Horvath [1], Korevaar [1], Maurin [1], Schwartz [1, 2, 3], Treves [1], Vladimirov [1]. The concept of distribution is a very natural one in applied mathematics. It can be used to represent an impulse, point mass, or point charge, for example. It also provides a natural method to use in obtaining fundamental solutions of partial differential equations, as we shall see in §1.2.

There are two definitions of distribution. One of these goes as follows. Let $\mathscr{D}$ be the space of *test functions* $f: \mathbb{R}^m \to \mathbb{C}$ such that f and all partials of all orders of f are continuous and vanish off a bounded set. Then a *distribution* T is a continuous linear functional $T: \mathscr{D} \to \mathbb{C}$. Here we need a topology or notion of convergence of sequences of test functions in order to say what we mean by the continuity of T. We say that a sequence of test functions converges if they all vanish off the same bounded set and all partials of the sequence converge uniformly to the corresponding partial of the limit function. Note: we do not ask that the convergence be simultaneously uniform for all partials. This rather intricate definition of convergence of test functions is designed to produce a very simple calculus of distributions. For example, all distributions will be infinitely differentiable.

Exercise 1 (Distributions Generalize Functions). Suppose that $f(x)$ is a *locally integrable function*; i.e., f is Lebesgue integrable over every bounded Lebesgue measurable subset of $\mathbb{R}^m$. Then f defines a distribution T_f via

$$T_f(g) = \int f(x)g(x)\,dx \qquad \text{for any } g \text{ in } \mathscr{D}.$$

Here $dx =$ the usual Lebesgue measure on $\mathbb{R}^m$. The exercise is to prove that T_f is indeed a distribution. You should also show that two locally integrable functions f_1 and f_2 define the same distribution if and only if $f_1 = f_2$ almost everywhere.

Exercise 2 (The Dirac Distribution). Define $\delta(g) = g(0)$ for each test function g in $\mathscr{D}$. Show that delta is indeed a distribution (i.e., continuous and linear).

1.1. Distributions or Generalized Functions

The second definition of distribution presents T as an equivalence class of Cauchy sequences of locally integrable functions. Here again convergence is defined with respect to test functions. That is, f_n converges to f means that $\int f_n g \, dx$ converges to $\int fg \, dx$ for every test function g. You can find more details in Korevaar [1], for example.

The Dirac delta distribution corresponds to the following type of sequence, for example.

Definition. A *Dirac sequence of positive type* $K_n : \mathbb{R}^m \to \mathbb{C}$ has the following properties:

(1) K_n is integrable over compact subsets of $\mathbb{R}^m$.
(2) K_n is nonnegative.
(3) $\int K_n \, dx = 1$.
(4) For every $\varepsilon, \delta > 0$, there exists $N \in \mathbb{Z}^+$ such that when $n \geq N$:

$$\int_{\|x\| \geq \delta} K_n(x) \, dx < \varepsilon.$$

It is easy to show (see Lang [1, pp. 211–213]) that if f is continuous on $\mathbb{R}^m$, then $K_n * f(x) = \int K_n(x - y) f(y) \, dy$ approaches $f(x)$ uniformly, as n goes to infinity, for x in any compact set.

Examples of Dirac Sequences for $\mathbb{R}^m = \mathbb{R}$

Example 1 (The Fejér Kernel).

$$K_n(x) = \begin{cases} \dfrac{1}{n} \left(\dfrac{\sin(n\pi x)}{\sin(\pi x)} \right)^2 & \text{for } |x| \leq \dfrac{1}{2} \\ 0 & \text{otherwise.} \end{cases}$$

This kernel can be used to show that the Fourier series of a continuous periodic function can be Cesàro summed to converge uniformly to the function (see §1.3 and Lang [1, p. 233]).

Example 2 (The Landau Kernel).

$$c_n = \left(\int_{-1}^{+1} (1 - x^2)^n \, dx \right)^{-1}.$$

$$L_n(x) = \begin{cases} c_n (1 - x^2)^n & \text{for } |x| \leq 1 \\ 0 & \text{otherwise.} \end{cases}$$

This kernel can be used to show that any continuous function on a finite interval can be uniformly approximated by polynomials since $L_n * f$ is a polynomial (see Lang [1, p. 214]).

Example 3 (The Gauss Kernel).

$$G_t(x) = (2\pi t)^{-1/2} \exp[-x^2/(2t)], \quad \text{as } t \text{ approaches zero from the right.}$$

This kernel is the fundamental solution of the heat equation in one space variable (see §1.2).

Exercise 3. (a) Check that the three preceding examples are Dirac sequences of positive type. Show that the convolution of any of these kernels with a continuous function $K_n * f$ approaches f uniformly on compact sets as n goes to infinity. For Example 3, the index n is replaced by the positive real number t which approaches 0 from the right.
(b) Then show that any continuous function of period 1 can be uniformly approximated by polynomials $P(x)$ as well as by trigonometric polynomials $P(\exp(2\pi i a x))$.

To obtain Dirac sequences for $\mathbb{R}^m$, $m > 1$, take $K_n(x) = K_n(x_1) \cdots K_n(x_m)$ for $x = (x_1, \ldots, x_m)$.

An example of a sequence approaching delta distributionally which is not a Dirac sequence of positive type is the Dirichlet kernel, used to show that the Fourier series of a continuously differentiable function converges to the function (see Exercise 1 of §1.3 and Lang [1, p. 237]).

It is clear that T_f in Exercise 1 corresponds to a sequence of functions. Take every element of the sequence to be f itself.

It is easy to define the restriction, sum, scalar multiple, translate, support, and derivative of distributions. For example, the *support of a function* f is the smallest closed set outside of which f vanishes. We say that a distribution T is zero on an open set U in $\mathbb{R}^m$ if $T(g) = 0$ for all test functions g with support in U. Then the *support of a distribution* is the smallest closed set outside of which the distribution is zero.

Exercise 4. Show that the support of the Dirac distribution is $\{0\}$.

Integration by parts says that if f' is locally integrable and g is a test function of one variable, then

$$T_{f'}(g) = \int f'g = -\int fg' = -T_f(g').$$

This suggests that the *partial derivative* $T_{x_i} = \partial T/\partial x_i$ of a distribution T should be defined by setting $T_{x_i}(g) = -T(g_{x_i})$ for any test function g. One must check that $-T(g_{x_i})$ has a continuous linear dependence on the test function g. This follows from the way that convergence of test functions was defined. Thus all distributions are infinitely differentiable and all mixed partials of distributions must be equal.

1.1. Distributions or Generalized Functions

Exercise 5 (The Distributional Derivative of the Heaviside Step Function is the Dirac Delta Distribution on $\mathbb{R}^m = \mathbb{R}$). Define the *Heaviside function* by

$$H(x) = \begin{cases} 0, & x < 0, \\ 1, & x > 0. \end{cases}$$

Show that $U' = \delta$, as distributions.

The following exercise is useful in the study of PDEs involving the Laplace operator Δ. Suppose that E is a fundamental solution of Laplace's equation (also known as the free space Green's function for the Laplacian); i.e. $\Delta E = \delta$ (a distributional differential equation). Then, if we are given a sufficiently nice function f and we seek to solve the PDE $\Delta u = f$ for the unknown function u, it will easily be seen that $u = f * E$ (cf. Theorem 1, which follows). It is harder to find the Green's functions satisfying various sorts of boundary conditions (see Garabedian [1]).

Exercise 6 (The Fundamental Solution of the Laplacian in $\mathbb{R}^m$, $m \geq 3$). Show that in the sense of distributions: if $\Delta = \partial^2/\partial x_1^2 + \cdots + \partial^2/\partial x_m^2$, then

$$\Delta(\|x\|^{2-m}) = (2-m)S_m\delta, \quad \text{where } S_m = \text{area of unit sphere in } \mathbb{R}^m.$$

Thus the fundamental solution for the Laplacian is $E(x) = c_m\|X\|^{2-m}$,

$$c_m = [S_m(2-m)]^{-1}.$$

Hint. Let T be the distribution of the function $\|x\|^{2-m}$ in the sense of Exercise 1. You must show that $(\Delta T)(g) = T(\Delta g) = g(0)$ for each test function g. Now $T(\Delta g) = \int \|x\|^{2-m}\Delta g\,dx$. To evaluate this integral, apply Green's theorem to the region obtained by removing a ball of radius r from $\mathbb{R}^m$. Then let r approach zero. In §1.2, we will learn another way to do this.

Exercise 7 (The Area of the Unit Sphere in $\mathbb{R}^m$). Show that

$$\text{area}\{x \in \mathbb{R}^m | \|x\| = 1\} = 2\Gamma(1/2)^m/\Gamma(m/2).$$

Hint. Consider the integral of $\exp(-\|x\|^2)$ in polar coordinates and recall Euler's formula for the gamma function (see Lebedev [1]).

It is useful to note *a few general facts about distributions* that we do not have the space to prove. There are three such results:

(1) Any distribution supported by $\{0\}$ is a finite linear combination of partial derivatives of delta.
(2) Any distribution with compact support extends to a continuous linear functional on the space $C^\infty(\mathbb{R}^m)$ of all infinitely differentiable functions on $\mathbb{R}^m$. Here convergence of a sequence of C^∞ functions means uniform convergence of partials on compacta.

(3) Every distribution is locally the result of applying a distributional differential operator (with constant coefficients) to a continuous function. Here locally means: when restricted to test functions with support in a given compact set.

REMARKS ON PROOFS.

(1) See Lang [2, p. 448].
(2) See F.G. Friedlander [1, p. 35].
(3) See Gelfand and Shilov [1, Ch. 2]. The proof can be sketched as follows. Now test functions with support in a compact set D are the intersection over p of functions which are continuously differentiable of order up to p on D. This dualizes to say that a p must exist so that T will be a continuous linear functional on functions continuously differentiable of order up to p. Then one uses the Hahn-Banach theorem, the Riesz representation theorem, and integration by parts. We shall use this result in the discussion of convolution and Fourier transform of distributions.

One defines *convergence* of a sequence T_n of distributions to a distribution T as n approaches infinity to mean $T_n(g)$ approaches $T(g)$ for every test function g. When thinking of functions as distributions, as in Exercise 1, this type of convergence is often called weak. The Lebesgue dominated convergence theorem gives an easy condition for weak convergence of sequences of functions. It is of course much easier for a sequence of functions to converge in the sense of distributional convergence than to converge pointwise. Note that if K_n is a Dirac sequence, then T_{K_n} approaches δ as $n \to \infty$.

It is easy to see that the operations of differentiation and passing to the limit can always be interchanged distributionally. This is false, of course, for sequences of functions. Just as easily, one can see that very weak conditions guarantee the convergence of series of distributions (for example, see Schwartz [1, p. 97]). In Theorem 3 of §1.3, we shall see some examples of Fourier series of distributions.

In general, it is impossible to define the *product of two distributions* without abandoning associativity, for we shall see that $x\delta = 0$. But then

$$\left(\frac{1}{x} x\right)\delta = \delta \quad \text{and} \quad \frac{1}{x}(x\delta) = 0.$$

Note also that if f is locally integrable, there is no reason for f^2 to be locally integrable; e.g., $f(x) = |x|^{-1/2}$. The nastier T is, the nicer S must be for TS to make sense. We shall confine ourselves to the following situation:

Definition. If T is a distribution and α is any infinitely differentiable function on $\mathbb{R}^m$, define the distribution αT by $(\alpha T)(g) = T(\alpha g)$ for test functions g.

Exercise 8 (Products of Infinitely Differentiable Functions α and Distributions T).

1.1. Distributions or Generalized Functions

(a) Show that
$$\frac{\partial}{\partial x_i}(\alpha T) = \frac{\partial \alpha}{\partial x_i} T + \alpha \frac{\partial T}{\partial x_i}.$$

(b) Show that $\alpha \delta = \alpha(0)\delta$.

One must similarly make some restrictions in order to define the convolution of two distributions. The idea of convolution is quite important. For example, solutions to partial (and ordinary) differential equations are often convolutions (cf. Exercise 6 and the next section). And in probability theory, the density function of the sum of two independent random variables is the convolution of the two density functions (see, for example, Apostol [1, Vol. II, p. 552]).

In order to figure out what the convolution of two distributions should be, we must recall the definition of *convolution $f * g$ of two functions $f, g : \mathbb{R}^m \to \mathbb{C}$*, assuming both of the functions are locally integrable and that at least one has bounded support:

$$f * g(x) = \int_{\mathbb{R}^m} f(x - y)g(y)\,dy = \int_{\mathbb{R}^m} f(y)g(x - y)\,dy. \tag{1.2}$$

The convolution of f, g in $L^1(\mathbb{R}^m)$ = the Lebesgue integrable functions also makes sense.

One can define the *convolution of two distributions* as the equivalence class of Cauchy sequences obtained by convolving the functions representing the two distributions, assuming that one of the distributions has bounded support. The definition as a continuous linear functional goes as follows.

Definition. Suppose that S and T are distributions and that S has bounded support. Then define the distribution $T*S$ as follows using the notation $T(g) = \langle T, g \rangle$:

$$\langle T*S, g \rangle = \langle T(y), \langle S(x), g(x + y) \rangle \rangle \qquad \text{for each test function } g.$$

Here we are using an abuse of notation, writing distributions as if they were functions in order to keep track of variables.

Exercise 9.

(a) Check that convolution of functions is associative and commutative, assuming that all the functions are in $L^1(\mathbb{R}^m)$. Show that convolution of a Lebesgue integrable function with a differentiable function produces a differentiable function.

(b) Suppose that f and g are locally integrable functions and that g has compact support. Show that $T_{f*g} = (T_f) * (T_g)$, using the notation of Exercise 1.

In order to see that the preceding definition of convolution of distributions makes sense, one must show that $\langle S(x), g(x+y)\rangle$ is a test function as a function of y. To see this, use the fact that S is the distributional derivative of a continuous function (locally for test functions vanishing off a given compact set).

Theorem 1 (Properties of Convolution).

(a) Regularization. *The convolution of a distribution of bounded support and a test function is a test function.*
(b) $T * \delta = T$, $T * (\delta_{x_j}) = T_{x_j}$, *with* T_{x_j} *denoting the partial derivative.*
(c) $T * S = S * T$ *if the support of* S *or of* T *is bounded.*
(d) $(T * S) * R = T * (S * R)$ *if the supports of any two of these distributions are bounded.*
(e) $(T * S)_{x_j} = (T_{x_j}) * S = T * (S_{x_j})$ *if the support of* S *or* T *is bounded.*
(f) *Suppose that* T_n *approaches* T *in the sense of distributions. Then* $T_n * S$ *approaches* $T * S$ *as* n *goes to infinity, provided that one of the following hold:*
   (i) *the* T_n *have uniformly bounded supports,*
   (ii) S *has bounded support.*
(g) *Define the translate of a distribution* T *by a vector* $a \in \mathbb{R}^m$ *as*

$$T_a(g) = T(g_{-a}) \quad \text{where} \quad g_a(x) = g(x+a).$$

Then $T * \delta_a = T_a$ *and* $\delta_a * \delta_b = \delta_{a+b}$.

Exercise 10. Prove Theorem 1.

It follows from Theorem 1 that every distribution is the limit of a sequence of test functions. To see this, note that there is a sequence of test functions K_n approaching the Dirac delta distribution. Therefore $T * K_n$ approaches T as n approaches infinity. For we can find K_n with uniformly bounded supports and apply part (f) of Theorem 1.

Exercise 11. Show that the support of $(S * T)$ is contained in the closure of the set of points $x + y$, with x in the support of S, and y in the support of T.

1.2. Fourier Integrals

> Euclidean methodology has developed a certain obligatory style of presentation. I shall refer to this as "deductivist style." This style starts with a painstakingly stated listed of *axioms, lemmas* and/or *definitions*. The axioms and definitions frequently look artificial and mystifyingly complicated. One is never told how these complications arose. The list of axioms and definitions is followed by the carefully worded *theorems*. These are loaded with heavy-going conditions; it seems impossible that anyone should ever have guessed them. The theorem is followed by the *proof*.
>
> ...

1.2. Fourier Integrals

> ...Deductivist style hides the struggle, hides the adventure. The whole story vanishes, the successive tentative formulations of the theorem in the course of the proof-procedure are doomed to oblivion while the end result is exalted into sacred infallibility.
>
> From Lakatos [1, p. 142].

We want to understand the Fourier transform of a distribution on $\mathbb{R}^m$, but it makes sense to start off with Fourier transforms of Schwartz functions. Our study of Fourier transforms on Euclidean space will be sketchy. For more details, the reader could consult any of the following references: Bochner [1], Choquet-Bruhat et al. [1], Dieudonné [1], Dym and McKean [1], Lang [1, 2], Maurin [1], Schwartz [1, 2], Stein and Weiss [1], Titchmarsh [1], Zygmund [1].

Definition. The *Schwartz space* $\mathscr{S}$ is the space of all infinitely differentiable functions $f: \mathbb{R}^m \to \mathbb{C}$ such that $|x^a D^b f|$ is bounded for all $a, b \in \mathbb{Z}^m$, with $a_j \geq 0$ and $b_j \geq 0$. Here we use the notation: $a = (a_1, \ldots, a_m)$,

$$x^a = x_1^{a_1} \cdots x_m^{a_m} \quad \text{and}$$

$$D^b = \frac{\partial^{|b|}}{\partial x_1^{b_1} \cdots \partial x_m^{b_m}}, \qquad |b| = b_1 + \cdots + b_m. \tag{1.3}$$

Definition (Fourier Transform of Schwartz Functions). Suppose $f \in \mathscr{S}$. Then define

$$\hat{f}(y) = \int_{\mathbb{R}^m} f(x) \exp(-2\pi i\, {}^t x y)\, dx.$$

Here we write $x \in \mathbb{R}^m$ as a column vector and ${}^t x$ as the transpose of that vector. Thus ${}^t x y$ is the inner product of x and y. Note that the above integral is easily seen to be convergent.

Theorem 1 (Properties of the Fourier Transform on the Schwartz Space).

(1) $f \in \mathscr{S}$ implies that $\hat{f} \in \mathscr{S}$.
(2) $D^a \hat{f} = \widehat{(-2\pi i x)^a f}$, using the notation (1.3) for differential operators.
(3) $\widehat{D^a f} = (2\pi i x)^a \hat{f}$.
(4) Convolution Theorem. $\widehat{(f * g)} = \hat{f} \cdot \hat{g}$.
(5) Translation. Set $f_a(x) = f(x + a)$ for a, x in $\mathbb{R}^m$. Then

$$\widehat{f_a}(x) = \exp(2\pi i\, {}^t a x) \hat{f}(x).$$

(6) A Function which is its own Fourier Transform. Let $f(x) = \exp(-\pi \|x\|^2)$. Then $\hat{f} = f$.
(7) Dilation. Let u be a positive real number and set ${}^u f(x) = f(ux)$ for x in $\mathbb{R}^m$. Then

$$\widehat{({}^u f)}(x) = u^{-m}\hat{f}(u^{-1}x).$$

(8) Multiplication Formula: $\int \hat{f}g = \int f\hat{g}$.
(9) Fourier Inversion. *Define* $f^-(x) = f(-x)$. *Then* $\hat{\hat{f}} = f^-$.

All of the functions in the preceding statements are assumed to be Schwartz functions.

PROOF. Everything is an exercise except part (9). To prove (9) note that the operation of replacing f by $\hat{\hat{f}}^-$ commutes with translation. Thus it suffices to prove (9) at $x = 0$. Then property (8) shows that it suffices to prove (9) for a Dirac sequence of test functions such as the Gauss kernel from §1.1. For we have, assuming that $K_n = K_n^- = \hat{\hat{K}}_n$,

$$\int \hat{\hat{f}} K_n = \int f \hat{\hat{K}}_n = \int f K_n.$$

The left-hand side of this equality approaches $\hat{\hat{f}}(0)$ and the right-hand side approaches $f(0)$ as n goes to infinity, since K_n approaches the Dirac delta distribution implies that $\hat{K}_n$ approaches 1. □

Exercise 1. Prove property (2) of the Fourier transform by differentiating under the integral sign. Then prove property (3) by integration by parts. Property (1) then follows easily from (2) and (3).

Exercise 2. Prove property (4) by changing the order of integration.

Exercise 3. Prove properties (5) and (7) by making the right substitutions.

Exercise 4 (Fourier Transforms of Gauss Kernels). Show that $f(x) = \exp(-\pi\|x\|^2)$, $x \in \mathbb{R}$, is its own Fourier transform by computing $\hat{f}'(x)$ using integration by parts and solving the resulting differential equation. Deduce property (6) of Theorem 1. Then let G_t be the Gauss kernel (Example 3 of §1.1). Show that $G_t = G_t^- = \hat{\hat{G}}_t$ using the 7th property of the Fourier transform.

Exercise 5 (More Properties of the Fourier Transform). Show that if f and g are Schwartz functions:

(a) $f \mapsto \hat{f}$ is one-to-one, linear from from $\mathscr{S}$ onto $\mathscr{S}$.
(b) $\widehat{(fg)} = \hat{f} * \hat{g}$.
(c) Define an inner product for f, g in $\mathscr{S}$ by $(f,g) = \int f\bar{g}$. Then $(f,g) = (\hat{f},\hat{g})$ (*Parseval's identity*). And, setting $\|f\|_2 = (f,f)^{1/2}$, we have the *Plancherel identity* $\|f\|_2 = \|\hat{f}\|_2$.

Mathematicians call the last statement of Exercise 5 the Plancherel theorem. And we will see many generalizations of it in the succeeding chapters. Plancherel proved it in 1910. However, Rayleigh used this result in his study of

blackbody radiation in 1889 (see Rayleigh [1]). Thus it might be more accurately called Rayleigh's theorem. In later chapters this result will be equivalent to the determination of the spectral or Plancherel measure in the inversion formula for the generalized Fourier transforms that we shall study.

Now we should turn to Fourier integrals of nastier functions than Schwartz functions. The space of all Schwartz functions is dense in the space $L^2(\mathbb{R}^m)$ of all Lebesgue square integrable functions on $\mathbb{R}^m$. See Apostol [2], Choquet-Bruhat et al. [1], Dieudonné [1, Vol. 2], Dym and McKean [1], Kolmogorov and Fomin [1], Korevaar [1], Lang [2], Maurin [1], or any of countless tests that treat the theory of the Lebesgue integral. The book of Dym and McKean is greatly to be recommended for this purpose. Now the continuous linear map of $\mathscr{S}$ onto $\mathscr{S}$ (using the L^2-topology of Exercise 5) given by $f \mapsto \hat{f}$, has a unique continuous extension to a map on $L^2(\mathbb{R}^m)$. The extended mapping will be one-to-one, onto and a linear Hilbert space isomorphism of $L^2(\mathbb{R}^m)$. It is easy to show that if $f \in L^2(\mathbb{R}^m)$, then $\hat{f}(y)$ is the L^2-limit, as n goes to infinity, of the finite Fourier transforms:

$$\int_{\|x\| \leq n} f(x) \exp(-2\pi i\,{}^t xy)\, dx.$$

Here you need the L^2 dominated convergence theorem. The inverse transform has a similar characterization.

Wiener showed in [1] that it is also possible to give an elegant description of the L^2 Fourier transform using Hermite polynomials. We will discuss this in §1.3.

If f is in $L^1(\mathbb{R}^m)$; i.e., if f is Lebesgue measurable and $\|f\|_1 - \int |f|$ is finite, then the Fourier transform $\hat{f}(y) = \int f(x) \exp(-2\pi i\,{}^t xy)\, dx$ exists as a Lebesgue integral. We could not say this for L^2 functions. Thus it appears more natural to discuss Fourier integrals for L^1 functions. However, there is a problem with L^1 functions. The Fourier transform of an L^1 function need not be L^1. A simple example on the real line is given by that in the following exercise.

Exercise 6 (A Function Whose Fourier Transform Is Not in $L^1(\mathbb{R})$). Define

$$\Pi(x) = \begin{cases} 0, & |x| \geq 1/2, \\ 1/2, & |x| = 1/2, \\ 1, & |x| \leq 1/2. \end{cases}$$

Show that $\hat{\Pi}(y) = \sin(\pi y)/(\pi y)$. Then show that $\hat{\Pi}$ is not in $L^1(\mathbb{R})$. In fact, this gives us an example of a function which has an improper Riemann integral on $\mathbb{R}$ but is not absolutely integrable and thus not Lebesgue integrable.

Theorem 2 (Properties of the Fourier Transform of L^1 Functions).

(1) $\|\hat{f}\|_\infty = \text{l.u.b.}\, |\hat{f}(y)| \leq \|f\|_1$ (l.u.b. = *least upper bound*).
(2) $\hat{f}\colon \mathbb{R}^m \to \mathbb{C}$ *is continuous.*
(3) $\widehat{(f * g)} = \hat{f} \cdot \hat{g}$.

(4) Riemann-Lebesgue Lemma: $\lim_{|y|\to\infty} \hat{f}(y) = 0$.
(5) $\hat{f} = 0 \Leftrightarrow f = 0$. *Here we mean* 0 *almost everywnere.*

Discussion. The proofs of these properties are not hard, using Lebesgue dominated convergence, Fubini's theorem, etc. To prove the Riemann-Lebesgue lemma, note that given an L^1 function f, one can find a Schwartz function g which is close to f in L^1 norm. Then $\hat{f}(x)$ and $\hat{g}(x)$ must be close for all x. Now use the fact that $\hat{g}$ approaches zero as x goes to infinity. More details can be found in Dym and McKean [1] or Stein and Weiss [1], for example.

In part (5) of Theorem 2, we are saying that the Fourier transform defines a 1–1 map of $L^1(\mathbb{R}^m)$ into the continuous functions on $\mathbb{R}^m$ which vanish at infinity. The actual image is a very complicated subalgebra, which has not yet been characterized.

You might well ask how to reclaim a function f in $L^1(\mathbb{R})$ from its Fourier transform. Since $\hat{f}$ need not be in L^1, it is necessary to use some kind of summability procedure. The time-honored method for piecewise smooth functions is outlined in the following exercise. Other methods can be found in Dym and McKean [1, pp. 103–106].

Exercise 7 (A Dirac Family Associated to Fourier Inversion). Suppose that $f \in L^1(\mathbb{R})$ and f is piecewise continuous (i.e., it can have at most a finite number of jump discontinuities in each finite interval). If the one-sided derivatives of f exist at x in $\mathbb{R}$, show that

$$\frac{1}{2}\Big(f(x+) + f(x-)\Big) = \lim_{r\to\infty}\left(\int_{|y|\le r} \hat{f}(y)\exp(2\pi i x y)\,dy\right)$$

$$= \lim_{r\to\infty}(f * W_r) \quad \text{with} \quad W_r(x) = \frac{\sin(2\pi r x)}{(\pi x)}.$$

Here

$$f(x+) = \lim_{w\to x,\, w>x} f(w) \quad \text{and} \quad f(x-) = \lim_{w\to x,\, w<x} f(w).$$

Hint (cf. Exercise 1 of §1.1). Write

$$\int_{|y|\le r} \hat{f}(y)\exp(2\pi i x y)\,dy = \int_{|y|\le r}\int_{u=-\infty}^{+\infty} \exp(2\pi i u(x-y))f(u)\,du\,dy.$$

Then interchange the order of integration.

To complete the solution, you need the Riemann-Lebesgue lemma (Thm. 2, Part 4) and Exercise 14:

$$\int_0^\infty \frac{\sin x}{x}\,dx = \frac{\pi}{2}.$$

Note. It is easy to extend Exercise 7 to $\mathbb{R}^m$, since

$$\int \cdots \int_{y\in\mathbb{R}^m,\, |y_j|\le r_j} f(y)\exp(2\pi i\,{}^t x y)\,dy = (f * W_r^m)(x),$$

where

$$W_r^m(x) = \prod_{j=1}^m W_{r_j}(x_j) \quad \text{for} \quad x \in \mathbb{R}^m, r \in (\mathbb{R}^+)^m.$$

Fourier analysis, however, did not originate in this textbook style of theorems and exercises. At the beginning, there were Fourier's experiments on heat diffusion. And there was much controversy! For example, Lagrange prevented the publication of Fourier's 1807 paper on the subject. See Grattan-Guinness and Ravetz [1] for the first published version of Fourier's paper, as well as an interesting discussion of Fourier's life and work. There was tremendous rivalry between mathematicians such as Fourier, Cauchy, and Poisson. Colossal arguments took place over the truth of the proposition that "any" function can be represented by its Fourier series. Mathematicians made insulting reviews of each other's papers. Out of real physical questions and equally real mathematical difficulties came modern analysis—the Riemann and Lebesgue integrals and, later, distributions. See also Riemann [1, pp. 227–271] for the fascinating history of Fourier analysis. Other historical references are Burkhardt [1] and Hilb and Riesz [1]. Another interesting reference is I. Lakatos [1, pp. 128–131].

Example 1 (Heat Flow on an Infinite Rod). The problem of heat flow on an infinite rod can be posed mathematically as follows, making many simplifying assumptions about the rod:

Find $u(x,t)$, $x \in \mathbb{R}$, $t > 0$, satisfying:

$$\begin{cases} \dfrac{\partial u}{\partial t} = a^2 \dfrac{\partial^2 u}{\partial x^2} & x \in \mathbb{R}, t > 0, \\ u(x,0) = f(x) = \text{initial heat distribution, assumed given.} \end{cases}$$

For a derivation see Dym and McKean [1, p. 61] or Vladimirov [1, p. 31], for example. Or you might like to read Fourier's original paper with its various approximations to the "correct" partial differential equation in Grattan-Guinness and Ravetz [1]. The constant a in the differential equation depends on the specific heat of the (uniform) material composing the rod.

To solve the problem, formally Fourier transform both sides of the partial differential equation with respect to x. Use the fact that the Fourier transform takes differentiation to multiplication by $2\pi i x$ (property 3 of Theorem 1). You obtain

$$\frac{\partial \hat{u}}{\partial t} = -4\pi^2 x^2 a^2 \hat{u}.$$

Therefore $\hat{u} = \hat{f} \exp(-4\pi^2 a^2 x^2 t)$.

Using Exercise 4, we see that the Fourier transform of $\exp[-(2\pi a x)^2 t]$ as a function of x is

$$(2a\sqrt{\pi t})^{-1} \exp[-x^2/(4a^2 t)] = G_v(x), \quad v = 2a^2 t, \quad G_v = \text{Gauss kernel}.$$

Now use the convolution property of the Fourier transform (Theorem 2, Part 3) and Fourier inversion to see that

$$u(x,t) = \hat{\hat{u}}(-x,t) = G_v * f = G_{2a^2 t}(x) * f(x).$$

Here the convolution is over the variable x.

The entire procedure must now be justified, since we did not know at the beginning that it was legal to use Fourier transform properties of $u(x,t)$, as it was an unknown function. Rather weak assumptions on f will suffice to make $G_{2a^2 t}(x) * f$ a solution of our problem, since the Gauss kernel is infinitely differentiable and a Schwartz function. Moreover, G_t is a Dirac sequence of positive type as t approaches 0 from the right. We are using Exercises 3 and 10 from §1.1.

We shall see at the end of this section that the Gauss kernel is also of central importance in probability and statistics.

Exercise 8 (D'Alembert's Solution of the Wave Equation). Consider the displacement $u(x,t)$ of an infinite homogeneous vibrating string at position x on the real axis and time t. Making enough simplifying assumptions, one finds that $u(x,t)$ must satisfy the wave equation (see, for example, Dym and McKean [1, p. 62]):

$$\begin{cases} \dfrac{\partial^2 u}{\partial t^2} = a^2 \dfrac{\partial^2 u}{\partial x^2} & x \in \mathbb{R}, t > 0, \\ u(x,0) = f(x), \quad \dfrac{\partial u}{\partial t}(x,0) = g(x), & x \in \mathbb{R}. \end{cases}$$

Apply similar methods to those that we used in the heat equation example to obtain d'Alembert's solution:

$$u(x,t) = \frac{1}{2}[f(x+at) + f(x-at)] + \frac{1}{2a}\int_{x-at}^{x+at} g(u)\,du.$$

Now that we have briefly reviewed the theory of the Fourier transform for rather nasty functions, it is time to describe the theory of the Fourier transform for distributions. The multiplication formula (property 8 of Theorem 1) suggests that the Fourier transform $\hat{T}$ of a distribution T should be defined by the equation $\hat{T}(g) = T(\hat{g})$ for all test functions g. The only problem is that when g and $\hat{g}$ are both test functions, then g must be identically zero, as we shall soon see. It is not, in fact, surprising that the support of $\hat{g}$ should be like that of $\exp(2\pi i x)$.

The moral is that we cannot easily define the Fourier transform of every distribution T. Instead we must restrict ourselves to tempered distributions. There is another way out of this dilemma which makes $\hat{T}$ a new sort of generalized function (see Gelfand and Shilov [1]).

Definition. A distribution T is said to be a *tempered distribution* if it extends to a continuous linear functional on the space $\mathscr{S}$ of Schwartz functions.

In order to understand what we mean by continuity in the definition above, we must define what we mean by convergence of a sequence of Schwartz functions g_n to a Schwartz function g. This means that all of the sequences $x^a D^b(g_n - g)$ converge to zero uniformly on $\mathbb{R}^m$ for all a, b in $(\mathbb{Z}^+)^m$, using notation (1) set up at the beginning of this section.

Exercise 9. Show that the Fourier transform defines a continuous linear map from $\mathscr{S}$ onto $\mathscr{S}$, using the preceding definition of continuous.

Definition. Suppose that T is a tempered distribution. Define the *Fourier transform* $\hat{T}$ by $\hat{T}(g) = T(\hat{g})$, for all Schwartz functions g.

Exercise 10 (Distributions Tempered and Not So Tempered).

(a) Show that the following distributions are all tempered:
  the distribution of a bounded or Lebesgue integrable function, any distribution with bounded support.

(b) Show that e^x does not define a tempered distribution on $\mathbb{R}$.

Theorem 3 (Properties of the Fourier Transform of Tempered Distributions).

(1) *The Fourier transform of a distribution T with bounded support is a function:*

$$V(s) = T(\exp(-2\pi i\,{}^t sx)).$$

Moreover, V may be continued to all complex numbers s as an entire function of exponential growth. The converse of this theorem is also true and is called the Paley-Wiener theorem.

(2) Fourier Transform Exchanges Differentiation and Multiplication by Polynomials. *Using the notation introduced in the definition of the Schwartz space, we have:*

$$D^a\hat{T} = (-2\pi ix)^a T \quad and \quad \widehat{(D^a T)} = (2\pi ix)^a \hat{T}.$$

(3) The Convolution Theorem. *Suppose that S and T are two distributions with bounded supports. Their Fourier transforms are functions and*

$$\widehat{(S * T)} = \hat{S}\hat{T}.$$

(4) Interchange of Unit for Convolution and Unit for Multiplication.

$$\hat{\delta} = 1.$$

(5) Fourier Inversion. *Suppose that T is a tempered distribution and g is a Schwartz function. Define T^- by $T^-(g) = T(g^-)$, where $g^-(x) = g(-x)$ for Schwartz functions g. Then*

$$T = (\hat{\hat{T}})^-.$$

A FEW PROOFS.

(1) Note that $V(s)$ makes sense because T has compact support and thus extends to the space of infinitely differentiable functions. For the same reason, there is a constant coefficient differential operator D with $T = DF$ for some continuous function F. Thus for Schwartz functions g, we have (setting $T(g) = \langle T, g\rangle$):

$$\langle \hat{T}, g\rangle = \langle T, \hat{g}\rangle = \int F\left(D_x \int \exp(-2\pi i\, {}^t s x) g(s)\, ds\right) dx$$

$$= \langle\langle T(x), \exp(-2\pi i\, {}^t s x)\rangle, g(s)\rangle,$$

upon interchanging orders of integrals. Proofs of the Paley-Wiener theorem can be found in Dym and McKean [1] and Schwartz [2, Ch. VI].

(3) $\langle S * T, g\rangle = \langle S * T, \hat{g}\rangle = \langle S(y), \langle T(x), \hat{g}(x + y)\rangle\rangle = \langle S(y), \langle T(x),$
$$\widehat{\exp(-2\pi i\, {}^t y x) g(x)}\rangle\rangle$$
$$= \langle \hat{T}(x), g(x) \langle S(y), \exp(-2\pi i\, {}^t y x)\rangle\rangle = \langle \hat{S}(x)\hat{T}(x), g(x)\rangle. \quad \square$$

Exercise 11. Prove parts (2), (4), (5) of Theorem 3.

It is clear from part (1) of Theorem 3 that if g and $\hat{g}$ are both test functions, then g must be zero everywhere. For a complex analytic function of one variable which is supported on a bounded infinite set must be identically zero (see Ahlfors [1, p. 127]).

Exercise 12. Verify the Table 1.1.

Table 1.1. A Short Table of Fourier Transforms of One Variable

$f(x)$, $x \in \mathbb{R}$	$\hat{f}(y) = \int f(x) \exp(-2\pi i x y)\, dy$
$\exp(-\pi x^2)$	$\exp(-\pi y^2)$
$\frac{1}{2}\exp(-\lvert x\rvert)$	$(1 + 4\pi^2 y^2)^{-1}$
$\Pi(x) = \begin{cases} 0, & \lvert x\rvert > 1/2 \\ \frac{1}{2}, & \lvert x\rvert = 1/2 \\ 1, & \lvert x\rvert < 1/2 \end{cases}$	$\dfrac{\sin(\pi y)}{(\pi y)}$
$\delta(x)$	1
$\frac{1}{2}[\delta(x - 1/2) + \delta(x + 1/2)]$	$\cos(\pi y)$
$(q^2 + x^2)^{-s}$, $\quad q \neq 0$, $\operatorname{Re} s > 0$	$\begin{cases} 2\pi^{1/2} \left\lvert \dfrac{\pi y}{q} \right\rvert^{s-1/2} \dfrac{K_{s-1/2}(2\pi \lvert yq\rvert)}{\Gamma(s)}, & y \neq 0 \\ \dfrac{\Gamma(\frac{1}{2})\Gamma(s - \frac{1}{2})}{\Gamma(s)} q^{1-s}, & y = 0 \end{cases}$

Hint. For the last formula, you need some properties of the gamma and K-Bessel functions (see Lebedev [1] or see Exercise 1 in §3.1).

If you should ever need a long table of Fourier transforms, try Erdélyi et al [2, Vol. 1] or Oberhettinger [3, 4].

Next let us consider some applications of Fourier transforms of distributions to partial differential equations. The general idea is the same as that which we used to solve the heat equation earlier. Suppose that D is a constant-coefficient partial differential operator and S is a tempered distribution. To solve the equation

$$DT = S$$

for T, try Fourier transforming the equation. You get

$$\widehat{DT} = M\hat{T} = \hat{S},$$

where M is multiplication by a polynomial. So you want to write

$$\hat{T} = \hat{S}/M.$$

Then Fourier inversion will find T. This makes perfect sense if the polynomial M never vanishes. But otherwise there is much work to do in order to legalize division by M. This has been done by Hörmander [1] and Łojasiewicz [1]. See F.G. Friedlander [1, p. 139ff] for a discussion of the theorem of Malgrange and Ehrenpreis to the effect that $DE = \delta$ can be solved for E.

In the examples that follow we will Fourier transform the *PDE* with respect to some subset of the variables. This will reduce the problem to an *ODE*. For more examples of this type, see Vladimirov [1].

Exercise 13 (Fundamental Solution of the Heat Equation). We want to solve

$$\frac{\partial E}{\partial t} - a^2 \Delta E = \delta(x, t), \quad x \in \mathbb{R}^m, t > 0, \quad \Delta = \frac{\partial^2}{\partial x_1^2} + \cdots + \frac{\partial^2}{\partial x_m^2}.$$

Claim. $E(x, t) = (4a^2 \pi t)^{-m/2} H(t) \exp[-\|x\|^2/(4a^2 t)]$, where

$$H(t) = \begin{cases} 1, & t \geq 0 \\ 0, & t < 0 \end{cases} = \text{Heaviside's step function.}$$

Hint. Imitate Example 1. Use $\hat{\delta} = 1$ and the fact that the distributional ordinary differential equation

$$S' + cS = \delta(t) = H'(t)$$

has the solution $S(t) = H(t)\exp(-ct)$. Prove this too.

Remarks. Note that $E(x, t)$ in Exercise 13 is positive for all positive t at every point x in space. Thus heat must be diffused with infinite velocity! This means that the heat equation does not appear to be a realistic model of heat transfer.

The transport equation appears to be better, but we shall not discuss it here (see Vladimirov [1]).

Example 2 (Fundamental Solution of Laplace's Equation in $\mathbb{R}^3$). We want to solve
$$\Delta E = \delta.$$

Claim. $E(x) = -(4\pi \|x\|)^{-1}$.

Discussion. Formally Fourier transform the *PDE* with respect to x. One obtains: $-4\pi^2 \|x\|^2 \hat{E} = 1$. It is possible to justify thinking of $\|x\|^{-2}$ as a tempered distribution since the behavior at 0 and ∞ is not really so bad. Thus $E(-x) =$ the Fourier transform of $(-4\pi^2 \|x\|^2)^{-1}$. Suppose that g is a Schwartz function and write, as usual, $T(g) = \langle T, g \rangle$. Then
$$\langle E(-x), g \rangle = \langle (-4\pi^2 \|x\|^2)^{-1}, \hat{g} \rangle$$
$$= \lim_{R \to \infty} \int_{|x| \le R} (-4\pi^2 \|x\|^2)^{-1} \int g(u) \exp(-2\pi i\, {}^t x u)\, du\, dx.$$

Interchange the order of integrals and use polar coordinates on the integral over the x variable. Then the x integral becomes:

$$-2\pi (4\pi^2)^{-1} \int_{r=0}^{R} \int_{\theta=0}^{\pi} \exp(-2\pi i \|u\| r \cos\theta) \sin\theta\, d\theta\, dr$$
$$= -(2\pi^2 \|u\|)^{-1} \int_0^{R'} \frac{\sin r}{r}\, dr. \qquad (R' = 2\pi \|u\| R)$$

By Exercise 14, this completes the proof that $\langle E(x), g \rangle = \langle -(4\pi \|x\|)^{-1}, g \rangle$.

Exercise 14 (An Improper Riemann Integral Which is Not Absolutely Convergent). Show that
$$\int_0^\infty \frac{\sin r}{r}\, dr = \frac{\pi}{2}.$$
Explain why the integral does not converge absolutely.

Exercise 15 (Fundamental Solution of the Wave Equation in $\mathbb{R}^3$). Define the simple surface layer distribution for a sphere
$$S_R = \{x \in \mathbb{R}^3 \mid \|x\| = R\},$$
by
$$\delta_{S_R}(g) = \int_{x \in S_R} g(x)\, dA, \qquad dA = \text{surface area element on } S_R.$$

Then show that the solution to the distributional equation $\square_a T = \delta(x,t)$, where $x \in \mathbb{R}^3$, $t \in \mathbb{R}$, and

$$\Box_a = \left(\frac{\partial^2}{\partial t^2} - a^2\left(\frac{\partial^2}{\partial x_1^2} + \frac{\partial^2}{\partial x_2^2} + \frac{\partial^2}{\partial x_3^3}\right)\right),$$

is $T = (4\pi a^2 t)^{-1} H(t)\delta_{S_{at}}(x)$.

Remarks. The last exercise can be used to derive Kirchoff's classical formula for the solution of the *Cauchy problem*:

$$\begin{cases} \Box_a u = h(x,t), & x \in \mathbb{R}^3,\ t > 0. \\ u(x,0) = f(x), & u_t(x,0) = g(x), \end{cases}$$

assuming that f, g, h are all sufficiently smooth (see Vladimirov [1]).

Huyghens' principle also follows from the last exercise. For a disturbance $\delta(x,t)$ propagates as a spherical wave along the spherical surface $\|x\| = at$, moving with speed a. After the wave passes a given point, there is no more disturbance. This principle fails in $\mathbb{R}$ and in even dimensional Euclidean space (see Exercise 8 and Vladimirov [1] or Dym and McKean [1]). Clearly Flatland would be a very noisy place!

It is possible to give many interpretations of the Fourier inversion formula that will aid us in our search for non-Euclidean analogues. The most important interpretation for this book is that Fourier inversion provides a spectral resolution of the Laplace operator on $\mathbb{R}^m$. That is, we are investigating *the spectral theorem for a very special unbounded self-adjoint operator* Δ on the Hilbert space $L^2(\mathbb{R}^m)$. For the Fourier inversion formula says that a nice function $f: \mathbb{R}^m \to \mathbb{C}$ has a representation as an integral of elementary eigenfunctions

$$e_a(x) = \exp(2\pi i\, {}^t a x), \qquad a, x \in \mathbb{R}^m, \tag{1.4}$$

of the Euclidean Laplacian:

$$\Delta = \frac{\partial^2}{\partial x_1^2} + \cdots + \frac{\partial^2}{\partial x_m^2}.$$

For if $f: \mathbb{R}^m \to \mathbb{C}$ is Schwartz, we have the spectral resolution

$$f(x) = \int_{\mathbb{R}^m} \hat{f}(y) e_y(x)\, dy \quad \text{with} \quad \hat{f}(y) = (f, e_y) = \int_{\mathbb{R}^m} f(u)\overline{e_y(u)}\, du. \tag{1.5}$$

Note that the eigenfunctions e_a are not in the L^2-space involved. This makes the elaboration of the spectral theorem rather intricate (see Maurin [2]). It is our aim to obtain analogues of Fourier inversion for symmetric spaces such as the non-Euclidean upper half-plane of Chapter 3. For a discussion of the general spectral theorem the reader should look at Maurin [2], Reed and Simon [1], or Yosida [1].

Another possible interpretation of the Fourier inversion formula comes from the theory of group representations. In this theory one views the eigenfunctions $e_a(x)$ in the preceding paragraph as irreducible unitary representations of the additive group $\mathbb{R}^m$, since $e_a(x + y) = e_a(x)e_a(y)$. We will not emphasize this aspect here, but if the reader is interested in group representations there are many good references, such as Dym and McKean [1], Hamermesh [1], Mackey [1, 2, 3], Talman [1], Vergne [1], and Vilenkin [1].

Exercise 16 (Heisenberg's Uncertainty Principle). Suppose that $\|f\|_2 = 1$. Then since $\|f\|_2 = \|\hat{f}\|_2$, both $|f(t)|^2$ and $|\hat{f}(w)|^2$ can be viewed as densities of probability distributions. Translation of f just leads to a phase shift in $\hat{f}$ (and similarly for translation of $\hat{f}$). So we can assume that the means vanish; i.e.,

$$\int t|f(t)|^2 \, dt = 0 \quad \text{and} \quad \int w|\hat{f}(w)|^2 \, dw = 0.$$

The variances are then

$$\sigma_t^2 = \int t^2 |f(t)|^2 \, dt \quad \text{and} \quad \sigma_w^2 = \int w^2 |\hat{f}(w)|^2 \, dw.$$

Then σ_t measures the time duration of the signal $f(t)$ while σ_w measures the frequency spread. The problem is to prove that

$$\sigma_t \sigma_w \geq 1/(4\pi).$$

And equality holds for $f(t) = C \exp(-at^2)$, with $a > 0$, C chosen to make $\|f\|_2 = 1$. You may assume that the functions involved are Schwartz functions. Then use Theorem 1, part (3) and Exercise 5, part (c) to see that it is equivalent to show that

$$\int |xf(x)|^2 \, dx \int |f'(x)|^2 \, dx \geq \frac{1}{4}.$$

See Dym and McKean [1, p. 119] for the details when f is only square integrable.

Many engineers prefer to use the Laplace transform rather than the Fourier transform.

Definition. The (one-sided) *Laplace transform* $\mathscr{L}f$ of a function $f: \mathbb{R}^+ \to \mathbb{C}$ is

$$(\mathscr{L}f)(s) = \int_0^\infty f(x) \exp(-sx) \, dx \quad \text{for } s \text{ in } \mathbb{C}.$$

One advantage is that $\mathscr{L}f$ tends to exist when $\hat{f}$ does not; for example, if f is merely bounded or even exponentially increasing, assuming that the real part of s is sufficiently large. Another advantage is that one can make use of the power of the theory of complex variables (see Exercise 20, for example). It is also possible to define Laplace transforms of functions of several positive real variables and of distributions. See Doetsch [1], Erdélyi et al. [2], Oberhettin-

1.2. Fourier Integrals

ger and Badii [1], Schwartz [1], Sneddon [1], or Widder [1] for more information about the Laplace transform.

Exercise 17 (Properties of the Laplace Transform).

(a) *Existence.* We shall say that a function $f: \mathbb{R}^+ \to \mathbb{C}$ is admissible if it is piecewise continuous on every finite interval and if there are constants a in $\mathbb{R}$ and M in $\mathbb{R}^+$ such that

$$|f(x)| \leq M \exp(ax) \quad \text{for all } x \text{ in } \mathbb{R}^+.$$

Show that if f is admissible, then $\mathscr{L}f(s)$ exists when $\operatorname{Re} s > a$.

(b) *Derivatives.* Suppose that f and f' are admissible. Then show that

$$\mathscr{L}(f') = s\mathscr{L}(f) - f(0) \quad \text{and} \quad (\mathscr{L}(f))' = \mathscr{L}(-xf(x)).$$

(c) *Integration.* Suppose that f is admissible and that $|f(x)/x| \leq Cx^{\varepsilon-1}$, for $\varepsilon > 0$. Then show that

$$\int_s^\infty \mathscr{L}f(p)\,dp = \mathscr{L}(f(x)/x) \quad \text{and} \quad \mathscr{L}\left(\int_0^x f(t)\,dt\right) = \frac{1}{s}\mathscr{L}(f).$$

(d) *Convolution.* Suppose that f, g, and fg are all admissible. Then prove that

$$(\mathscr{L}f)(s)(\mathscr{L}g)(s) = \mathscr{L}\left(\int_0^x f(x-t)g(t)\,dt\right),$$

$$\mathscr{L}(fg)(p) = \frac{1}{2\pi i}\int_{\operatorname{Re} s = c}(\mathscr{L}f)(s)(\mathscr{L}g)(p-s)\,ds.$$

Exercise 18 (Inversion of the Laplace Transform). Suppose that $e^{-cx}f(x)$ lies in $L^1(\mathscr{R})$ and $f(x)$ vanishes for negative x. Assume also that $f(x)$ is piecewise differentiable. Use Exercise 7 to show that

$$\frac{1}{2}\bigl(f(x+) + f(x-)\bigr) = \lim_{r\to\infty}\frac{1}{2\pi i}\int_{c-ir}^{c+ir} e^{sx}\mathscr{L}f(s)\,ds.$$

Exercise 19. Compute the Laplace transform of

(a) x^a, $\operatorname{Re} a > -1$.
(b) $\delta(x-a)$, $a > 0$,
(c) the square wave of period a which has the values

$$f(t) = \begin{cases} 1, & 0 < t < a/2 \\ 0, & a/2 < t < a \end{cases}, \quad f(t+a) = f(t).$$

Exercise 20 (Fourier Series via Laplace Inversion).

(a) If $f(x)$ is periodic, with period $a > 0$; i.e., $f(x+a) = f(x)$, show that

$$(\mathscr{L}f)(s) = (1 - e^{-as})^{-1}\int_0^a f(x)\exp(-sx)\,dx.$$

(b) Use (a) to obtain the Fourier series expression for a periodic function f, assuming that the function $f(x)$ is sufficiently nice.

Hint. See Sneddon [1, pp. 166–169]. For part (b) use Cauchy's residue theorem to see that the Laplace inversion integral gives a Fourier series via part (a).

The Laplace transform can be used in the same sort of way as the Fourier transform to solve differential equations. It also has seen much use in the study of the asymptotics of functions. For example, it is useful in number theory where one wants to study asymptotic properties of the sequence of prime numbers. Similarly, as we shall see, it allows one to study the asymptotic properties of eigenvalues of the Laplace operator, by making use of solutions of the heat equation. The Laplace transform results needed here are called Tauberian theorems. Such theorems involve Laplace-Stieltjes transforms:

$$\int_0^\infty \exp(-st)\,d\alpha(t),$$

with α of bounded variation and *normalized* to make $\alpha(0) = 0$ and $\alpha(x) = (\alpha(x+) + \alpha(x-))/2$. See Apostol [2] for a treatment of Riemann-Stieltjes integrals. Before thinking about Tauberian theorems, however, one should consider the following.

Theorem 4 (An Abelian Theorem). *Suppose that $f(s) = \int_0^\infty \exp(-st)\,d\alpha(t)$ for $s > 0$, where the integral is a normalized function of bounded variation and*

$$\alpha(t) \sim At^c/\Gamma(c+1) \quad as \quad t \to \infty \ (or\ t \to 0+).$$

Then

$$f(s) \sim As^{-c} \quad as \quad s \to 0+ \ (or\ s \to \infty).$$

The Abelian theorem is not hard to prove (e.g., see Widder [1, p. 182]). Its converse is not true without extra hypotheses. Such a converse follows.

Theorem 5 (A Tauberian Theorem). *Suppose that $\alpha(t)$ is a nondecreasing, normalized function of bounded variation. And suppose that*

$$f(s) = \int_0^\infty \exp(-st)\,d\alpha(t)$$

converges for all $s > 0$. If for some nonnegative number c

$$f(s) \sim As^{-c} \quad as \quad s \to 0+ \ (or\ s \to \infty),$$

then

$$\alpha(t) \sim At^c/\Gamma(c+1) \quad as \quad t \to \infty \ (or\ t \to 0+).$$

There is a proof in Widder [1, p. 192].

1.2. Fourier Integrals

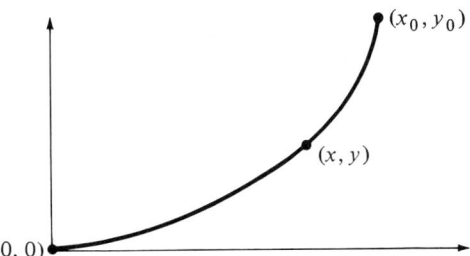

Figure 1.1. A particle sliding along a curve under the influence of gravity.

We will see an application of this result in the next section.

Exercise 21 (Abel's Integral Equation. The Tautachrone). Consider a particle of mass m sliding along a curve under the influence of gravity (with no friction), starting at (x_0, y_0) and ending at $(0, 0)$, as pictured in Figure 1.1. Let $T(y_0)$ be the time for the particle to fall from (x_0, y_0) to $(0, 0)$, assuming the shape of the curve is given by some function $y = f(x)$. Then conservation of energy says

$$\sqrt{2g}\,T(y_0) = \int_{y=0}^{y_0} \varphi'(y)(y_0 - y)^{-1/2}\,dy, \qquad \varphi'(y) = \left(1 + \left(\frac{dx}{dy}\right)^2\right)^{1/2},$$

where g is the acceleration of gravity and the curve is assumed to look like that in Figure 1.1. Derive this integral equation.

Now assume that $T = T_0 = $ constant; i.e., that the time of descent is independent of the starting point. The tautachrone problem is to find the curve $y = f(x)$ under this hypothesis. Solve this problem by taking the Laplace transform of both sides of the integral equation. Then use the convolution property of the Laplace transform plus the fact that $\mathscr{L}(y^{-1/2}) = (\pi/s)^{1/2}$. The result is that $\varphi'(y) = (2gT_0^2\pi^{-2}y^{-1})^{1/2}$. This gives a differential equation for $x(y)$ which has a cycloid for its solution.

Application to Probability and Statistics: The Central Limit Theorem

The Fourier transform can also be used to derive the central limit theorem, which is one of the most fundamental results in probability and statistics. Special cases of this theorem go back to DeMoivre and Laplace. The general result was proved by Lindeberg in 1922. Previously Liapunov, Chebychev, and Markov had proved the theorem under more restrictive hypotheses. The proof using Fourier transforms was developed from 1925 to 1940 using P. Lévy's proofs of properties of Fourier transforms of probabilistic density functions called characteristic functions. References for probability and statistics are Cramér [1], Feller [1], Kolmogorov [2], and Loève [1]. Collections of papers on applications to physics are Bharucha-Reid [1] and Wax [1].

Let us give a brief summary of the main definitions. For a historical perspective on the subject, see M. Kac [2]. For example, Kac recalls that in 1936 "independent random variables were to me (and others, including my teacher Steinhaus) shadowy and not really well-defined objects."

Let S be a measurable set in some Euclidean space, Let P be a *probability measure* on the Borel sets in S. This means that $P(S) = 1$. A *random variable* $X : S \to \mathbb{R}$ is a measurable function. The *distribution function* F_X of X is defined by

$$F_X(t) = P(\{w \in S \mid X(w) \le t\}) = P(X \le t).$$

If

$$F_X(t) - F_X(a) = \int_a^t f_X(u)\,du$$

for some nonnegative measurable function f_X, we call f_X the *density function* for X. The density function can be obtained via the Radon-Nikodym theorem provided that the probability distribution on $\mathbb{R}$ is absolutely continuous with respect to Lebesgue measure. Otherwise one can use the Lebesgue decomposition theorem to split the probability distribution on $\mathbb{R}$ into a sum of an absolutely continuous and a singular part (see Lang [2], Feller [1, Vol. II, pp. 138–143], or Loève [1]). Given two random variables X, Y, their *joint distribution* is

$$F(a,b) = P\{w \in S \mid X(w) \le a \quad \text{and} \quad Y(w) \le b\} = P(X \le a, Y \le b).$$

We say that X and Y are *independent* if $F(a,b) = F_X(a) F_Y(b)$.

If X is a random variable, we say that the *expectation* or *mean value* of X is

$$\mu = E(X) = \int_{-\infty}^{+\infty} t f_X(t)\,dt = \int_{-\infty}^{+\infty} t\,dF_X(t).$$

The *variance* of X is

$$\sigma^2 = \mathrm{Var}(X) = E((X-\mu)^2) = \int_{-\infty}^{+\infty} (t-\mu)^2 f_X(t)\,dt,$$

where σ is the *standard deviation*.

Exercise 22. Suppose that X and Y are independent random variables with density functions f_X and f_Y. Prove that the density function of the random variable $X + Y$ is the convolution $f_X * f_Y$.

The *characteristic function* φ_X of a random variable X is the Fourier transform

$$\varphi_X(t) = E(e^{itX}) = \int_{-\infty}^{+\infty} e^{itu} f_X(u)\,du = \hat{f}_X(-t/2\pi).$$

1.2. Fourier Integrals

Theorem 6. (Properties of Characteristic Functions). *Suppose that X is a random variable with density function $f_X = f$. Then $f \in L^1(\mathbb{R})$, $f \geq 0$, $\int f(x)\,dx = 1$ and the characteristic function of X, which is $\varphi(t) = \hat{f}(-t/2\pi)$, has the following properties.*

(1) *The function φ is continuous, $\varphi(0) = 1$, and $|\varphi(t)| \leq 1$ for all t.*
(2) *The random variable $aX + b$, for $a, b \in \mathbb{R}$, has the characteristic function*

$$\varphi_{aX+b}(t) = e^{ibt}\varphi_X(at).$$

(3) *If X and Y are two independent random variables, then the characteristic function of the sum $X + Y$ is*

$$\varphi_{X+Y}(t) = \varphi_X(t)\varphi_Y(t).$$

(4) *Fourier inversion. Suppose that φ is the characteristic function of the random variable X and suppose that $\varphi \in L^1(\mathbb{R})$. Then X has a bounded continuous density function f given by*

$$f(x) = \frac{1}{2\pi}\int_{-\infty}^{+\infty} e^{-itx}\varphi(t)\,dt.$$

Exercise 23. Prove Theorem 6.

Hint. See Feller [1, Vol. II, Ch. XV].

Exercise 24. Suppose that X and Y are two independent random variables. Show that $\mathrm{Var}(X + Y) = \mathrm{Var}(X) + \mathrm{Var}(Y)$.

A random variable X is said to be *normally distributed with parameters m and σ or normal (m, σ)* if the density function of X is

$$\sigma^{-1}(2\pi)^{-1/2}\exp[-(x-m)^2/(2\sigma^2)].$$

We have graphed some of these density curves in Fig. 1.2.

Exercise 25 (Properties of Normal Distributions). Assume that X is normal (m, σ).

(a) Show that $E(X) = m$ and $\mathrm{Var}(X) = \sigma^2$.
(b) Prove that

$$P(|X - m| > \lambda\sigma) = 2(2\pi)^{-1/2}\int_{\lambda}^{\infty}\exp(-x^2/2)\,dx.$$

(c) Show that the characteristic function of X is $\varphi_X(t) = \exp[imt - (\sigma t)^2/2]$.
(d) Show that the sum of any two independent normally distributed random variables is itself normally distributed.

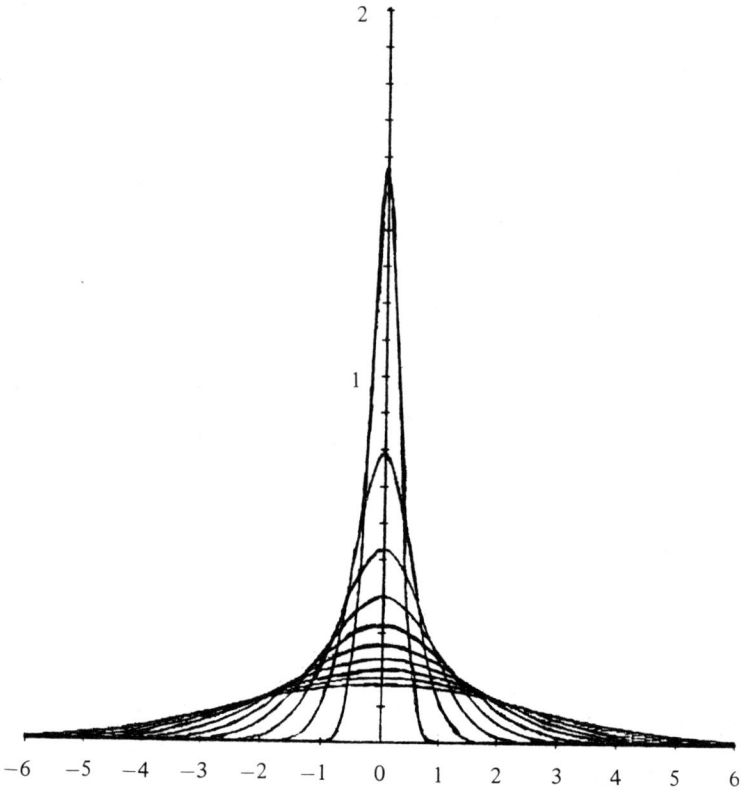

Figure 1.2. Graphs of normal density curves. $m = 0, \sigma = I/4, I = 1, 2, \ldots, 10$.

Theorem 7 (The Central Limit Theorem). *Suppose that $\{X_n\}_{n\geq 1}$ is a sequence of independent random variables, each having the same density function $f(x)$. Suppose that the mean is 0 and the standard deviation is 1; i.e.,*

$$\int f(x)\,dx = 1, \qquad \int xf(x)\,dx = 0, \qquad \text{and} \qquad \int x^2 f(x)\,dx = 1.$$

Then $(X_1 + \cdots + X_n)/\sqrt{n}$ is nearly Gaussian for large n; i.e.,

$$P\left(a \leq \left(\frac{X_1 + \cdots + X_n}{\sqrt{n}}\right) \leq b\right) \sim (2\pi)^{-1/2} \int_a^b \exp(-x^2/2)\,dx \qquad \text{as } n \to \infty.$$

This means that

$$\int_{a\sqrt{n}}^{b\sqrt{n}} (\underbrace{f * \cdots * f}_{n \text{ times}})(x)\,dx \sim (2\pi)^{-1/2} \int_a^b \exp(-x^2/2)\,dx \qquad \text{as } n \to \infty.$$

Proof. Using Exercise 22 and Theorem 6, one sees that the Fourier transform of the asymptotic relation in the central limit theorem is

1.2. Fourier Integrals

$$\hat{f}(s/\sqrt{n})^n \sim e^{-2\pi^2 s^2} \quad \text{as } n \to \infty.$$

Recall that $\hat{f}(s) = \int e^{-2\pi i s x} f(x)\,dx$. So the hypotheses of the theorem say that

$$\hat{f}(0) = 1, \quad \frac{d}{ds}\hat{f}(0) = 0, \quad \frac{d^2}{ds^2}\hat{f}(0) = -4\pi^2.$$

It follows, using the Taylor formula, that since $\hat{f}$ has a continuous second derivative,

$$\hat{f}(s/\sqrt{n}) = 1 + s(\hat{f})'(0)/\sqrt{n} + (\hat{f})''(0)s^2/(2n) + \cdots \sim (1 - 2\pi^2 s^2/n).$$

Therefore $\hat{f}(s/\sqrt{n})^n \sim e^{-2\pi^2 s^2}$, as $n \to \infty$.

To go from the Fourier transform of the central limit theorem back to the result we want, one must know a "continuity theorem" for characteristic functions due to Lévy and Cramér (see Cramér [1, pp. 97–98] and Loève [1]). This is analogous to a Tauberian theorem (see Theorem 5). Instead we choose to follow the discussion in Dym and McKean [8, p. 116].

We have proved that

$$\lim_{n \to \infty} \varphi_{(X_1 + \cdots + X_n)/\sqrt{n}}(2\pi s) = \lim_{n \to \infty} \int_\mathbb{R} f_n(x) \exp(2\pi i x/\sqrt{n})\,dx$$

$$= \exp[-2\pi^2 s^2],$$

where

$$f_n(x) = (f * \cdots * f)(x)$$

if the number of f's in the convolution is n.

Let d_n be the density for the random variable $(X_1 + \cdots + X_n)/\sqrt{n}$. Then we have for Schwartz functions k:

$$\lim_{n \to \infty} \int_\mathbb{R} d_n(x)k(x)\,dx = \lim_{n \to \infty} \int_\mathbb{R} \hat{d}_n(s)\hat{k}(s)\,ds$$

$$= \int_\mathbb{R} \exp(-2\pi^2 s^2)\hat{k}(s)\,ds = \int_\mathbb{R} (2\pi)^{-1/2} \exp(-x^2/2)k(x)\,dx,$$

using the Plancherel theorem, dominated convergence, and part (1) of Theorem 2.

Exercise 26. Complete the proof of the central limit theorem by approximating

$$\chi_{[a,b]}(x) = \begin{cases} 0, & x \notin [a,b] \\ 1, & x \in [a,b] \end{cases}$$

by Schwartz functions k. □

It is possible to relax the hypotheses in the central limit theorem and one can also obtain error estimates. See the references for the details.

There is also a multivariate central limit theorem for random vectors in $\mathbb{R}^m$. This can be quickly deduced from the case $m = 1$ (see Anderson [1, pp. 74–75]).

In this section, we have seen the Gaussian or normal distribution in two seemingly different contexts—as the fundamental solution of the heat equation (see Example 1) and as the limiting density function for the normalized sum of a sequence of independent identically distributed random variables. The Gaussian density also appears in a third context closely related to the previous two; namely, in Einstein's work on Brownian motion. The latter is the motion—visible with a microscope—of tiny particles suspended in a liquid. Such motion had been observed by Robert Brown in 1827, as well as many others before him. Einstein proved in 1905 that if S_t denotes the displacement after t minutes of a particle in Brownian motion then S_t has density function

$$(4\pi Dt)^{-1/2} \exp(-x^2/(4Dt)),$$

where D depends in a very explicit way on the temperature and friction coefficient of the medium. See the references mentioned above for more information on Brownian motion. Another interesting reference is Nelson [1].

We will find in Chapter 3 that we can generalize the central limit theorem to the non-Euclidean case, using the non-Euclidean Fourier transform. The proof will be basically the same.

1.3. Fourier Series and the Poisson Summation Formula

Jétais d'abord parvenu à plusieurs de ces équations par des éliminations très-laborieuses, mais j'emploie maintenant une règle beaucoup plus générale et très-expéditive pour résoudre une fonction arbitraire quelconque en série de sinus ou de cosinus d'arcs multiples. Ces résultats confirment pleinement l'opinion de Daniel Bernoulli.

Les développements dont il s'agit ont donc cela de commun avec les équations aux différences partielles qu'ils peuvent exprimer les propriétés des fonctions entièrement arbitraires et discontinues; c'est pour cela qu'ils se présentent naturellement pour l'intégration de ces dernières équations, et leur application offre des facilitiés singulières dans les questions des lignes || mouvements des fluides, de la propagation du son, des vibrations des corps élastiques, et donne un moyen aisé de déterminer les mouvements avec toute la généralité que l'on obtiendrait de l'emploi des fonctions arbitraires. J'en ai fait une application plus particulière à la question de la propagation de la chaleur et l'on parvient ainsi à reconnaître

distinctement comment elle se propage par ondes successives dans l'intérieur des corps.*

From Fourier's 1805 draft of his work on the theory of heat propagation in Grattan-Guinness and Ravetz [1, pp. 185–186].

Als Fourier in einer seiner ersten Arbeiten über die Wärme, welche er der französischen Akademie vorlegte, (21. Dec. 1807) zuerst den Satz aussprach, dass eine ganz willkürlich (graphisch) gegebene Function sich durch eine trigonometrische Reihe ausdrücken lasse, war diese Behauptung dem greisen Lagrange so unerwartet, dass er ihr auf das Entschiedenste entgegentrat. Es soll sich hierüber noch ein Schriftstuck im Archiv der Pariser Akademie befinden. Dessenungeachtet verweist Poisson überall, wo er sich der trigonometrischen Reihen zur Darstellung willkürlicher Functionen bedient, auf eine Stelle in Lagrange's Arbeiten über die schwingenden Saiten, wo sich diese Darstellungsweise finden soll. Um diese Behauptung, die sich nur aus der bekannten Rivalität zwischen Fourier aund Poisson erklären lässt, zu widerlegen, sehen wir uns genöthigt, noch einmal auf die Abhandlung Lagrange's zurückzukommen; denn über jenen Vorgang in der Akademie findet sich nichts veröffentlicht.

... [Here Riemann discusses a formula in Lagrange's work on the vibrating string].

... Hätte Lagrange in dieser Formel n unendlich gross werden lassen, so wäre er allerdings zu dem Fourier'schen Resultat gelangt. Wenn man aber seine Abhandlung durchliest, so sieht man, dass er weit davon entfernt ist zu glauben, eine ganz wirkürliche Function lasse sich wirklich durch eine unendliche Sinusreihe darstellen.**

From Riemann [1, pp. 232–233].

* At first I arrived at several of these equations by very laborious eliminations, but I now use a much more general and expeditious rule to resolve any arbitrary function in a series of sines or cosines of [integral] multiples of angles. These results fully confirm the opinion of Daniel Bernoulli.

As with partial differential equations, the expansions in question can express the properties of entirely arbitrary and discontinuous functions. It is for that reason that they naturally present themselves for the integration of these last equations and their application offers singular facility for questions of fluid motion, sound propagation, vibration of elastic bodies, and gives an easy method for determining the motion in all the generality that can be obtained through the use of arbitrary functions. I have made a more particular application to the question of the propagation of heat and one thus succeeds in recognizing distinctly how it propagates by successive waves in the interior of a body.

** Fourier, in his first papers on heat submitted to the French Academy (December 21, 1807) was the first to formulate the theorem that an arbitrarily (graphically) given function could be represented by a trigonometric series. This statement was so unexpected to old Lagrange that he refuted it vehemently. It is claimed that the Archive of the Paris Academy contains such a document. Nevertheless, whenever Poisson uses trigonometric series for the representation of arbitrary functions he refers to a passage in Lagrange's papers on the vibrating string where such a representation is supposed to be found. In order to refute this statement which can only be explained by the well-known rivalry between Fourier and Poisson we have to come back to Lagrange's paper because nothing has been published about Lagrange's opposition in the Academy.

If Lagrange had considered these formulas for infinite n he would have obtained Fourier's result. However, when one goes through his paper one sees that the notion of representing an arbitrary function by an infinite trigonometric series is very foreign to Lagrange.

Historically the study of Fourier series of periodic functions preceded the study of Fourier integrals. We have reversed the order because it reverses itself naturally in later chapters of this book. Electrical engineers have sometimes argued that one should look solely at Fourier integrals, since nothing in nature is really periodic (see Bracewell [1, Introduction]). However, Fourier series are of great importance, both in pure and applied mathematics. They are so essential that people have designed machines to compute the coefficients of many specialized sorts of functions. In 1903, for example, Michelson had a machine which would find the Fourier coefficients of a sound using vertical rods, each vibrating at a different frequency. Today people use computers and the fast Fourier transform to analyze all sorts of phenomena, as we shall see later in this section. At this point, let us just quote an article on weather prediction that appeared in *Science* (Vol. 220 (1983), p. 40): "Fourier transform methods solved that problem [of making the computer time necessary for the calculations affordable] in the early 1970's, allowing [weather] forecasters to transform the grid mesh into a wave representation, perform the nonlinear calculations, and transform the results back to a grid with no substantial penalty in computing time."

Suppose that $f: \mathbb{R}^m/\mathbb{Z}^m \to \mathbb{C}$. This means that f has period one in each variable. We can also think of f as a function on the unit cube $[0, 1]^m$ in $\mathbb{R}^m$, with opposite sides identified. We say that the unit cube is a fundamental domain for $\mathbb{R}^m/\mathbb{Z}^m$.

Definition. The *Fourier series* of $f: \mathbb{R}^m/\mathbb{Z}^m \to \mathbb{C}$ is

$$\sum_{a \in \mathbb{Z}^m} (f, e_a) e_a(x),$$

where $e_a(x) = \exp(2\pi i\, {}^t a x)$, for $x \in \mathbb{R}^m$, $a \in \mathbb{Z}^m$, and

$$(f, e_a) = \int_{[0,1]^m} f(y)\overline{e_a(y)}\, dy.$$

Note that the functions $e_a(x)$, $a \in \mathbb{Z}^m$ are eigenfunctions of the Laplacian

$$\Delta = \frac{\partial^2}{\partial x_1^2} + \cdots + \frac{\partial^2}{\partial x_m^2}$$

which have period one in each variable. The first two statements of the theorem below say that nice functions $f(x)$ are represented by their Fourier series. This can be interpreted as the spectral resolution of Δ on $\mathbb{R}^m/\mathbb{Z}^m$. This parallels the view of Fourier integrals given by formulas (2) and (3) of §1.2. It is this point of view that we will seek to generalize in later chapters, where we replace $\mathbb{Z}^m$ by discrete groups of matrices.

Theorem 1 (Properties of Fourier Series).

(1) *Suppose that f is in $L^2(\mathbb{R}^m/\mathbb{Z}^m)$. Then f is the L^2-limit of the partial sums of its Fourier series. Moreover, we have the* **Parseval** *equality:*

1.3. Fourier Series and the Poisson Summation Formula

$$\sum_{a \in \mathbb{Z}^m} |(f, e_a)|^2 = \|f\|_2^2.$$

(2) *Suppose that $f \in L^1(\mathbb{R}^m/\mathbb{Z}^m)$ and that*

$$\sum_{a \in \mathbb{Z}^m} |(f, e_a)| < \infty.$$

Then you can change f on a set of measure 0 to make f continuous on $\mathbb{R}^m/\mathbb{Z}^m$ and such that

$$f(x) = \sum_{a \in \mathbb{Z}^m} (f, e_a) e_a(x).$$

(3) *Suppose that f has continuous partial derivatives of all orders less than or equal to k. If $k > m/2$, then the Fourier series of f converges uniformly and absolutely to f.*

(4) *Riemann-Lebesgue Lemma. Suppose that $f \in L^1(\mathbb{R}^m/\mathbb{Z}^m)$. Then (f, e_a) approaches zero, as $\|a\|$ approaches infinity.*

PROOF.

(1) For example, the Stone-Weierstrass theorem (see Lang [1, p. 148]) says that the exponentials $e_a(x)$, $a \in \mathbb{Z}^m$, form a complete orthornomal set in the Hilbert space $L^2(\mathbb{R}^m/\mathbb{Z}^m)$. You can also deduce this easily from Exercise 3 of §1.1. The machinery of Hilbert spaces then completes the proof easily.

(2) The difference of $f(x)$ and the sum of its Fourier series is a function with zero Fourier coefficients and is therefore orthogonal to all trigonometric polynomials and thus to all continuous functions. Since continuous functions are dense in L^1, the difference of f and its Fourier series must be zero except on a set of measure zero.

(3) Note that, if we use the notation (1) from §1.2, we have

$$(D^a f, e_b) = (2\pi i b)^a (f, e_b).$$

Then $D^a f \in L^2(\mathbb{R}^m/\mathbb{Z}^m)$ implies that

$$\sum_{|a|=k} \sum_{b \in \mathbb{Z}^m} |(f, e_b)|^2 [(2\pi b)^a]^2 < \infty.$$

Now there is a constant $c > 0$ such that

$$\sum_{|a|=k} [(2\pi b)^a]^2 \geq c \|b\|^{2k}.$$

So the Cauchy-Schwarz inequality enables us to compare the series of absolute values of Fourier coefficients and the series $\Sigma \|b\|^{-2k}$, which is an Epstein zeta function. We will see that this latter series converges for $k > n/2$ before the end of this section. See Theorem 5 of this section and Exercise 5 of §1.4.

(4) This is clear for a dense set of L^1 functions; e.g., functions in $L^1 \cap L^2$, or infinitely differentiable functions. □

It follows from part (3) of Theorem 1 that the speed of convergence of the Fourier series increases with the smoothness of the function. This sort of theorem goes back to Dirichlet in 1829 (see Grattan-Guinness and Ravetz [1, pp. 471–474]). Fourier had not quite managed to show that any piecewise-smooth function can be represented by its Fourier series, taking the average value at the jump discontinuities. For we can, in fact, allow the function $f(x)$ to have a jump discontinuity at x, setting

$$f(x+) = \lim_{w \to x,\, w > x} f(w) \quad \text{and} \quad f(x-) = \lim_{w \to x,\, w < x} f(x).$$

The convergence of Fourier series of piecewise-smooth functions is discussed in the next Exercise.

Exercise 1 (Jump Discontinuities and Fourier Series).

(a) Suppose that $f(x)$ is a piecewise-continuous function on $[0, 1)$. Assume that both one-sided derivatives of f exist at x in $[0, 1)$. Define the *Dirichlet kernel* $D_n(x)$ by

$$D_n(x) = \sum_{|k| \leq n} \exp(2\pi i k x) = \frac{\sin[\pi(2n+1)x]}{\sin(\pi x)}.$$

Then the nth partial sum of the Fourier series of f is

$$s_n f(x) = \sum_{|k| \leq n} (f, e_a) e_a(x) = (f * D_n)(x) = \int_0^1 f(y) D_n(x - y)\, dy.$$

Show that $(f * D_n)(x)$ approaches $\frac{1}{2}(f(x+) + f(x-))$ as n approaches infinity.

(b) Generalize the results of part (a) to functions of several variables. Assume that the function has sufficiently many continuous derivatives. Use the Dirichlet kernel:

$$D_a(x) = \prod_{j=1}^{m} D_{a_j}(x_j).$$

Then $f * D_a$ is the partial sum of the Fourier series of f corresponding to the sum over $b \in \mathbb{Z}^m$ with $|b_j| \leq a_j, j = 1, \ldots, m$. Show that $D_a * f$ approaches f as all the a_j approach infinity, assuming that $f(x)$ is sufficiently smooth.

Although we can allow the function $f: \mathbb{R}/\mathbb{Z} \to \mathbb{C}$ to have jump discontinuities, the partial sums of the Fourier series of such functions will always overshoot the mark by about 9% at the jumps. This is called the *Gibbs phenomenon* because it was pointed out by Gibbs in a letter to *Nature* in 1899. The letter was a reply to Michelson, who was angry that his machine for computing Fourier series failed badly at jumps. Actually the phenomenon was first observed by Wilbraham in 1848. There are many methods for improving the convergence of Fourier series and these are quite important for filter design (see Hamming [1]). The Gibbs phenomenon also occurs for Fourier integrals.

1.3. Fourier Series and the Poisson Summation Formula

Exercise 2 (The Gibbs Phenomenon and Smoothing in the One-Variable Case).

(a) Set
$$f(x) = \begin{cases} -1 & \text{for } -\tfrac{1}{2} \le x < 0 \\ +1 & \text{for } 0 \le x \le +\tfrac{1}{2}. \end{cases}$$

Show that for large n and small $|x|$, the partial sums $S_n f(x)$ of the Fourier series of f, defined in Exercise 1, will overshoot the mark by about 9%. This is the *Gibbs phenomenon*. Either graph the result by computer or follow the theoretical approach in Dym and McKean [1, p. 43].

(b) Show that it is possible to smooth out the horns of the Gibbs phenomenon in part (a) by averaging the first n partial sums (*Cesàro summation*). Show that, in fact,

$$K_n * f(x) = \frac{1}{n+1} \sum_{j=0}^{n} S_n f(x),$$

with K_n = the *Fejér kernel* in Example 1 of §1.1. Graph the result by computer showing that Fejér smoothing eliminates the oscillations of the partial sums but it has a slow "rise time."

(c) Graph the result of the more popular smoothing method called *Lanczos smoothing*, which is defined by

$$h_n f(x) = \frac{n}{2\pi} \int_{t-\pi/n}^{t+\pi/n} S_n f(u)\, du.$$

See Hamming [1] for more details on smoothing and filters. Your answers should resemble the graph of Fig. 1.3.
See Exercise 14 of Section 2.1 for some work on an analogue of the Gibbs phenomenon in a higher-dimensional setting.

Carleson [1] proves that the Fourier series of any L^2 function on $\mathbb{R}/\mathbb{Z}$ converges almost everywhere. For $\mathbb{R}^2/\mathbb{Z}^2$, the work of C. Fefferman and other, shows that the answers to the convergence question depend on the type of partial sums used; e.g., sums over squares, rectangles, circles (see J.M. Ash [1]). In fact, Ash notes: "At the present time, the passage from 2 to 3 dimensions seems far more substantial and nontrivial than that from 1 to 2 or than that from 3 to more."

There are examples of continuous functions of one variable whose Fourier series diverge at uncountably many points. And there is an L^1 function whose Fourier series diverges everywhere (see Kolmogorov [1]). References for such results can be found in the collection of articles edited by J.M. Ash mentioned in the preceding paragraph. See also the work of Zygmund [1]. For some of the older history of Fourier series, see Burkhardt [1], Grattan-Guinness and Ravetz [1], Hilb and Riesz [1], and Riemann [1, pp. 227–271]. Theorem 4 will show, however, that if all you seek is distributional convergence, then you always get what you want.

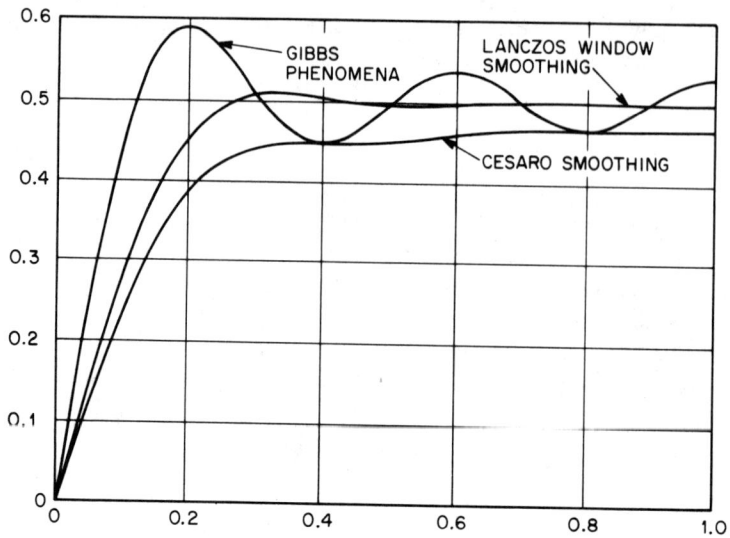

Figure 1.3. Comparison of various smoothing techniques. (From R.W. Hamming, *Digital Filters*, © 1977, p. 74. Reprinted by permission of Prentice-Hall, Inc., Englewood Cliffs, N.J.)

Exercise 3 (Weyl's Ergodic Theorem or Weyl's Criterion for Uniform Distribution). For any irrational number a and any continuous function $f(x)$ of period one, show that

$$\lim_{N\to\infty} \frac{1}{N} \sum_{n=1}^{N} f(na) = \int_0^1 f(x)\,dx.$$

Hint. First prove it for trigonometric polynomials.

Exercise 4 (The Shannon Sampling Theorem). This result was proved by Shannon in [1]. A function whose Fourier transform is zero outside of some interval is called *band limited*. We seek to reconstruct such a function from samples taken at equally spaced points. Suppose that $f: \mathbb{R} \to \mathbb{C}$ and $\hat{f}(y) = 0$ for all y with $|y| > y_c$. Show that

$$f(x) = \sum_{n \in \mathbb{Z}} f(n/(2y_c)) \frac{\sin[\pi(2y_c x + n)]}{\pi(2y_c x + n)}.$$

Hint. First expand $\hat{f}(y)$ in a Fourier series over the interval $[-y_c, +y_c]$:

$$\hat{f}(y) = \sum_{n \in \mathbb{Z}} A_n \exp(\pi i n y / y_c) \quad \text{for} \quad y \in [-y_c, +y_c].$$

Let $p_{y_c}(x)$ be 1 if $x \in [-y_c, +y_c]$ and 0 otherwise. Then, for any $y \in \mathbb{R}$,

$$\hat{f}(y) = p_{y_c}(y) \sum_{n \in \mathbb{Z}} A_n \exp(\pi i n y / y_c).$$

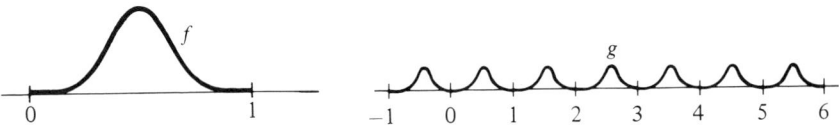

Figure 1.4. Periodization of a bump function.

Use the Fourier inversion theorem, the translation property of the Fourier transform, and the formula from Exercise 6 of §1.2 giving the Fourier transform of p_{y_c}, and you will find the solution of the problem.

Now we want to investigate the connection between Fourier series and Fourier integrals. This is the Poisson summation formula. Suppose that $f : \mathbb{R}^m \to \mathbb{C}$ is a Schwartz function. Form the periodic function (of period one in each variable),

$$g(x) = \sum_{a \in \mathbb{Z}^m} f(x + a). \tag{1.6}$$

Exercise 5. Show that the series (1.6) converges uniformly for all $x \in \mathbb{R}^m$, assuming that f is a Schwartz function.

When $m = 1$ and the support of f lies inside $(0, 1)$, the graphs of f and g may resemble those in Fig. 1.4. If the support of f is larger, the graph of g will be more complicated.

Theorem 2 (The Poisson Summation Formula). *If $f : \mathbb{R}^m \to \mathbb{C}$ is a Schwartz function, then*

$$g(x) = \sum_{a \in \mathbb{Z}^m} f(x + a) = \sum_{a \in \mathbb{Z}^m} \hat{f}(a) \exp(2\pi i\, {}^t a x).$$

PROOF. The Poisson summation formula simply says that the periodic function is represented by its Fourier series. For $\hat{f}(a)$ is really the ath Fourier coefficient of g, by the following calculation:

$$\begin{aligned}
\hat{f}(a) &= \int_{\mathbb{R}^m} f(y) \exp(-2\pi i\, {}^t a y)\, dy \\
&= \int_{[0,1]^m} \sum_{b \in \mathbb{Z}^m} f(y + b) \exp[-2\pi i\, {}^t a (y + b)]\, dy \\
&= \int_{[0,1]^m} g(y) \exp(-2\pi i\, {}^t a y)\, dy.
\end{aligned}$$

Thus Theorem 1 implies Theorem 2. □

In fact, Poisson is not the only name attached to Theorem 2, for Gauss and Cauchy also found the formula (see Burkhardt [1, p. 1338]). However Jacobi, who needed the result in his work on theta functions, attributes it to Poisson (see Jacobi [1, p. 307]).

The hypotheses on the function $f(x)$ in Theorem 2 should be weakened (see Stein and Weiss [1, pp. 250–257]). There is also an interesting discussion of the Poisson summation formula in Feller [1, Vol. II, p. 632]. Exercise 19 is a cautionary example on the failure of Poisson summation. See Hejhal [5] for a discussion of a related summation formula due to Voronoi.

It would be useful to be able to derive these results on Fourier series directly from the inversion formula for the Fourier transform in §1.1. Bracewell [1, pp. 204ff] claims to do this, but there seems to be a gap in his argument. Exercise 20 of §1.2 gives a way of deriving the representation of certain functions by Fourier series, using the inversion of Laplace transforms plus Cauchy's integral theorem (see also Titchmarsh [1, pp. 4–6]). There is also a short distribution-theoretic proof of the Poisson sum formula which we will describe after Theorem 4. We mention this because it would be nice to be able to create similar arguments in later chapters. There is another distribution-theoretic discussion of the formula in Lighthill [1, pp. 67–68].

On the other hand, it is possible to derive Fourier integrals from Fourier series of functions of period P, by letting P approach infinity. This is done by Titchmarsh in [1, pp. 70–73], for example. To sketch the formal argument, suppose that $f(x)$ is a Schwartz function on $\mathbb{R}$. Let

$$f_P(x) = \sum_{n \in \mathbb{Z}} f(x + nP).$$

The Fourier series of this function f_P of period P is easily seen to be

$$f_P(x) = \frac{1}{P} \sum_{k \in \mathbb{Z}} \left(\int_{-P/2}^{P/2} f(y) \exp(-2\pi i k y/P) \, dy \right) \exp(2\pi i k x/P)$$

$$= \frac{1}{P} \sum_{k \in \mathbb{Z}} \int_{-P/2}^{+P/2} f(y) \exp[2\pi i (x - y) k / P] \, dy.$$

Then argue that, as P approaches infinity, this last sum approaches the Riemann sum for

$$\int_{w \in \mathbb{R}} \int_{y \in \mathbb{R}} f(y) \exp[2\pi i (x - y) w] \, dy \, dw.$$

A simpler way of comparing Fourier series with Fourier integrals comes from comparing the kernels that result from approximating the Fourier series by its nth partial sum in Exercise 1 and approximating the Fourier integral by that on a finite interval $[-r, r]$, as in Exercise 7 of §1.2. These kernels are

$$D_n^P(x) = \frac{\sin[\pi(2n+1)x/P]}{P \sin(\pi x/P)}$$

and

$$W_r(x) = \frac{\sin(2\pi r x)}{\pi x}.$$

1.3. Fourier Series and the Poisson Summation Formula

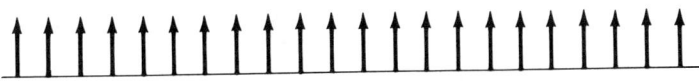

Figure 1.5. The Dirac comb or Shah functional.

Now

$$D_n^P(x) = \frac{\sin[\pi(2n)x/P]\cos(\pi x/P)}{P\sin(\pi x/P)} + \frac{\cos(\pi 2nx/P)\sin(\pi x/P)}{P\sin(\pi x/P)},$$

which approaches $W_{n/P}(x)$ as P approaches infinity (for fixed x).

Next consider Fourier series of periodic distributions. The space of *test functions* $\tilde{\mathcal{D}}$ consists of all infinitely differentiable functions on $\mathbb{R}^m/\mathbb{Z}^m$, which should be thought of as a product of circles (or torus). Convergence of a sequence of test functions means uniform convergence of all derivatives. A *distribution* T on $\mathbb{R}^m/\mathbb{Z}^m$ is a continuous linear function $T : \tilde{\mathcal{D}} \to \mathbb{C}$. It is possible to identify T with a periodic distribution $\tilde{T}$ on $\mathbb{R}^m$ with $\tilde{T}_a = \tilde{T}$ for all a in $\mathbb{Z}^m$, using the definition of $\tilde{T}_a$ in Theorem 1(g) of §1.1. To see this, suppose that g is a test function on $\mathbb{R}^m$ and set $\tilde{g} =$ the $\mathbb{Z}^m$-*periodization of* g:

$$\tilde{g}(x) = \sum_{a \in \mathbb{Z}^m} g(x+a).$$

Then if T is a distribution on $\mathbb{R}^m/\mathbb{Z}^m$, define the periodic distribution $\tilde{T}$ of period 1 on $\mathbb{R}^m$ by

$$T(\tilde{g}) = \tilde{T}(g).$$

For example, if δ is the Dirac delta distribution then

$$\tilde{\delta} = \sum_{n \in \mathbb{Z}^m} \delta_a \tag{1.7}$$

Some engineers call the distribution $\tilde{\delta}$ on $\mathbb{R}^m$ the *shah functional*, for the Russian letter Ш. Note that if f and g are related by formula (1.6), then $\tilde{\delta}f =$ Ш$f = g$. One can think of $\tilde{\delta}$ as an infinite "Dirac comb" of impulses pictured in Fig. 1.5 (see the article of Sakai in Vanasse [1, p. 7]). The Fourier transform of δ_a is the function $\exp(2\pi i\,^t ax)$. Thus the Fourier transform of the shah functional is

$$\hat{Ш}(x) = \sum_{a \in \mathbb{Z}^m} \exp(-2\pi i\,^t ax). \tag{1.8}$$

The Poisson summation formula is exactly the statement that the shah functional is its own Fourier transform; i.e., that Ш $=$ $\hat{Ш}$.

Theorem 3 (A Criterion for Distributional Convergence of a Trigonometric Series). *A trigonometric series*

$$\sum_{n \in \mathbb{Z}^m} c_n \exp(2\pi i\,^t nx)$$

converges in the distributional sense (see §1.1) if there are positive constants C, a such that

$$|c_n| \leq C \|n\|^a \quad \text{for all } n \in \mathbb{Z}^m.$$

PROOF. Set

$$f(x) = \sum_{n \in \mathbb{Z}^m - 0} \|2\pi i n\|^{-2k} c_n \exp(2\pi i\,^t n x).$$

This series converges uniformly to a continuous function $f(x)$ provided that k is sufficiently large (by the proof of Theorem 1). Now apply the Laplace operator k times to obtain $\sum c_n \exp(2\pi i\,^t n x)$, which must therefore converge as a distribution. □

Bochner [2] notes that Riemann had already considered this sort of convergence long before Schwartz' book [2].

Definition. Let T be a distribution on $\mathbb{R}^m/\mathbb{Z}^m$. The *Fourier series* of T is $\sum c_n(T) \exp(2\pi i\,^t n x)$, with $c_n(T) = \langle T(x), \exp(-2\pi i\,^t n x) \rangle$.

Now we can lay to rest all worries about convergence of Fourier series in the sense of distributions.

Theorem 4. *The Fourier series of any distribution T on $\mathbb{R}^m/\mathbb{Z}^m$ converges to T.*

PROOF. First note that the nth Fourier coefficient of δ on $\mathbb{R}^m/\mathbb{Z}^m$ is 1, since $\delta(e_n) = 1$. And the Fourier series of δ converges to δ by Theorem 1 or Theorem 2. For this just says that the Fourier series of any test function converges to the function when everything is evaluated at $x = 0$.

Now let T be any distribution on $\mathbb{R}^m/\mathbb{Z}^m$. Then

$$T = T * \delta = T * \left(\sum_{n \in \mathbb{Z}^m} e_n \right) = \sum_{n \in \mathbb{Z}^m} (T * e_n).$$

Here all convolutions are over $\mathbb{R}^m/\mathbb{Z}^m$. Finally

$$(T * e_n)(x) = \langle T(y), \exp(2\pi i\,^t n(x - y)) \rangle = \exp(2\pi i\,^t n x) \langle T(y), \exp(-2\pi i\,^t n y) \rangle.$$

This completes the proof of Theorem 4. □

Now let us discuss a distribution-theoretic proof of the Poisson sum formula for $\mathbb{R}/\mathbb{Z}$ from Donoghue [1, p. 162] (cf. also Friedlander [1, p. 104]). Let T be a periodic tempered distribution. Then, since $T_a = T$ for all $a \in \mathbb{Z}$, we have

$$(e^{2\pi i a s} - 1) \hat{T}(s) = 0.$$

Thus $\hat{T}(s)$ must be supported on the zeros of $e^{2\pi i a x} - 1$; i.e., on $\mathbb{Z}$. Then, using the first general fact about distributions that was stated in §1.1, it follows that $\hat{T}$ must be a sum over a of linear combinations of derivatives of δ_a, $a \in \mathbb{Z}$. Now

suppose that $T = \text{Ш}$, the shah functional. Then the fact that shah is its own Fourier transform follows fairly easily.

Next we consider some applications of Poisson's summation formula. The first exercise can also be viewed as a example of the method of images (see Sommerfeld [1, pp. 72–74] and Exercise 20). We will see many applications of this result throughout these notes.

Exercise 6 (Jacobi's Transformation Formula for the Theta Function). Show that for $x \in \mathbb{R}^m$ and $t > 0$ the following equation is a consequence of Theorem 2:

$$\sum_{a \in \mathbb{Z}^m} \exp[-2\pi i\,{}^t a x - 4\pi^2 c\|a\|^2 t] = (4\pi c t)^{-m/2} \sum_{a \in \mathbb{Z}^m} \exp[-\|a + x\|^2/(4ct)].$$

In the special case $m = 1$, $x = 0$, the theta function of Exercise 6 is the partition function for the planar rigid rotator in quantum statistical mechanics. And it approaches the high-temperature limit as $t \to 0+$, a limit that can be computed from the transformation formula in Exercise 6 (see Hurt [1, p. 10]).

Exercise 7 (Heat Diffusion on a Circle). Consider the problem

$$\begin{cases} \dfrac{\partial u}{\partial t} = c\dfrac{\partial^2 u}{\partial t^2} & x \in \mathbb{R}/\mathbb{Z},\ t > 0, \\ u(x, 0) = f(x) & x \in \mathbb{R}/\mathbb{Z}. \end{cases}$$

Claim. The solution is the same as that of Example 1 in §1.2. The exercise is to start out writing the solution as a Fourier series in x. The Fourier coefficients will be functions of t satisfying an ordinary differential equation that is easily solved. This allows you to write the solution $u(x, t)$ as a convolution over $[0, 1]$ of f with the theta function in Exercise 6. Then use that last exercise to see that the solution is really the same as that of Example 1 in §1.2.

Exercise 8. Derive some more results from the Poisson summation formula, such as the following result for $t > 0$:

$$\sum_{n \in \mathbb{Z}} \exp(-|n|t) = \frac{2}{t} \sum_{n \in \mathbb{Z}} (1 + (2\pi n/t)^2)^{-1}.$$

Justify the formula, even though $\exp(-|x|)$ is not a Schwartz function.

A more general result of this type involves (for $t > 0$)

$$\sum_{n \in \mathbb{Z}} (t + n^2)^{-s}$$

when $\operatorname{Re} s > \frac{1}{2}$. The K-Bessel function will enter into the last formula. And this result will appear in Chapter 3 as the Fourier expansion of Epstein's zeta function.

Theorem 5 (Asymptotics of the Eigenvalues of Δ on $L^2(\mathbb{R}^m/\mathbb{Z}^m)$). *Let $N(x)$ be the number of eigenvalues λ of Δ such that $|\lambda| \leq x$ and $\Delta h = \lambda h$ for $h \neq 0$ and $h \in L^2(\mathbb{R}^m/\mathbb{Z}^m)$. The eigenvalues are counted with multiplicity. Then*

$$N(x) \sim (4\pi)^{-m/2}\Gamma(1 + m/2)^{-1}x^{m/2} \qquad \text{as } x \to \infty.$$

PROOF. Clearly a complete orthonormal set of eigenfunctions of Δ on $\mathbb{R}^m/\mathbb{Z}^m$ consists of the exponentials $e_a(x) = \exp(2\pi i\, {}^t a x)$, $a \in \mathbb{Z}^m$. Now

$$\Delta e_a = -4\pi^2 \|a\|^2 e_a.$$

It follows that

$$N(x) = \#\{a \in \mathbb{Z}^m |\ 4\pi^2 \|a\|^2 \leq x\}.$$

When $m = 2$ this is the number of lattice points inside of a circle of radius $\sqrt{x}/(2\pi)$. One could give a simple geometrical argument to prove Theorem 3 (cf. Hardy and Wright [1, pp. 270–272] or Courant and Hilbert [1, Vol. I, p. 430]). But we choose instead to use the Tauberian theorem which was Theorem 5 of §1.2.

So we look at the Laplace transform of $N(x)$. This Stieltjes integral is really the sum

$$\sum_{a \in \mathbb{Z}^m} \exp[-4\pi^2 \|a\|^2 t] = (4\pi t)^{-m/2} \sum_{n \in \mathbb{Z}^m} \exp[-\|a\|^2/(4t)].$$

Here we have used Exercise 6, which is a simple application of the Poisson summation formula. The right side of the last equality is asymptotic to $(4\pi t)^{-m/2}$ as $t \to 0+$. Thus the Tauberian Theorem (Theorem 5 of §1.2) finishes the proof. □

Finding the order of magnitude of the error term in the asymptotic expression of Theorem 5 is called the circle problem in number theory (when $m = 2$). It appears to be quite difficult. A history of results on the subject can be found in Hejhal [2, p. 450].

Theorem 5 is a special case of Weyl's result on the asymptotic distribution of the eigenvalues of the Laplacian on a compact domain (see Weyl [1, Vol. I, pp. 393–430; Vol. IV, pp. 432–456]).

Next we want to consider an example of Fourier analysis in the real world of smog, earthquakes, etc.

Application. Spectroscopy and the Search for Hidden Periodicities

Spectroscopy was created by Michelson and Morley in 1887 to develop a length standard based on the wavelength of an emission line of an element. In 1892 the wavelength of the red line of cadmium became the international length standard. This lasted until 1960 when the orange line of krypton-86 replaced the red line of cadmium. And Michelson was awarded a Nobel prize in 1907 for this work, which has had an enormous variety of applications— from the identification of air pollutants in Los Angeles smog to the components of the Venusian atmosphere. Fourier analysis was always involved in the theory of spectroscopy. And the theory of spectroscopy has certainly been one of the motivating factors leading to interesting developments in harmonic analysis (see Wiener [2, pp. 119–260]). But Fourier transform spectrometers were not built until the 1970s because they require fast computers and the fast Fourier transform of Cooley and Tukey [1] (see Brigham [1]). The graph of observations of spectra of Venus (Fig. 1.6) from the article of Bell in Vanasse [1, p. 138] shows that Fourier transform spectroscopy has greatly improved resolving power (from 8 cm^{-1} in 1962 to 0.05 cm^{-1} in 1973).

Here we shall only be able to give a brief glimpse into the fascinating theory of spectroscopy. The interested reader could consult any of the following references for more details: Bousquet [1], Griffiths [1], Harris and Bertolucci [1], Vanasse [1]. Some of the background in crystallography will be discussed in §1.4. There are also many connections with time-series analysis as Wiener notes in [2, pp. 119–260]. More information on time-series analysis can be found in Bloomfield [1], Jenkins and Watts [1], Kanasewich [1], and Mackey [2, §18], for example. Time-series analysis studies phenomena $f(t)$ which are neither periodic nor decaying as t approaches infinity. Thus classical Fourier analysis does not apply, although the theory of distributions does allow one to treat such functions. For example, given the variable star data below from Bloomfield [1], one asks: What are the hidden periodicities? This question goes back to Schuster [1]. We will consider this question briefly after discussing spectroscopy.

Now let us review the basic ideas of spectroscopy. This is the study of the interaction of electromagnetic radiation and matter. Quantum mechanics says that radiation emitted at frequency v can move an atom from energy level E_1 to E_2 according to Planck's law: $E_2 - E_1 = hv$. Here $v = c/\lambda_v$, where c = the speed of light, λ_v = wavelength. Since c is known to less accuracy, spectroscopists use $\sigma = 1/\lambda_v$ = the *wave number*, and not v. Figure 1.7 from Harris and Bertolucci [1, p. 2] gives an idea of the wave numbers of various sorts of radiation.

In order to understand what is going on with spectral lines, you really need to know some quantum mechanics, chemistry, crystallography, and group representations. We will touch on this in §1.4 and Chapter 2.

A Fourier spectrometer uses a Michelson interferometer to divide the beam from some source into two separate beams of equal strength. After the two beams travel different paths, they are recombined. A signal is obtained which ultimately becomes the *interferogram function*

$$F(x) = \int_0^\infty B(\sigma) \cos(2\pi\sigma x) \, d\sigma,$$

where $B(\sigma)$ is the *source spectral density* at wave number σ. Then Fourier

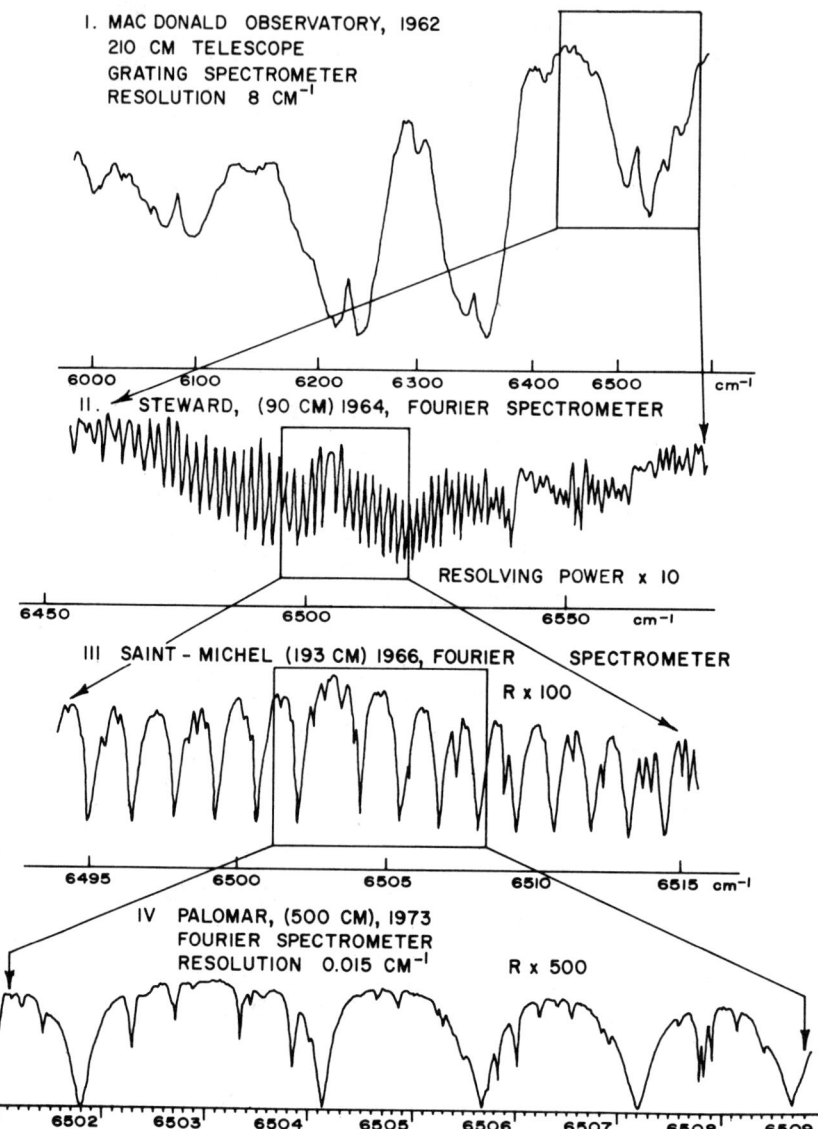

Figure 1.6. *Improvements in the near-IR Venus spectrum due to Fourier spectroscopy*: same type detectors (cooled PbS) with almost the same NEP used throughout. Curve I by Kuiper (1962); Curve II from Connes and Connes (1966); Curve III from Larson and Fink (1975); and Curve IV from Connes and Michel (1975). Four strong CO_2 Venusian bands are shown in I; the rotational structure is resolved in II; III shows lines from much weaker overlapping banks; IV gives a good approximation of the true line profile, together with ever fainter line. Trace IV presents approximately 1/800th of the actual spectral range available from the magnetic tape—general-purpose computer output (parts of which are obscured by H_2O) (after Connes and Michel, 1975). (From Bell's article in Vanasse [1, p. 138]. Reprinted by permission of Academic Press.)

1.3. Fourier Series and the Poisson Summation Formula

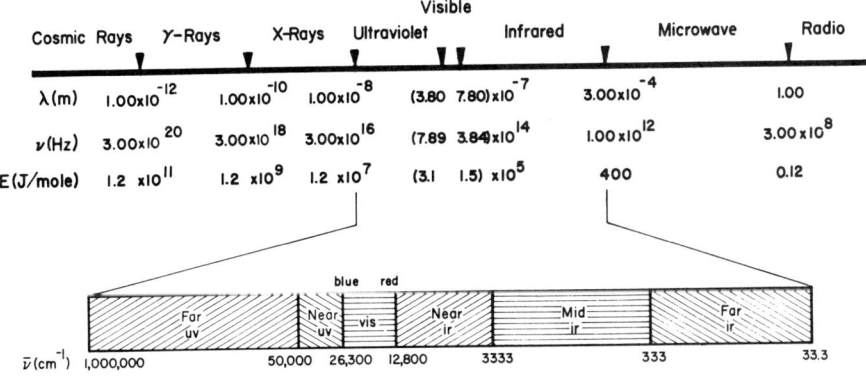

Figure 1.7. The electromagnetic spectrum. The wavelength λ, is given in units of meters; the frequency, ν, is given in units of hertz (1 Hz = 1 oscillation per second); and the energy, E, carried by a *mole* of photons is given in joules (4.184 J = 1 calorie). The wave number, $\bar{\nu}$, is expressed in units of cm^{-1} (read "reciprocal centimeters" or "wave numbers"). (From Harris and Bertolucci [1, p. 2]. Reprinted by permission of Oxford University Press.)

inversion gives the *equation of spectral recovery*:

$$B(\sigma) = 4 \int_0^\infty F(x) \cos(2\pi\sigma x) \, dx.$$

We view $F(x)$ as an even function on the real line here. Since all functions actually live on finite intervals, we have to look at

$$F_X = F \cdot \chi_{[-X,X]},$$

where

$$\chi_{[-X,X]}(x) = \begin{cases} 1 & \text{if } |x| \le X \\ 0 & \text{if } |x| > X. \end{cases}$$

Then the real source spectral density (using the convolution theorem for the Fourier transform and Exercise 12 of §1.2) is

$$B_X(\sigma) = B(\sigma) * (\sin(2\pi\sigma X)/\pi\sigma).$$

So spectroscopists call $\sin(2\pi\sigma X)/\pi\sigma$ the *instrument function*. To avoid the Gibbs phenomenon, one may modify the instrument function using methods analogous to those mentioned in Exercise 2. The spectroscopy literature usually does this by replacing $\chi_{[-X,X]}(x)$ by $\chi_{[-X,X]}(x)(1-|x|/X)$. This replaces B_X by $B * (\sin(\pi\sigma X)/\pi\sigma X)^2$. Physically this "apodization" can be obtained by putting on suitable apertures.

In fact, one can only sample F to reconstruct B and thus Shannon sampling (Exercise 4) enters the picture. Suppose that the samples of F are taken at nd, $n = 0, 1, 2, 3, \ldots$. Then if $B(\sigma) = \hat{F}(\sigma)$ vanishes for $|\sigma| > \sigma_m$, we need $d = 1/2\sigma_m$

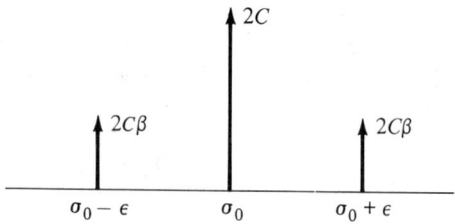

Figure 1.8. Satellites caused by sampling errors.

from the sampling theorem. One says that a sampling distance $d \geq 1/2\sigma_m$ makes the spectrum free of aliasing.

Periodic sampling errors can lead to "satellites" near strong spectral lines, as in Fig. 1.8. Such errors come from defects in the equipment. Let us consider a theoretical explanation.

Figure 1.8 means $B(\sigma) = 2C\delta(\sigma - \sigma_0)$, with C a constant. Then $F(x) = C\cos(2\pi\sigma_0 x)$. Suppose that F is sampled with periodic errors. Then x would be replaced by x' with $x' = x + \beta\cos(2\pi\varepsilon x)$. If β is small then $F(x') \cong C\{\cos(2\pi\sigma_0 x) - \pi\beta\sigma_0 \sin[2\pi(\sigma_0 + \varepsilon)x] - \pi\beta\sigma_0 \sin[2\pi(\sigma_0 - \varepsilon)x]\}$. This gives two satellites pictured as in Fig. 1.8.

Finally let us make a few connections with time-series analysis. The remarks go back to Wiener [2, pp. 119–260]. If $E(t)$ is the electric field and δ the path difference of the interfering beams in the spectrometer, then the intensity or mean power is, for $\tau = \delta/c$,

$$a_E(\tau) = \lim_{T\to\infty} \frac{1}{2T} \int_{-T}^{+T} \overline{E(t)} E(t+\tau) \, dt.$$

This expression is called the *autocovariance* or *autocorrelation* of E and plays a large role in time-series analysis. Strictly speaking, it should be normalized to $a(\tau)/a(0)$. It is the Fourier transform of the power spectrum by the *Wiener-Khintchine formula*. To see this formally, suppose $\chi_{[-T,T]}$ is as above. Then set $E_T = E \cdot \chi_{[-T,T]}$. Then (under appropriate hypotheses on E)

$$a_E(\tau) = \lim_{T\to\infty} \frac{1}{2T} \int_{-T}^{+T} \overline{E(t)} E(t+\tau) \, dt$$

$$= \lim_{T\to\infty} \frac{1}{2T} \int_{-T}^{+T} \overline{E_T(t)} \int_{-\infty}^{+\infty} \hat{E}_T(w) \exp[2\pi i w(t+\tau)] \, dw \, dt.$$

$$= \lim_{T\to\infty} \frac{1}{2T} \int_{-\infty}^{+\infty} \exp(2\pi i w \tau) |\hat{E}_T(w)|^2 \, dw.$$

This is the desired formula since *the power spectrum* is

$$P(w) = \lim_{T\to\infty} |\hat{E}_T(w)|^2 / 2T.$$

1.3. Fourier Series and the Poisson Summation Formula

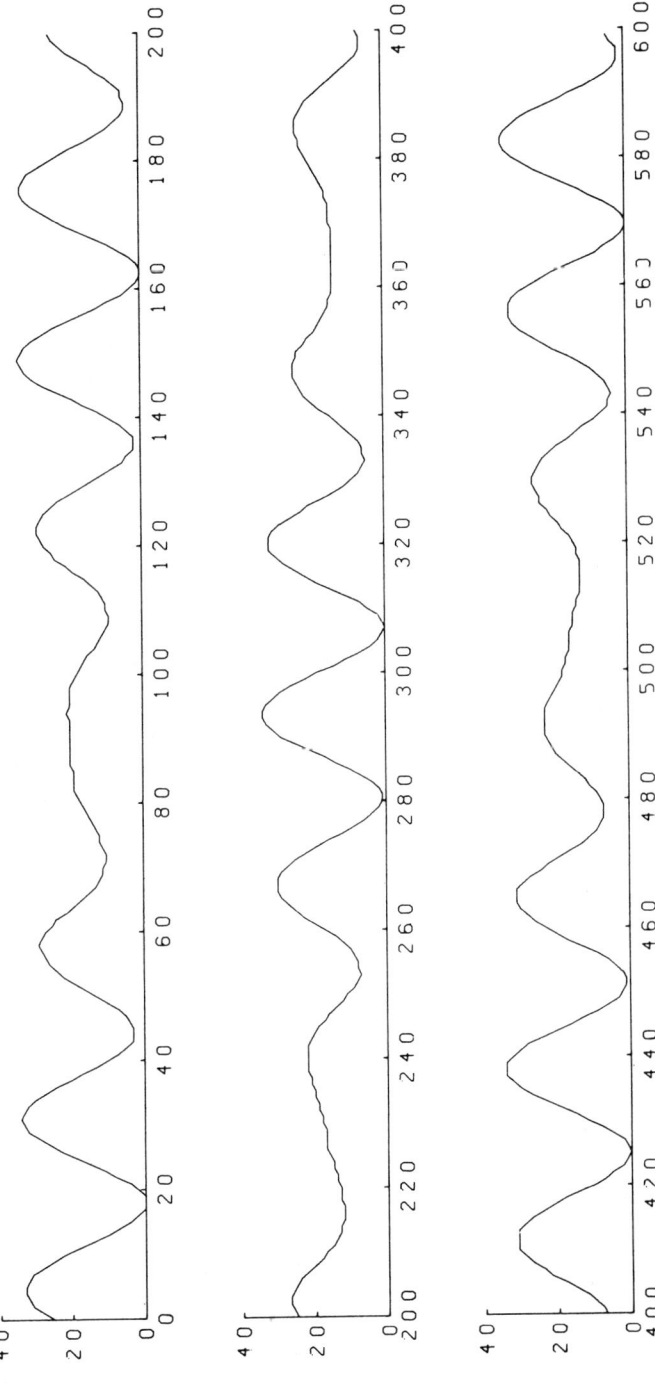

Figure 1.9. Magnitude of a variable star at midnight on 600 successive nights. (From Bloomfield [1, p. 3]. Reprinted by permission of John Wiley & Sons.)

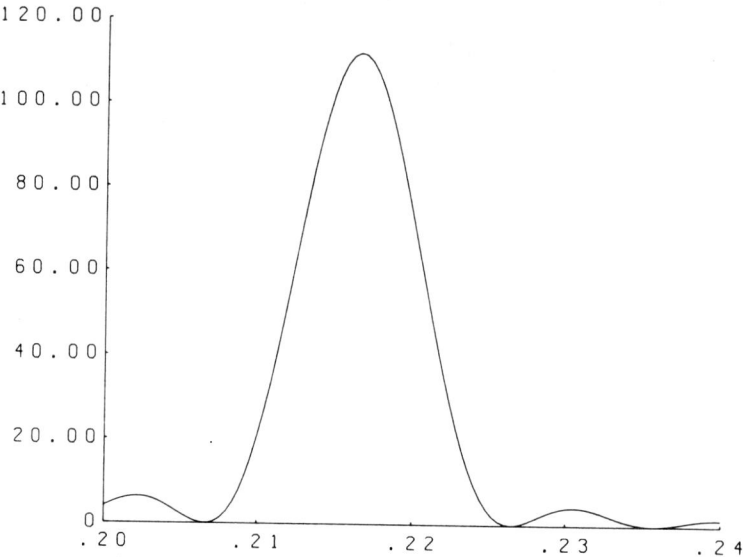

Figure 1.10. Periodogram of the variable-star data for frequencies ω, $0.20 < \omega < 0.24$. (From Bloomfield [1, p. 19]. Reprinted by permission of John Wiley & Sons.)

But what does all this have to do with the hidden periodicities in the variable star data of Fig. 1.9 from Bloomfield [1, p. 3]? Bloomfield [1, Ch. 2] explains this quite well. The most natural method to try in searching for preciodicity is to model the 600 data values x_t, for $t = 0, 1, \ldots, 599$ by $(\mu + A\cos\omega t + B\sin\omega t)$ or by a sume of such terms. You could, for example, guess ω, and then choose μ, A, B to minimize

$$S(\mu, A, B) = \sum_{t=0}^{599} (x_t - \mu - A\cos\omega t - B\sin\omega t)^2.$$

This is the *method of least squares*. If you count the 21 peaks in the data, you could guess the period to be $600/21 \cong 28.6$. This leads you to take $\omega \cong 2\pi/28.6 \cong 0.220$. The method of least squares then gives, if you ignore certain terms,

$$\mu = \bar{x} = \frac{1}{n}(x_0 + \cdots + x_{n-1}), \quad n = 599,$$

$$A(\omega) = \frac{2}{n}\sum(x_t - \bar{x})\cos\omega t,$$

$$B(\omega) = \frac{2}{n}\sum(x_t - \bar{x})\sin\omega t.$$

So we are approximately looking for the discrete Fourier transform.

1.3. Fourier Series and the Poisson Summation Formula

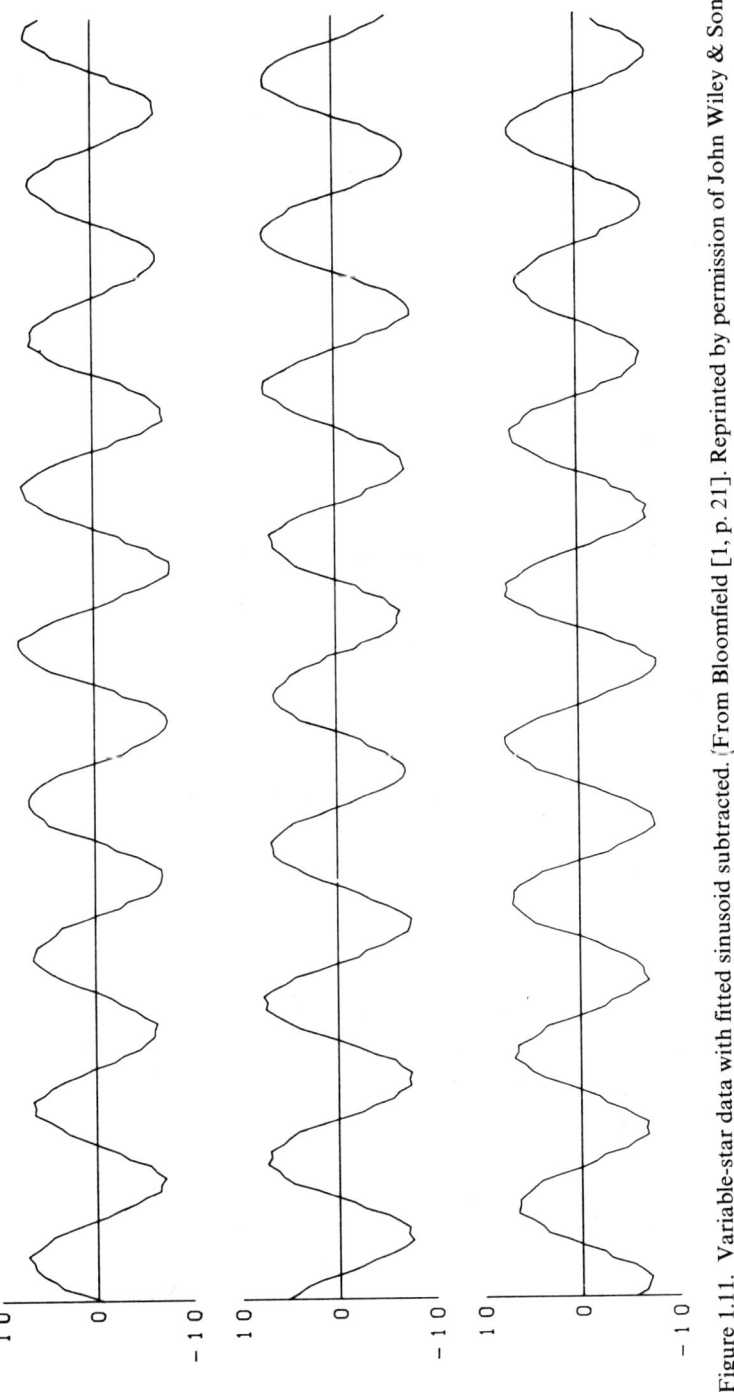

Figure 1.11. Variable-star data with fitted sinusoid subtracted. (From Bloomfield [1, p. 21]. Reprinted by permission of John Wiley & Sons.)

In order to obtain further information on the periods, one forms the *periodogram* $n(A^2 + B^2)/2$. The graph of this function is shown in Fig. 1.10 (from Bloomfield [1, p. 19]). The maximum occurs at $\omega \cong 0.21644$. If one then does a least squares analysis using this value of ω, one obtains the graph of Fig. 1.11 for $x_t - (\mu + A\cos\omega t + B\sin\omega t)$ (from Bloomfield [1, p. 21]). The graph has an obvious period of 24 days. The new ω is thus about 0.262. Thus one concludes that the original variable star data had two periodic components.

It should be noted that the Shannon sampling theorem affects the choice of ω. If the sampling interval is $d = 1$ day, then $0 \le \omega \le \pi/d$. Every frequency not in the range has an alias in the range.

In Chapter 5 of Bloomfield [1] filtering methods are used to show that x_{300} in the variable star data is in error, with $x_{300} = 18$ and not 19. This rather impressive deduction does not appear to be in the realm of possibility for other examples considered by Bloomfield; e.g., European wheat prices for 1500–1869.

Poisson's Sum Formula as a Trace Formula

Our next goal is to obtain an interpretation of Poisson's summation formula in terms of traces of integral operators. In order to do this we shall briefly sketch the theory of integral operators on $L^2(D)$, where D is a measurable subset of $\mathbb{R}^m$. References for this subject include Courant and Hilbert [1], Dieudonné [1], Lang [2], Maurin [2], Stakgold [1, 2], and Yosida [1].

Definition. Let $K \in L^2(D \times D)$. Then the integral operator L_K defined by

$$L_K f(x) = \int_D K(x,y) f(y) \, dy \tag{1.9}$$

is called a *Hilbert-Schmidt operator with kernel K*. Actually, more general Hilbert-Schmidt operators are considered in the references.

Clearly L_K maps functions f in $L^2(D)$ to functions $L_K f$ in $L^2(D)$. And just as clearly, the map L_K is linear. So we call it an "operator," which just means linear map.

Definition. An operator $L: H \to H$, on a Hilbert space H, is said to be *self-adjoint* if $(Lf, g) = (f, Lg)$, for all f, g in H. Here (f, g) denotes the inner product in the Hilbert space H.

Note that the Hilbert-Schmidt operator (1.9) is self-adjoint if and only if

$$K(x, y) = \overline{K(y, x)} \quad \text{for almost all } x, y \text{ in } D. \tag{1.10}$$

Definition. An operator $L: H \to H$, on a separable Hilbert space H, is *compact* if for every bounded sequence x_n in H, the sequence Lx_n has a convergent subsequence.

1.3. Fourier Series and the Poisson Summation Formula

Theorem 6. *A Hilbert-Schmidt operator on a compact domain is a compact operator.*

For a proof, see Yosida [1, pp. 277–278] or any of the references above some of which consider only special cases.

Exercise 9 (Examples of Kernels).

(a) Let $D = (0, 1)$ and $K(x, y) = |x - y|^{-a}$, $0 < a < \frac{1}{2}$. Show that this is the kernel of a Hilbert-Schmidt operator on $L^2(D)$. And show that, if instead $\frac{1}{2} \le a < 1$, then the operator is no longer Hilbert-Schmidt, although it is compact.

(b) Show that the kernel $\exp(-x^2 - y^2)$ generates a Hilbert-Schmidt operator on $L^2(\mathbb{R})$, while the kernel $\exp(-|x - y|)$ generates an operator that is not even compact although it is bounded on $L^2(\mathbb{R})$.

Hint. See Stakgold [1, pp. 353–354].

Theorem 7 (The Spectral Theorem for Compact Self-Adjoint Operators). *Suppose that $L: H \to H$ is a compact self-adjoint operator on a separable Hilbert space H. Then H has a complete orthonormal set of eigenvectors $\{v_n\}$ of L with $Lv_n = \lambda_n v_n$, $\lambda_n \in \mathbb{R}$, $n = 1, 2, 3, \ldots$.*

Here we explicitly allow 0 to be an eigenvalue. And every vector v in H has a generalized Fourier series representation

$$v = \sum_{n \ge 1} (v, v_n) v_n.$$

For a proof of the spectral theorem see the references above. The idea is that one can obtain the eigenvalues of L by finding maxima of the quadratic form (Lv, v) for v in H with $\|v\| = 1$. This method can actually be put on a computer, as the Rayleigh-Ritz and finite element methods. See Chapter 3 for an example of the finite element method in practice, or Strang and Fix [1] for a more complete story of the theory.

Exercise 10 (Expansion of the Kernel of a Self-Adjoint Hilbert-Schmidt Operator). Suppose that $K(x, y)$ is the kernel of a self-adjoint Hilbert-Schmidt operator L_K given by (1.9). Let $\{v_n\}$ be a complete orthonormal set of eigenvectors of L_K as provided by Theorem 2, with $L_K v_n = \lambda_n v_n$. Show that the series

$$\sum \lambda_n v_n(x) \overline{v_n(y)}$$

converges to $K(x, y)$ in the L^2 norm.

If the trace of the self-adjoint integral operator L_K in (1.9) has any meaning, it must be the infinite sum $\Sigma \lambda_n$ of all the eigenvalues λ_n of L_K. Let us suppose that the operator L_K is *positive*; i.e., that $(L_K f, f)$ is positive for all $f \ne 0$ in H.

This means that all the eigenvalues of L_K are positive. Then we have the following theorem.

Theorem 8 (Mercer's Theorem). *Suppose that L_K is a positive self-adjoint Hilbert-Schmidt operator (1.9) with a continuous kernel K on a compact set in $\mathbb{R}^m$. Then*

$$K(x, y) = \sum \lambda_n v_n(x) \overline{v_n(y)},$$

where the v_n and λ_n are as in Theorem 7, and the convergence of the series is absolute and uniform.

This is proved in the references.

Exercise 11 (The Trace of a Positive Self-Adjoint Hilbert-Schmidt Operator). Suppose that the Hilbert-Schmidt operator L_K satisfies the hypotheses of Theorem 8. Show that

$$\operatorname{Trace} L_K = \sum_{n \geq 1} \lambda_n = \int_D K(x, x)\, dx.$$

We want to apply this last exercise to Poisson's summation formula. Suppose that $f : \mathbb{R}^m \to \mathbb{C}$ is a Schwartz function. Define an integral operator on functions f in $L^2(\mathbb{R}^m/\mathbb{Z}^m)$ by

$$L_f g(x) = (f * g)(x) = \int_{\mathbb{R}^m} f(x - y) g(y)\, dy. \tag{1.11}$$

Then the functions $e_n(x) = \exp(2\pi i {}^t n x)$, $n \in \mathbb{Z}^m$, give a complete orthonormal set of eigenfunctions of L_f on $L^2(\mathbb{R}^m/\mathbb{Z}^m)$. To see this, note that

$$\begin{aligned} L_f e_n(x) = (f * e_n)(x) &= \int_{\mathbb{R}^m} f(x - y) \exp(2\pi i {}^t n y)\, dy \\ &= \int_{\mathbb{R}^m} f(v) \exp(2\pi i {}^t n (x - v))\, dv = \hat{f}(n) e_n(x). \end{aligned} \tag{1.12}$$

On the other hand we have

$$L_f g(x) = \int_{\mathbb{R}^m/\mathbb{Z}^m} K_f(x, y) g(y)\, dy,$$

with

$$K_f(x, y) = \sum_{n \in \mathbb{Z}^m} f(x - y - n). \tag{1.13}$$

Making sufficiently stringent assumptions on $f(x)$, we can use Exercise 11 to say that

$$\operatorname{Trace} L_f = \sum_{n \in \mathbb{Z}^m} \hat{f}(n) = \int_{\mathbb{R}^m/\mathbb{Z}^m} K_f(x, x)\, dx. \tag{1.14}$$

But this last integral is easily seen to be $\Sigma f(n)$, by formula (1.13). Thus (1.14) is really Poisson's summation formula of Theorem 2. This discussion comes from a seminar talk of Larry Verner in 1973. though it was certainly well known to Selberg as early as 1951. In Chapter 3, we will begin consideration of Selberg's trace formula with an analogous analysis. Our main interest will be in integral operators on noncompact domains, however.

Exercise 12 (Band and Time-Limited Functions—Another Look at Uncertainty). Slepian and Pollak [1, 2] note that Exercise 16 of §1.2 does not really tell you "just how close one can come to simultaneously limiting in both time and frequency." Consider f in $L^2(\mathbb{R})$. Set $F(w) = \hat{f}(w/2\pi)$. If $F(w) = 0$ for $|w| > B$, we shall say that f is *band-limited*, and write $f \in \mathscr{B}_B$. Show that the function f in $\mathscr{B}_B$ with $\|f\|_2 = 1$ and

$$\int_{-T}^{+T} |f(t)|^2 \, dt = \text{maximum}$$

is the eigenfunction corresponding to the largest eigenvalue of the integral operator

$$Lf(t) = \int_{-T}^{T} f(s) \frac{\sin[B(t-s)]}{\pi(t-s)} \, ds \qquad \text{for } |t| \leq T.$$

Hint. Consider the operators

$$\mathscr{B}f(t) = \frac{1}{2\pi} \int_{-B}^{+B} \hat{f}(w/2\pi) \exp(itw) \, dw,$$

and

$$\mathscr{T}f(t) = \begin{cases} f(t) & \text{if } |t| \leq T, \\ 0 & \text{if } |t| > T. \end{cases}$$

Show that $\mathscr{B}\mathscr{T}f = Lf$.

Note. Slepian and Pollak calculate the eigenvalues of L by showing that the corresponding eigenfunctions must be spheroidal wave functions. These results have been generalized to some symmetric spaces by Grünbaum et al. [1]. Slepian and Pollak give the following table of values of $\lambda_n(c)$ (Table 1.2) where $c = BT/2$.

Some of the most interesting applications of the spectral theorem for integral operators are in partial differential equations. This occurs because one can often write the inverse operator of a differential operator D (or of $D - uI$, I = identity, $u \in \mathbb{C}$) as an integral operator. The kernel of this integral operator is called the Green's function. For example, the *Green's function* $G(x,y)$ *for* $-\Delta$ *on a region D in $\mathbb{R}^3$* with the boundary condition that f vanish on the boundary ∂D satisfies

Table 1.2. Values of $\lambda_n(c) = L_n(c) \times 10^{-p_n(c)}$

	c = 0.5		c = 1.0		c = 2.0		c = 4.0		c = 8.0	
n	L	p	L	p	L	p	L	p	L	p
0	3.0969	1	5.7258	1	8.8056	1	9.9589	1	1.0000	0
1	8.5811	3	6.2791	2	3.5564	1	9.1211	1	9.9988	1
2	3.9175	5	1.2375	3	3.5868	2	5.1905	1	9.9700	1
3	7.2114	8	9.2010	6	1.1522	3	1.1021	1	9.6055	1
4	7.2714	11	3.7179	8	1.8882	5	8.8279	3	7.4790	1
5	4.6378	14	9.4914	11	1.9359	7	3.8129	4	3.2028	1
6	2.0413	17	1.6716	13	1.3661	9	1.0951	5	6.0784	2
7	6.5766	21	2.1544	16	7.0489	12	2.2786	7	6.1263	3
8	1.6183	24	2.1207	19	2.7768	14	3.6066	9	4.1825	4

From Slepian and Pollak [1, p. 4].

$$(Tf)(x) = \int_D G(x, y) f(y) \, dy,$$

$$-\Delta Tf = f \quad \text{for all } f \text{ in } L^2(D), \tag{1.15}$$

$-T\Delta f = f \quad$ and f such that both f and Δf are in $L^2(D)$ and such that f vanishes on the boundary ∂D.

Equations (1.15) say that

$$-\Delta_x G(x, y) = \delta(x - y) \quad \text{for } x, y \text{ in } D$$
$$G(x, y) = 0 \quad \text{if } x \text{ is in } \partial D \tag{1.16}$$

Thus Exercise 6 of §1.1 implies that

$$G(x, y) = (4\pi \|x - y\|)^{-1} + h(x, y),$$

where $h(x, y)$ has no singularities and is chosen to make $G(x, y)$ vanish if x lies on the boundary of D. Finding $G(x, y)$ explicitly is difficult or impossible unless the region D is nice. For example, sharp spikes are not allowed in the region D. The following exercise gives a simple example of a Green's function for the Laplace operator on a ball of radius r. You can find more examples and discussions of Green's functions in §2.2 and in the following references: Courant and Hilbert [1], Dunford and Schwartz [1, Vol. II, Ch. XIII, where Green's functions are called resolvent kernels], Garabedian [1], Morse and Feshbach [1, Vol. II, Ch. 10], and Stakgold [1, 2].

Exercise 13 (Green's Function for $-\Delta$ on a Ball of Radius r by the Method of Images). William Thompson developed this method and applied it to problems in electro- and magnetostatics. The idea is to use the map of $\mathbb{R}^3$ which sends a point $x \in \mathbb{R}^3$ to its inverse point x^* with respect to the sphere of radius r. By this, we mean that x^* lies on the same radial line from the origin as x and

$\|x\| \|x^*\| = 1$. The problem is to show that the Green's function for the ball of radius r in $\mathbb{R}^3$ is

$$G(x, y) = (4\pi)^{-1}(\|x - y\|^{-1} - r\|y\|^{-1}\|x - y^*\|^{-1}).$$

Hint. To show that $G(x, y) = 0$ when $\|x\| = r$, you must use the fact that

$$\frac{\|x - y^*\|}{\|x - y\|} = \frac{r}{\|y\|} \quad \text{if } \|x\| = r.$$

If the Green's function $G(x, y)$ exists for the region D, then the operator T in (1.15) will be a Hilbert-Schmidt operator which means that both T and $-\Delta$ will have a complete orthonormal set of eigenfunctions v_n in $L^2(D)$ with $-\Delta v_n = \lambda_n v_n$ and

$$G(x, y) = \sum_{n \geq 1} \lambda_n^{-1} v_n(x) \overline{v_n(y)}. \tag{1.17}$$

In general, the convergence of this series is only in the L^2 norm. This gives another method for finding the Green's function. However, even in the simple example below, one finds a series which does not converge absolutely.

Exercise 14 (Green's Function for $-\Delta$ on a Rectangular Solid). Consider the Dirichlet eigenvalue problem

$$\begin{cases} -\Delta v = \lambda v & 0 < x < a, \, 0 < y < b, \, 0 < z < c \\ v = 0 & \text{on the boundary of the rectangular solid.} \end{cases}$$

Find a complete orthonormal set of eigenfunctions and consider formula (1.17) for the Green's function. Show that the series does not converge absolutely, using information about Epstein's zeta function (Theorem 2 of §1.4). You can find a discussion of the method of images for this problem in Courant and Hilbert [1, Vol. 1, pp. 378–384].

As a last topic in this discussion of spectral theory, we want to consider *a Schrödinger eigenvalue problem in one dimension*:

$$y''(x) - 4\pi^2 x^2 y(x) = \lambda y(x), \quad x \in \mathbb{R}$$
$$y(x) \in L^2(\mathbb{R}). \tag{1.18}$$

This is a good, though simplistic, model of a quantum-mechanical problem (see Messiah [1, Vol. I, Ch. XII]), which indicates quite clearly that a continuous problem can sometimes have a only a discrete sequence of possible eigenvalues. In quantum theory this is thought of as saying that the system is quantized and can only be in a certain discrete list of states. The eigenvalues of the differential operator in (1.18) correspond to the energy levels of the system, once everything has been normalized in the proper way for physics. We have chosen our normalization to connect with the theory of Fourier integrals. If the problem were on a finite interval, then we could find a Green's function and use Hilbert-Schmidt theory to see that the spectrum is discrete. However, this problem is singular, because it is on an infinite interval. Thus the spectrum has no a priori reason to be discrete. And certainly there are other Schrödinger eigenvalue problems with mixtures of continuous and discrete spectra. An example is the eigenvalue problem associated with the hydrogen atom which is considered in §2.1. Another such eigenvalue problem will be discussed in Chapter 3.

We could approach the eigenvalue problem (1.18) by substituting $y(x) = w(x)\exp(-\pi x^2)$. The resulting differential equation for $w(x)$ can be seen to have a polynomial solution only for special values of λ. However, we shall take a different approach—*the factorization method of Infeld and Hull* [1] (see Talman [1]). Set $D = d/dx$. Then we can factor the operator

$$H = D^2 - (2\pi x)^2 = (D - 2\pi x)(D + 2\pi x) - 2\pi = (D + 2\pi x)(D - 2\pi x) + 2\pi.$$

Define the raising operator $A = D - 2\pi x$ and the lowering operator $B = D + 2\pi x$. Then $H = AB - 2\pi = BA + 2\pi$.

Exercise 15 (Hermite Functions are Eigenfunctions of H and the Fourier Transform).

(a) Show that if we define $y_0 = \exp(-\pi x^2)$, $y_n = Ay_{n-1}$, then $Hy_n = \lambda_n y_n$, where the eigenvalues $\lambda_n = -(4n + 2)\pi$.
(b) Show that if $\hat{f} =$ the Fourier transform of f, then $\hat{y}_n = (-i)^n y_n$.
(c) Show that $y_n(x) = (D - 2\pi x)^n \exp(-\pi x^2) = \exp(-\pi x^2)D^n \exp(-2\pi x^2)$.
(d) Show that $y_n(x)$ is the product of $\exp(-\pi x^2)$ and a polynomial of degree n.

Exercise 16 (Spectral Decomposition of the Schrödinger Operator H). Set $v_n = \|y_n\|_2^{-1/2} y_n$, with y_n as in Exercise 15. Show that v_n gives a complete orthonormal set for $L^2(\mathbb{R})$ and thus the spectral decomposition of the differential operator $H = D^2 - (2\pi x)^2$.

Hint. The functions v_n are pairwise orthogonal since they are eigenfunctions of the self-adjoint operator H corresponding to distinct eigenvalues. To see that they form a complete orthonormal set, you must show that if $f \in L^2(\mathbb{R})$ and $(f, v_n) = 0$ for all n, then f is zero (almost everywhere). One way to do this is to look at

$$\widehat{f \exp(-\pi x^2)} = \int f(y)\exp(-2\pi i xy - \pi y^2)\,dy$$

$$= \sum_{n\geq 0} \frac{(-2\pi i x)^n}{n!} \int y^n f(y)\exp(-\pi y^2)\,dy.$$

If $(f, v_n) = 0$ for all n, this must vanish. Why?

Exercise 17 (Connection With the Classical Hermite Polynomials). Define $H_n(x) = (-1)^n \exp(x^2)D^n \exp(-x^2)$, $D = d/dx$. The $H_n(x)$ are the classical Hermite polynomials (see Courant and Hilbert [1, Vol. I, pp. 91–97]). Find the formula connecting y_n and H_n.

Exercise 18 (A New Expression for the Fourier Transform on $L^2(\mathbb{R})$). This idea comes from Wiener [1, pp. 51–71]. Suppose that the functions v_n are as in Exercise 16. Then any function f in $L^2(\mathbb{R})$ has an expression $f = \Sigma (f, v_n) v_n$. What is the corresponding expansion for the Fourier transform of f?

Exercise 19. Find an example of $f \in L^1(\mathbb{R})$ such that $\hat{f} \in L^1(\mathbb{R})$, and $f(n) = 0$ for all $n \in \mathbb{Z}$; $\hat{f}(0) = 1$, $\hat{f}(n) = 0$, if $n \in \mathbb{Z}$, $n \neq 0$. This shows that the Poisson summation formula is not always valid when you might expect it to be.

Hint. See Katznelson [1, Ex. 15, pp. 130–131].

Exercise 20 (Periodic Green's Function for the Helmholtz Equation in $\mathbb{R}^3$ by the Method of Images).

(a) Show that $g_k(x, y) = \exp(ik\|x - y\|)/4\pi \|x - y\|$ is a Green's function for the Helmholtz operator $(\Delta - k^2)$ on $\mathbb{R}^3$. Other choices for $g_k(x, y)$ are $\exp(-ik\|x - y\|)/4\pi \|x - y\|$, $\cos(k\|x - y\|)/4\pi \|x - y\|$, etc. Our choice of $g_k(x, y)$ dies off exponentially at infinity, if $\operatorname{Im} k > 0$. It can be viewed as corresponding to outgoing waves in the wave equation.

(b) Use the method of images to obtain the Green's function for $(\Delta - k^2)$ on the unit cube $\mathbb{R}^3/\mathbb{Z}^3$ corresponding to periodic boundary conditions.

Answer.
$$G_k(x, y) = \frac{1}{4\pi} \sum_{n \in \mathbb{Z}^3} \frac{\exp(ik\|x + n - y\|)}{\|x + n - y\|}.$$

For what values of k does this series converge?

(c) Show that, as in formula (1.17), we can represent the Green's function in part (b) by the following series converging in the L^2-sense:
$$G_k(x, y) = \sum_{n \in \mathbb{Z}^3} \frac{\exp(2\pi i {}^t n(x - y))}{\|2\pi n\|^2 - k^2}.$$

For absolute convergence, show that one can look at $G_k - G_m$, since if $R_k = (\Delta - k^2)^{-1}$, then $R_k - R_m = (k^2 - m^2) R_k R_m$.

(d) What does Poisson summation do to the formulas in parts (b) and (c)?

1.4. Mellin Transforms, Epstein and Dedekind Zeta Functions

It is remarkable that the deepest ideas of number theory reveal a far-reaching resemblance to the ideas of modern theoretical physics. Like quantum mechanics, the theory of numbers furnishes completely nonobvious patterns of relationship between the continuous and the discrete (the technique of Dirichlet series and trigonometric sums, p-adic numbers, nonarchimedean analysis) and emphasizes the role of hidden symmetries (classfield theory, which describes the relationship between prime numbers and the Galois groups of algebraic number fields). One would like to hope that this resemblance is no accident, and that we are already hearing new words about the World in which we live, but we do not yet understand their meaning.

From Yu. I. Manin, as translated by Ann and Neal Koblitz, *Mathematics and Physics*, Birkhãuser, Boston, 1981.

In this section we shall consider a method used both by number theorists and physicists to obtain analytic continuations of Dirichlet series. Riemann used this procedure in [1, pp. 145ff] to obtain the analytic continuation of the *Riemann zeta function*:

$$\zeta(s) = \sum_{n \geq 1} n^{-s}, \qquad \operatorname{Re} s > 1 \tag{1.19}$$

to the entire complex s-plane. Ewald used the same method to compute potentials of crystal lattices (see Born and Huang [1, p. 389]). Riemann's method can also be used to compute Green's functions for a rectangular parallelepiped (see Courant and Hilbert [1, Vol. 1, pp. 378–384] and Exercise 20 of §1.3). We shall discuss these examples in this section, as well as the Dedekind zeta function of an algebraic number field.

Before considering Riemann's method we must understand another important integral transform.

Definition. The *Mellin transform* of $f: \mathbb{R}^+ \to \mathbb{C}$ is

$$Mf(s) = \int_0^\infty f(y) y^{s-1}\, dy.$$

The Mellin transform is as well suited to the study of the multiplicative properties of numbers as the Fourier transform is well suited to the study of the additive properties of numbers. In fact, the change of variables $y = e^x$ allows you to go from one transform to the other. You could also consider the Mellin transform as a two-sided Laplace transform.

Exercise 1 (Inversion of the Mellin Transform). Make the change of variables $y = e^x$ to see that $Mf(s) = \widehat{F}(t/2\pi)$, where $s = \sigma + it$, with σ, t real, and $F(x) = f(e^x)e^{\sigma x}$. Suppose that $F(x) \in L^1(\mathbb{R})$ and that $f(y)$ is piecewise continuous with one-sided derivatives always existing. Use Exercise 7 of §1.2 to show that

$$\frac{1}{2}\Big(f(y+) + f(y-)\Big) = \frac{1}{2\pi i} \lim_{r \to \infty} \int_{c-ir}^{c+ir} y^{-s} Mf(s)\, ds.$$

1.4. Mellin Transforms, Epstein and Dedekind Zeta Functions

Exercise 2 (Properties of the Mellin Transform). Prove the following properties, assuming that the functions which are Mellin transformed satisfy the conditions of Exercise 1, for example.

(a) Set $D = y\,d/dy$. Then $MDf(s) = -sMf(s)$.

(b) $\dfrac{d}{ds} Mf(s) = M(f(y)\log y)$.

(c) $M(fg)(s) = \dfrac{1}{2\pi i} \displaystyle\int_{\operatorname{Re} z = c} Mf(z) Mg(s-z)\,dz$.

(d) For $f, g : \mathbb{R}^+ \to \mathbb{C}$, define $(f * g)(y) = \int_0^\infty f(y/u)g(u)u^{-1}\,du$. Then
$$M(f * g)(s) = Mf(s)Mg(s).$$

(e) Set $f_a(y) = f(ay)$, for a and y in $\mathbb{R}^+$. Then $M(f_a)(s) = a^{-s} Mf(s)$.

(f) $\int_{\operatorname{Re} s = c} Mf(s)g(s)\,ds = \int_{y=0}^\infty f(y) M^{-1} g(y) y^{-1}\,dy$.

Exercise 3. Verify Table 1.3.

Table 1.3. A Short Table of Mellin Transforms

$f(y)$, $y > 0$	$Mf(s)$
$\exp(-ay)$, $y > 0$	$a^{-s}\Gamma(s)$, $\operatorname{Re} s > 0$
$\begin{cases}\exp(-y), & y > a \\ 0, & 0 < y < a\end{cases}$	$\Gamma(s, a)$
$\exp\left[\dfrac{a}{2}\left(y + \dfrac{1}{y}\right)\right]$, $a > 0$	$2K_s(a)$
$\exp(-zy)(1+y)^{a-1}$, $\operatorname{Re} z > 0$	$\Gamma(s)\Psi(s, a+s, z)$, $\operatorname{Re} s > 0$

Hint. Review the properties of the gamma, incomplete gamma, K-Bessel, and confluent hypergeometric functions. See, for example, Lebedev [1, Chapters 1, 5, 9]. A longer table of Mellin transforms can be found in Erdélyi et al. [2] or Oberhettinger [1].

More information on Mellin transforms can be found in Sneddon [1] and Titchmarsh [1]. You could, of course, also deal with Mellin transforms of distributions.

It is possible to view the Mellin inversion formula in Exercise 1 as *the spectral resolution of the singular differential operator* $(y\,d/dy)^2$ acting on functions in the space

$$L^2(\mathbb{R}^+, y^{-1}\,dy) = \left\{ f : \mathbb{R}^+ \to \mathbb{C} \text{ measurable} \,\bigg|\, \int_0^\infty |f(y)|^2 y^{-1}\,dy < \infty \right\}.$$

The functions y^s are all eigenfunctions of this differential operator. See Stakgold [1, pp. 465–466] for an exercise on the Mellin inversion formula from this point of view. In Chapters 3 and 4 we will generalize the Mellin inversion formula to spaces of positive $n \times n$ matrices. This is, in fact, one of the main objectives of this book. Note that the operator $(y\,d/dy)^2$ and the measure $y^{-1}\,dy$ are invariant under the action change of variables $w = ay$ for a $\mathbb{R}^+$. Thus we are really viewing a spectral decomposition which is intimately related to the multiplicative group of positive real numbers when we study the Mellin inversion formula. The generalizations of Mellin inversion to be considered in Chapters 3 and 4 are directly related to the general linear group of $n \times n$ nonsingular real matrices.

All of the analytic continuations of zeta functions that we want to consider are related to the analytic continuation of Epstein's zeta function. The latter lives on the symmetric space $\mathscr{P}_n$ of positive definite symmetric $n \times n$ real matrices and it will appear in the study of harmonic analysis on $\mathscr{P}_n$ in Chapters 3 and 4.

Definition. *Epstein's zeta function of* $Y \in \mathscr{P}_n$ $s \in \mathbb{C}$, *with* $\operatorname{Re} s > n/2$ *is*

$$Z(Y, s) = \frac{1}{2} \sum_{a \in \mathbb{Z}^n - 0} Y[a]^{-s}.$$

Here $Y[a] = {}^t a Y a$, thinking of a as a column vector, with ${}^t a$ = transpose of a. Thus $Y[a]$ is the quadratic form

$$Y[a] = \sum_{i,j=1}^{n} y_{ij} a_i a_j \qquad \text{if } Y = (y_{ij}),\, a = {}^t(a_1, \ldots, a_n).$$

In the special case that $m = 1$, then $Y = y \in \mathbb{R}^+$, and $Z(y, s) = y^{-s}\zeta(2s)$, where ζ = Riemann's zeta function. See Edwards [1] for a fascinating treatment of the work on that zeta function, whose main importance for number theory comes from the following formula:

$$\zeta(s) = \prod_{p = \text{prime}} (1 - p^{-s})^{-1}, \qquad \operatorname{Re} s > 1. \tag{1.20}$$

Exercise 4. Prove formula (1.20) which is called the *Euler product* for Riemann's zeta function. You need to know that every natural number $n \in \mathbb{Z}^+$ can be factored uniquely (up to order) as a product of primes.

Many other special cases of Epstein's zeta function arose in number theory before Epstein's papers (see Epstein [1, 2]) in 1903 and 1907. Most of these cases arise from the fact that the Dedekind zeta function of an algebraic number field can be written as a finite sum of integrals of Epstein zeta functions (Theorem 2). The sad story of the end of Epstein's life and the horrors of Hitler's Germany can be found in Siegel [5, Vol. III, pp. 464–470] or the translation [7].

There are many applications of Epstein zeta functions in statistical and solid-state physics. We will discuss some of these at the end of this section. An interesting reference which gives applications of Epstein's zeta function in quantum statistical mechanics is Hurt and Hermann [1, Ch. 8].

1.4. Mellin Transforms, Epstein and Dedekind Zeta Functions

Exercise 5 (Convergence of Epstein's Zeta Function).

(a) Use Theorem 5 of §1.3 to show that Epstein's zeta function does indeed converge absolutely and uniformly on compact subsets of the half-plane $\operatorname{Re} s > n/2$.

(b) Develop an integral test in several variables to do the convergence proof in another way. We will return to this topic in Chapter 4.

Our next goal is to analytically continue Epstein's zeta function to the whole s-plane as a meromorphic function of s, having its unique pole at $s = n/2$. The method goes back to Riemann [1, pp. 147ff] and this paper is translated and discussed in Edwards [1]. Riemann did not mention incomplete gamma functions, but he did, in fact, obtain this expansion.

Theorem 1 (The Analytic Continuation of Epstein's Zeta Function). *Let $\Gamma(s, x)$ be the incomplete gamma function as in Exercise 3 and set*

$$G(s, x) = x^{-s}\Gamma(s, x) = \int_1^\infty y^{s-1} \exp(-xy)\, dy.$$

Then Epstein's zeta function can be analytically continued to all $s \in \mathbb{C}$ with its only pole a simple one at $s = n/2$ having residue $\frac{1}{2}\pi^{n/2}|Y|^{-1/2}\Gamma(n/2)^{-1}$. Here $|Y|$ denotes the determinant of Y.

The analytic continuation comes from the incomplete gamma expansion:

$$\Lambda(Y, s) = \pi^{-s}\Gamma(s)Z(Y, s)$$

$$= \frac{|Y|^{-1/2}}{2s - n} - \frac{1}{2s} + \frac{1}{2} \sum_{a \in \mathbb{Z}^n - 0} \left(G(s, \pi Y[a]) + |Y|^{-1/2}G\left(\frac{n}{2} - s, \pi Y^{-1}[a]\right) \right).$$

Thus Epstein's zeta function satisfies the functional equation

$$\Lambda(Y, s) = |Y|^{-1/2}\Lambda(Y^{-1}, n/2 - s).$$

Furthermore, Epstein's zeta function takes on the values

$$Z(Y, 0) = -\tfrac{1}{2}, \qquad Z(Y, -k) = 0, \ k = 1, 2, 3, \ldots.$$

PROOF. This demonstration will be broken up into a sequence of exercises. □

Exercise 6 (The Transformation Formula of a Theta Function). For Y in $\mathscr{P}_n$, $t \in \mathbb{R}^+$, define

$$\theta(Y, t) = \sum_{a \in \mathbb{Z}^n} \exp(-\pi Y[a]t).$$

Show that the series converges and that

$$\theta(Y, t) = |Y|^{-1/2} t^{-n/2} \theta(Y^{-1}, t^{-1}) \qquad \text{with } |Y| = \text{determinant of } Y.$$

Hint. Use the same method as in Exercise 6 of §1.3. To do this you must write $tY = B^2$, with $B \in \mathscr{P}_n$. Then set $f(x) = g(Bx)$. The Fourier transforms of f and g are related by

$$\hat{f}(x) = |B|^{-1}\hat{g}({}^tB^{-1}x).$$

Note. The series for theta converges quickly for large t. When t is near 0, the transformation formula allows one to replace t by t^{-1} and obtain a quickly convergent series again. We used this fact already in Theorem 5 of §1.3. It will be used over and over again in these notes.

Exercise 7 (Riemann's Trick).

(a) Show that for $\operatorname{Re} s > n/2$:

$$\Lambda(Y,s) = \frac{1}{2}\int_0^\infty t^{s-1}[\theta(Y,t) - 1]\,dt.$$

You can justify the interchange of summation and integration by Fubini's theorem (see Lang [2, p. 295]).

(b) Obtain the analytic continuation of Epstein's zeta function by writing

$$\int_0^\infty = \int_0^1 + \int_1^\infty.$$

Then replace t by $1/t$ in the first integral and use Exercise 6. This should lead you to the formula

$$\Lambda(Y,s) = \int_1^\infty t^{s-1} w(Y,t)\,dt + |Y|^{-1/2}\int_1^\infty t^{n/2-s-1} w(Y^{-1},t)\,dt$$

$$+ \frac{1}{2}\int_1^\infty (|Y|^{-1/2}t^{n/2} - 1)t^{-s-1}\,dt,$$

where $w(Y,t) = \frac{1}{2}(\theta(Y,t) - 1)$. Then interchange sum and integral in the first two integrals and evaluate the last integral.

(c) Show that the incomplete gamma expansion of $Z(Y,s)$ converges exponentially faster than the original Dirichlet series defining $Z(Y,s)$.

Hint. The incomplete gamma function has the asymptotic expansion

$$G(s,x) \sim x^{-1}e^{-x}\left(1 + \frac{s-1}{x} + \frac{(s-1)(s-2)}{x^2} + \cdots\right) \quad \text{as } x \to \infty.$$

This completes the proof of Theorem 1. Much use has been made of such incomplete gamma expansions in number theory (see Montgomery [1] and Lavrik [1]). Our next goal is to apply the result to study the Riemann zeta function in the interval (0, 1). We do this in the next set of exercises, which go back to Riemann again (see Edwards [1, pp. 16–18]). Selberg generalized these exercises to the symmetric space $\mathscr{P}_n$ in order to obtain the analytic continu-

ation of generalizations of Epstein's zeta function known as Eisenstein series. We will consider this in Chapter 4.

Exercise 8 (Multiplication Invariant Differential Operators on The Positive Reals).

(a) Let L be a differential operator acting on functions $f: \mathbb{R}^+ \to \mathbb{C}$. Suppose $u: \mathbb{R}^+ \to \mathbb{R}^+$ is differentiable with a differentiable inverse. Define L^u by $L^u f = L(f \circ u) \circ u^{-1}$. Let $a \in \mathbb{R}$ and $D_a = y^a(d/dy)y^{1-a}$. Then define $c(y) = cy$ for $c \in \mathbb{R}^+$. Show that $D_a^c = D_a$; i.e., D_a is multiplication invariant.

(b) Suppose that L is a multiplication-invariant differential operator on $\mathbb{R}^+$. Define its formal adjoint L^* by

$$\int_{\mathbb{R}^+} (Lf)(y)\overline{g(y)}\, y^{-1}\, dy = \int_{\mathbb{R}^+} f(y)\overline{L^* g(y)}\, y^{-1}\, dy.$$

Note. dy/y is the multiplication invariant or Haar measure on the group of positive real numbers under multiplication. Show that if $u(y) = 1/y$, then $L^u = \overline{L^*}$.

(c) Use integration by parts to prove that $D_a^u = -D_{2-a}$.

Exercise 9 (Transformation Formula of a Differentiated Theta Function). Set $L = D_{1/2}D_1$, using the notation of Exercise 8. Show that if $\theta(y,t)$ is the theta function of Exercise 6, when $n = 1$, and $t, y \in \mathbb{R}^+$, then

$$L_t\theta(y,t) = y^{-1/2}t^{-1/2}L_t\theta(y^{-1},t^{-1}).$$

Exercise 10 (Power Series for $\zeta(2s)$ around $s = \frac{1}{4}$). If $L = D_{1/2}D_1$, then $L^* = D_1 D_{3/2}$ and $L^* t^s = s(s - \frac{1}{2})t^s$. Then replace θ by $L_t\theta$ in the proof of Theorem 1 to show that

$$2s(s - \tfrac{1}{2})\pi^{-s}\Gamma(s)\zeta(2s) = \sum_{n\geq 1} \int_{t=1}^{\infty} (t^s + t^{1/2-s})\pi t n^2(2\pi t n^2 - 3) \times \exp(-\pi t n^2)t^{-1}\, dt.$$

Then show that we can expand the right-hand side as a power series in even powers of $s - \frac{1}{4}$ with positive coefficients. Deduce that $\zeta(s) < 0$ for $0 \leq s < 1$.

The Riemann hypothesis says that $\zeta(s)$ has its only zeros at

$$s = -2, -4, -6, \ldots \qquad \text{(the trivial zeros)}$$

and at points on the line $\operatorname{Re} s = \frac{1}{2}$—a line which is fixed by the functional equation of the zeta function. The hypothesis has been checked by computer for the first many million zeros (see Brent [1] and Edwards [1]). However, the proof or disproof of this hypothesis has so far skillfully eluded its many pursuers.

Epstein's zeta function behaves very differently from Riemann's in some ways. For example, there are many matrices Y in $\mathscr{P}_2$ such that $Z(Y,s)$ vanishes for s with $\operatorname{Re} s > 1$ (see Davenport and Heilbronn [1]). And the following Exercise shows that the behavior of $Z(Y,s)$ for s in the interval $(0,1)$ is very different from that of Riemann's zeta function when Y has a small minimum $m_Y = \min\{Y[a] = {}^t a Y a | a \in \mathbb{Z}^2 - 0\}$.

Exercise 11. Use the incomplete gamma expansion of $Z(Y,s)$, $Y \in \mathscr{P}_2$, in Theorem 1 to graph the function $\Lambda(Y,s)$ for x in $(0,1)$ and

$$Y = \begin{pmatrix} t & 0 \\ 0 & 1/t \end{pmatrix}, \quad t = 1, .1, .01, .001.$$

Hint. You can compute the incomplete gamma functions using the ALGOL procedure below devised by Riho Terras in [1] and [2].

ALGOL PROCEDURE TO COMPUTE INCOMPLETE GAMMA FUNCTION G(S,X)

```
100   REAL PROCEDURE G(S,X);
200   COMMENT THIS FINDS INCOMPLETE GAMMA G(S,X)
      FOR 0<S<1 AND X>0;
300   VALUE S,X; REAL S,X;
400   IF X GEQ 0.5 THEN BEGIN REAL W; INTEGER K;
500   FOR K:=ENTIER(8+50/X) STEP -1 UNTIL 1 DO
600   W:=X+W*(K-S)/(W+K);
700   G:=EXP(-X)/W;
800   END ELSE BEGIN
900   FOR K:=ENTIER(17+X+X) STEP -1 UNTIL 0 DO
1000  W:=(1+X*W)/(S+K);
1100  G:=X**(-S)*GAMMA(S)-EXP(-X)*W;
1200  END;
1300  END OF PROCEDURE G;
```

The procedure G is based on two formulas. When $0 < s < 1$ and $x > 0.5$, it uses the *continued fraction* for $G(s, x)$:

$$G(s,x) = e^{-x} \left(\cfrac{1}{x + \cfrac{1-s}{1 + \cfrac{1}{x + \cfrac{2-s}{1 + \cfrac{2}{x + \cdots}}}}} \right)$$

$$\equiv e^{-x} \left(\frac{1}{x+} \frac{1-s}{1+} \frac{1}{x+} \frac{2-s}{1+} \frac{2}{x+} \cdots \right) \quad \text{for } \operatorname{Re} s > 0. \qquad (1.21)$$

This is due to Legendre. See Exercise 12 for hints on a derivation. You can find a discussion of continued fractions in Henrici [1, Vol. II, Ch. 12, especially p. 629]. The ALGOL procedure quoted above views the continued fraction (1.21) as a composition of fractional linear transformations:

$$T_k(w) = x + w(k - s)/(w + k).$$

That is, the continued fraction should be viewed as

$$T_1 T_2 \cdots T_k(w) = (e^x G(s,x))^{-1}, \tag{1.22}$$

for the correct choice of w. In the ALGOL procedure, we replace w by zero. This is analogous to using the first k terms of a series. The recursions sending w to $T_k(w)$ are error correcting when $0 < s < 1$ and $x > 0$. The value of k used to compute $G(s,x)$ by truncating the continued fraction via (1.22) is $k = [8 + 50/x]$. Clearly this is very large if x is very near 0. In fact, the continued fraction is also bad when $s \geq 1$, because then the T_k start to magnify rather than correct errors. See R. Terras [1, 2] and Henrici [2] for discussions of error-correcting recursions.

When $0 < s < 1$ and $0 < x < 0.5$, the ALGOL procedure above uses a power series:

$$G(s,x) = x^{-s}\Gamma(s,x) = x^{-s}\Gamma(s) - x^{-s}\gamma(s,x),$$

$$\text{when } \gamma(s,x) = \int_0^x t^{s-1}e^{-t}\,dt = x^s e^{-x} \sum_{n\geq 0} \frac{x^n}{(s)_{n+1}}, \tag{1.23}$$

$$(s)_0 = 1, \text{ and } (s)_n = \Gamma(s+n)/\Gamma(s) - s(s+1)\cdots(s+n-1).$$

Exercise 12 (Properties of the Incomplete Gamma Function).

(a) Show that if $\varphi(s,x) = e^x G(s,x)$, then $\varphi(s+1,x) = 1 + s\varphi(s,x)/x$.
(b) Prove formula (1.21) above. It can be done using part (a). Or use recursions for the confluent hypergeometric function.
(c) Prove formula (1.23) above.
(d) Use part (a) to prove the asymptotic expansion for $G(s,x)$ in part (c) of Exercise 7.

Your answer to Exercise 11 should resemble the graphs in Fig. 1.12, which were produced by a Burroughs 6700 computer at the University of California (San Diego) using the ALGOL program for incomplete gamma functions above.

Exercise 13 (Epstein Zeta Functions for Positive Definite Matrices with Small Minima over the Integer Lattice). Let $0 < u < 1$. Suppose Y is in $\mathscr{P}_n$ with $\det Y = |Y| = 1$. Let

$$m_Y = \min\{Y[a] = {}^t a Y a \mid a \in \mathbb{Z}^n - 0\}.$$

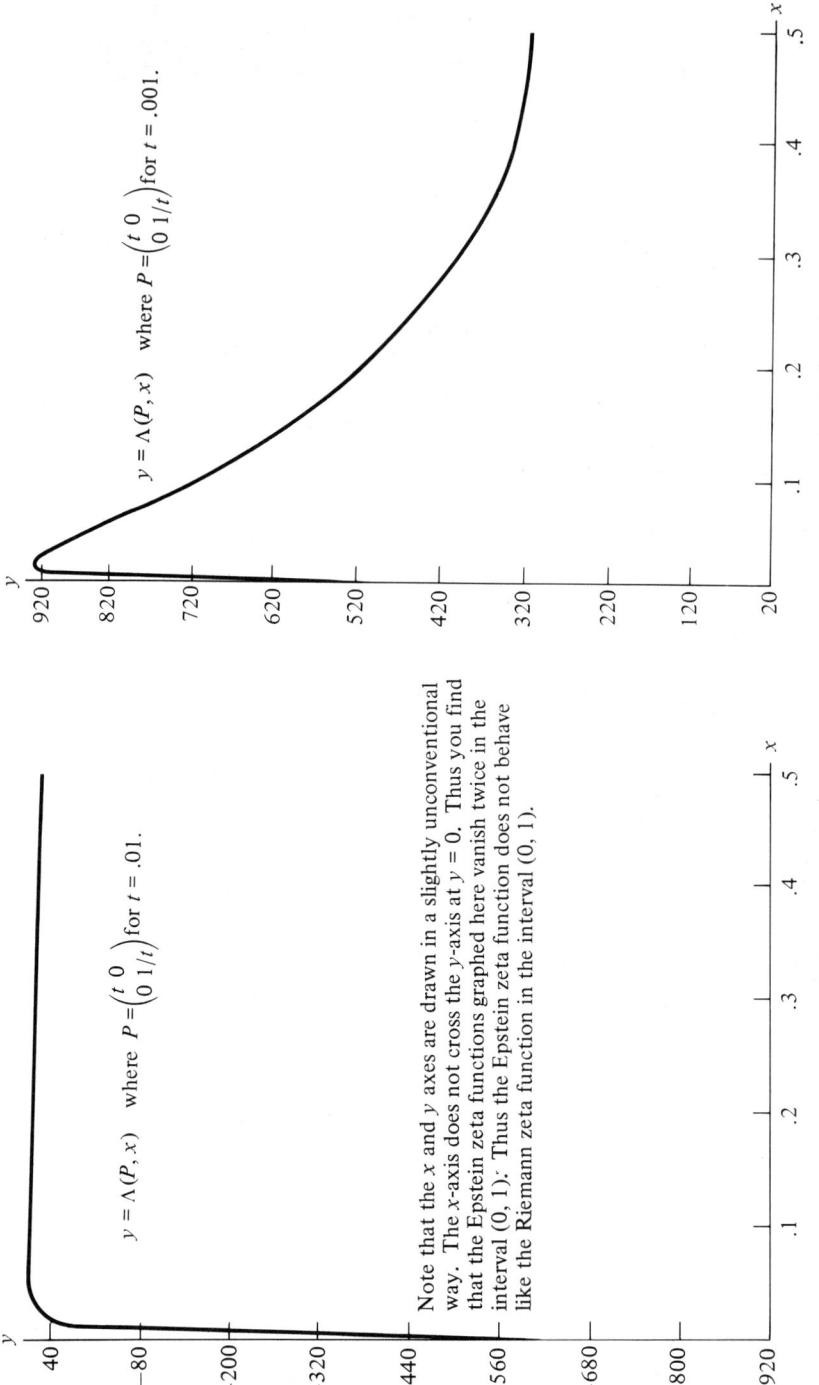

Figure 1.12. Graphs of the product of the Epstein zeta function with its gamma factors in the interval (0, 1).

Suppose that $m_Y \le nu/2\pi e$ or $m_{Y^{-1}} \le n(1-u)/2\pi e$. Show that if n is sufficiently large (dependent on u), $Z(Y, nu/2) > 0$. Conclude that $Z(Y,s)$ vanishes in $(nu/2, n/2)$.

Hint. Use the fact that $G(s,a) > a^{-s}\Gamma(s) - 1/s$. Or see A. Terras [3].

The size of m_Y in Exercise 12 is connected with part of Hilbert's 18th problem (see Milnor [1]) and we shall discuss it in more detail in Chapter 4. It should be noted here that there are forms Y in $\mathscr{P}_n$ with $|Y| = 1$ and m_Y as small as you like. Moreover, there are forms with $m_Y > n/(2\pi e)$, if n is large, as we shall prove in Chapter 4. In fact, we shall show in that same Chapter that given s in $(0, n/2)$, there exist Y in $\mathscr{P}_n$ with $Z(Y, s)$ positive, negative, or zero.

Next we would like to consider some applications of Theorem 1 in algebraic number theory and crystallography. We shall discuss algebraic number theory first. This will require a brief survey of the essentials of that subject. More details can be found in the books of Hecke [3], Lang [3], Narkiewicz [1], and Samuel [1], for example.

Application 1. Algebraic Number Theory

Algebraic number theory originated with work of Kummer, Dedekind and others in the 1800s. They wanted to prove Fermat's conjecture that for $n = 3, 4, 5, \ldots$, the equation $x^n + y^n = z^n$ has no integer solutions x, y, z with $xyz \ne 0$. It was found that if one knows that the ring of cyclotomic integers $O_n = \mathbb{Z}[\exp(2\pi i/n)]$ has unique factorization into primes, then Fermat's conjecture follows. Unfortunately O_n is only a unique factorization domain for 29 of these rings (see Masley and Montgomery [1] and Table 1.3).

Another question which might appear to have nothing to do with cyclotomic integers is that of deciding whether an ordinary integer is prime. Recently this has become interesting to the CIA and the NSA because public key cryptography involves two huge primes and Fermat's little theorem (see Simmons [1]). The current fastest algorithm for distinguishing prime numbers from composite numbers involves algebraic number theory in cyclotomic fields (see Adleman, Pomerance, and Rumely [1]).

Thus one is lead to the study of *algebraic number fields* $K = \mathbb{Q}(\omega)$, ω algebraic; i.e., ω is a solution of an equation

$$f(\omega) = \omega^n + a_{n-1}\omega^{n-1} + \cdots + a_0 = 0, \qquad a_i \in \mathbb{Q}.$$

Assuming that the polynomial $f(x)$ is irreducible, we say that n is the *degree* of K over $\mathbb{Q}$. Let $\omega^{(j)}, j = 1, 2, \ldots, n$, denote the roots of the irreducible polynomial $f(x)$ with $\omega^{(j)} \in \mathbb{R}$, for $j = 1, 2, \ldots, r_1$ and $\omega^{(j)} \notin \mathbb{R}$, for $j = r_1 + 1, \ldots, n$. Furthermore, we shall order the roots so that the nonreal roots come in pairs of complex conjugates as follows:

$$\omega^{(j)} = \overline{\omega^{(j+r_2)}}, \qquad j = r_1 + 1, \ldots, r_1 + r_2.$$

We always use an overbar to denote complex conjugate. The map from ω to $\omega^{(j)}$ extends to a field isomorphism $x \to x^{(j)}$ of K into $\mathbb{C}$, fixing points of $\mathbb{Q}$. We shall call these isomorphisms the *conjugations* of K.

Problems involving the ordinary ring of integers $\mathbb{Z}$ in the rational number field $\mathbb{Q}$ lead directly to the study of the ring O_K of *algebraic integers* α in the number field K; i.e., solutions of equations

$$\alpha^n + a_{n-1}\alpha^{n-1} + \cdots + a_0 = 0, \quad \text{with } a_i \text{ in } \mathbb{Z}, \text{ all } i.$$

Number fields generalize ordinary arithmetic since $O_\mathbb{Q} = \mathbb{Z}$. And the arithmetic of these generalized integers is quite fascinating. It turns out that, just as in physics, analysis is needed to study these seemingly purely algebraic objects. A direct way to obtain algebraic information is to study the Dedekind zeta function of K and other related L-functions. Since O_K is not, in general, a unique factorization domain (UFD), one has to replace integers by ideals. An *ideal* $\mathfrak{A}$ in O_K is an abelian group under addition such that $O_K \mathfrak{A} \subset \mathfrak{A}$ and $\mathfrak{A} \neq \{0\}$. A *principal ideal* has the form $\alpha O_K = (\alpha)$, for some α in O_K. One can show that every ideal can be factored uniquely (up to order) into a product of prime ideals. Here the product of two ideals $\mathfrak{A}$ and $\mathfrak{B}$ is defined by

$$\mathfrak{A}\mathfrak{B} = \left\{\sum_{i=1}^{k} a_i b_i \, | \, a_i \in \mathfrak{A}, b_i \in \mathfrak{B}\right\}.$$

Using elementary divisor theory (which is the theory lying behind the fundamental theorem of abelian groups), one can show that every ideal $\mathfrak{A}$ has a $\mathbb{Z}$-basis or integral basis; i.e.,

$$\mathfrak{A} = \mathbb{Z}\alpha_1 \oplus \mathbb{Z}\alpha_2 \oplus \cdots \oplus \mathbb{Z}\alpha_n, \quad \text{for some } \alpha_1, \alpha_2, \ldots, \alpha_n \in \mathfrak{A}.$$

Thus we can view $\mathfrak{A}$ as an analogue of a crystal lattice. Moreover, one finds that the quoteint $O_K/\mathfrak{A}$ is a finite ring. We will say that the *norm of the ideal* $\mathfrak{A}$ is

$$\text{Norm } \mathfrak{A} = N\mathfrak{A} = \#(O_K/\mathfrak{A}).$$

In particular, $O_K = \mathbb{Z}\alpha_1 \oplus \cdots \oplus \mathbb{Z}\alpha_n$ and one defines the *discriminant*

$$d_K = \det(\alpha_i^{(j)})^2_{1 \leq i,j \leq n}.$$

In general, it is hard to find a $\mathbb{Z}$-basis for O_K. We give a table of examples at the end of our brief review of number theory. The divisors of d_K are the primes p in $\mathbb{Z}$ such that the ideal pO_K factors into prime ideals in K with squares of prime ideals appearing. This is Dedekind's discriminant theorem (see Samuel [1, p. 75]).

As we said, one wants to know if O_K is a UFD for many applications. To measure how far O_K deviates from unique factorization, Dedekind defined the *ideal class group* I_K of ideals modulo the equivalence relation

$$\mathfrak{A} \sim \mathfrak{B} \Leftrightarrow \gamma\mathfrak{A} = \mathfrak{B}, \quad \text{for some } \gamma \neq 0 \text{ in } K.$$

Multiplication of ideals induces a group operation on I_K which makes I_k a finite group whose order h_K is called the *class number* of K. The ring O_K is a

1.4. Mellin Transforms, Epstein and Dedekind Zeta Functions

UFD if and only if $h_K = 1$. We have only considered ideals which are subsets of O_K. But it is also useful to enlarge this definition. A *fractional ideal* $\mathfrak{A}$ is a subset of K forming an abelian group under addition such that $O_K \mathfrak{A} \subset \mathfrak{A}$ and $\alpha \mathfrak{A} \subset O_K$ for some $\alpha \neq 0$ in O_K. The arithmetic of fractional ideals is very like that of the rational numbers $\mathbb{Q}$. It is usually fairly hard to tell whether the class number of a given field is one. For imaginary quadratic fields $\mathbb{Q}(\sqrt{d})$ the answer was obtained during the last 20 years or so (see Stark [2]). The imaginary quadratic fields $\mathbb{Q}(\sqrt{d})$ with class number one have discriminant $d = -3, -4, -7, -8, -11, -19, -43, -67, -163$. Proofs of this statement involve analysis, using either Epstein zeta functions or modular forms. The theorem was suspected for a long time because it is easy to produce big tables of class numbers for those fields. We will show how to do this in Chapter 3. For real quadratic fields, the situation is very different. Tables of class numbers lead one to believe that there should be an infinite number of real quadratic fields with class number one, but no one can prove it (see Williams and Broere [1]). Recently Masley and Montgomery in [1] found all the cyclotomic fields with class number one. Previously Siegel had reduced the problem of determining the prime cyclotomic fields of class number one to a finite problem (see Siegel [5, Vol. III, p. 442]).

Note that there is a problem computing with ideals, since they are sets of numbers. Thus one often wants to go up to a field $E \supset K$ such that the ideals of K become principal in E. Then one can compute in E. Hecke had to do this many times (see Hecke [1, pp. 255ff] for example). Class field theory shows that E exists (see Cassels and Fröhlich [1, Exercise 3, pp. 355–357]).

There is a more surprising invariant of a number field waiting to be described—the *group of units* $U_K = \{u \in O_K | u^{-1} \in O_K\}$. This group is not very interesting if $K = \mathbb{Q}$, since it then consists of $+1$ and -1. It is also not very interesting for an imaginary quadratic field since if m is square-free and negative then $U_{\mathbb{Q}(\sqrt{m})} = \{+1, -1\}$ unless $m = -1$ or -3. If $m = -1$, you get the 4th roots of unity. If $m = -3$, you get the 6th roots of unity. However, the real quadratic fields contain infinite cyclic groups of units as well as $+1$ and -1. The generator of the infinite cyclic group of units is called the fundamental unit for the real quadratic field. It can be found using the continued fraction expansion for $\sqrt{d}$.

In general, Dirichlet's unit theorem says that the group of units $U_K = \langle \varepsilon_1 \rangle \times \cdots \times \langle \varepsilon_r \rangle \times W$, where $\langle \varepsilon_j \rangle$ denotes the infinite cyclic group generated by ε_j, W is the group of roots of unity in K (a finite cyclic group), $r = r_1 + r_2 - 1$, with r_i as defined at the very beginning of the discussion of algebraic number fields. The ε_i are called *fundamental units* of K. The measure of the unit group is given to a certain extent by the *regulator* defined by

$$R_K = \det\left(\log |\varepsilon_i^{(j)}|^{e_j}\right)_{1 \leq i,j \leq r},$$

with

$$e_j = \begin{cases} 1, & j = 1, \ldots, r_1 \\ 2, & j = r_1 + 1, \ldots, r. \end{cases}$$

Since the norm of a unit is one, it does not matter which embedding $K \to \mathbb{C}$ is left out in the definition of the regulator.

Minkowski's lemma in the geometry of numbers (see Chapter 4 or Samuel [1, p. 55]) gives the standard proof of the finiteness of the class number as well as Dirichlet's unit theorem. One can also deduce these things from the convergence of Dedekind's zeta function (see Theorem 2 and Siegel [3, Vol. II, pp. 93–94]).

Note that if $r = 1$, the regulator is the logarithm of the fundamental unit. This is a transcendental number (see Baker [1, p. 6]), as was proved by Lindemann in 1882. Thus one might expect the same of the regulator itself, but no one knows how to prove that.

The search for fundamental units is a problem that we will be able to say something about, since there are many connections with the tools developed in later chapters of this book. The connection between algebraic number theory and that of the general linear group of $n \times n$ nonsingular real matrices is basic to much recent work (see T. Shintani [1]). It is that connection that we shall begin to analyze in this section. The really interesting work can only begin after we have Chapters 3 and 4 behind us.

Conjectures of H. Stark allow one to construct units of class fields of number fields from values of derivatives of L-functions. The units are not fundamental, in general, but the conjecture allows one to find the units explicitly in many interesting cases. This also relates to Hilbert's 12th Problem (see Stark [3, 4]).

One of the major problems of algebraic number theory is that of describing the growth of the product of the class number times the regulator as the discriminant goes to infinity. We shall see why one looks at the class number times the regulator. In fact, it is hard to separate the two with the available tools. The *Brauer-Siegel theorem* says that under certain hypotheses on the sequence of fields (e.g., fixed degree over $\mathbb{Q}$), one has

$$\log(hR) \sim \log|d|^{1/2} \quad \text{as } |d| \to \infty.$$

Thus, in particular, for imaginary quadratic fields (where $R = 1$), there can be only a finite number of fields with a given class number. Siegel did the quadratic version of the theorem. Brauer did the more general result using the theory of representations of finite groups and Artin L-functions. The whole proof would be trivial if one knew that the Dedekind zeta function has no real zeros near 1 (see Lang [3]). We are about to define this function. Unfortunately its behavior near $s = 1$ is still a mystery. The possible zeros are called "Siegel zeros." References for such things are Stark [1], Purdy et al. [1], and Siegel [3].

Table 1.4 gives some examples of number fields and their invariants.

The *Dedekind zeta function* of a number field K is defined to be

$$\zeta_K(s) = \sum_{\text{ideals } \mathfrak{A} \subset O_K} N\mathfrak{A}^{-s} = \prod_{\text{prime ideals } \mathfrak{P}} (1 - N\mathfrak{P}^{-s})^{-1}$$

for $\operatorname{Re} s > 1$. Unique factorization of ideals into prime ideals gives the Euler

1.4. Mellin Transforms, Epstein and Dedekind Zeta Functions

Table 1.4. Examples of Number Fields

K	O_K	d_K	U_K	h_K
$\mathbb{Q}(\sqrt{m})$ m square-free negative	$\mathbb{Z}[\sqrt{m}]$ if $m \not\equiv 1 \pmod 4$	m or $4m$	$\{+1, -1\}$ if $m \neq -1$ or -3	$= 1$ for only 9 fields
$\mathbb{Q}(\sqrt{m})$ m square-free positive	$\mathbb{Z}[\sqrt{m}]$ if $m \not\equiv 1 \pmod 4$	m or $4m$	$\{+1, -1\} \times \langle \varepsilon \rangle$ find ε by continued fractions	Tables lead to unproved conjecture that it is 1 infinitely often
$\mathbb{Q}(\sqrt[3]{m})$ m cube-free	Sometimes $\mathbb{Z}[\sqrt[3]{m}]$	Sometimes a power of 3 times m	$\{\pm 1\} \times \langle \varepsilon \rangle$ = fundamental unit computed by Voronoi's algorithm	Tables lead to unproved conjecture that it is 1 infinitely often
$\mathbb{Q}(\exp(2\pi i/m))$	$\mathbb{Z}[\exp(2\pi i/m)]$	For $m = p =$ prime it is $\pm p^{p-2}$	$r = \dfrac{\varphi(m)}{2} - 1$ I haven't seen a good algorithm to produce them $\varphi =$ Euler's function	1 for only 29 distinct fields

References for these tables are Narkiewicz [1], Stark [1], Williams and Broere [1], Barrucand, Williams, and Baniuk [1], and Masley and Montgomery [1].

product above just as in Exercise 4 for the case $K = \mathbb{Q}$. For $\zeta_\mathbb{Q}(s)$ is nothing other than Riemann's zeta function.

The convergence of Dedekind's zeta function can be proved algebraically using a fact from algebraic number theory. One needs to know that if p is a prime in $\mathbb{Z}$, then

$$pO_K = \mathfrak{P}_1^{e_1} \cdots \mathfrak{P}_g^{e_g}, \text{ with prime ideals } \mathfrak{P}_i \text{ in } O_K,$$

such that $N\mathfrak{P}_i = p^{f_i}$. Now $p^n = N(p) = p^{\sum f_i e_i}$ and it follows that

$$n = \sum_{i=1}^{g} f_i e_i.$$

It is also clear that if $\mathfrak{P}$ is a prime ideal in K then $\mathfrak{P} \cap \mathbb{Z} = (p)$, where p is a prime number in $\mathbb{Z}$. This shows (using the Euler product for ζ_K) that

$$\log \zeta_K(s) = \sum_{\mathfrak{P}} \sum_{k \geq 1} k^{-1} N\mathfrak{P}^{-ks} \leq n \sum_p \sum_{k \geq 1} k^{-1} p^{-ks} = n \log \zeta(s).$$

So we have

$$\zeta_K(s) \leq (\zeta(s))^n \quad \text{when } s > 1. \tag{1.24}$$

In 1916 Hecke figured out how to continue $\zeta_K(s)$ to the whole complex s-plane as a meromorphic function with a simple pole at $s = 1$ with residue

$$\operatorname*{Res}_{s=1} \zeta_K(s) = 2^{r_1}(2\pi)^{r_2} hRw^{-1}|d|^{-1/2},$$

where r_1, r_2 are the numbers of real and pairs of complex conjugations of K, h is the class number, R the regulator, d the discriminant, and w the number of roots of unity in K. This clearly relates class number-regulator problems to the Dedekind zeta function. Moreover, Hecke's proof can easily be seen to connect these problems with the behavior of the Dedekind zeta function in the interval $(0, 1)$.

Landau used Hecke's result (plus Hadamard factorization and Cauchy's integral theorem) to prove the *prime ideal theorem*:

$$\# \{\text{prime ideals } \mathfrak{P} \text{ in } K | N\mathfrak{P} \leq x\} \sim x/\log x \qquad \text{as } x \to \infty.$$

There is a proof in Goldstein [1]. See also Lagarias and Odlyzko [2].

We want to relate $\zeta_K(s)$ and Epstein's zeta function $Z(Y, s)$ using a method devised by Hecke [pp. 198–207]. This method is the essence of the relation between the general linear group and algebraic number fields. It will tell us all the properties of the Dedekind zeta function.

Theorem 2 (Hecke's Relation Between Dedekind and Epstein Zeta Functions). *Let $\varepsilon_1, \ldots, \varepsilon_r$ be a system of fundamental units for the algebraic number field K. For $x \in \mathbb{R}^r$, set*

$$\tau_j = \prod_{i=1}^{r} |\varepsilon_i^{(j)}|^{2x_i}, \qquad j = 1, 2, \ldots, n.$$

Let C be an ideal class in the ideal class group I_K of K. For any ideal $\mathfrak{A}$ in C, let $\mathfrak{A} = \mathbb{Z}\omega_1 \oplus \mathbb{Z}\omega_2 \oplus \cdots \oplus \mathbb{Z}\omega_n$. Then define the matrix $Y(C, x) \in \mathcal{P}_n$ by

$$Y(C, x) = \begin{pmatrix} \tau_1 & & 0 \\ & \ddots & \\ 0 & & \tau_n \end{pmatrix} \left\{ \begin{pmatrix} \omega_1^{(1)} \cdots \omega_n^{(1)} \\ \omega_1^{(n)} \cdots \omega_n^{(n)} \end{pmatrix} \right\},$$

using the notation $Y\{A\} = {}^t\bar{A} Y A$. If $\zeta_K(s)$ is Dedekind's zeta function, set

$$\Lambda_K(s) = (2^{-r_2}|d_K|^{1/2}\pi^{-n/2})^s \Gamma(s/2)^{r_1} \Gamma(s)^{r_2} \zeta_K(s).$$

Here $n = [K : \mathbb{Q}]$, r_1 is the number of real conjugate fields of K, and r_2 is the number of pairs of complex conjugate fields of K. Let $Z(Y, s)$ be Epstein's zeta function and

$$\Lambda(Y, s) = \pi^{-s} \Gamma(s) Z(Y, s).$$

We also use the notation $Y^0 = |Y|^{-1/n} Y$ for Y in $\mathcal{P}_n$. Clearly $|Y^0| = 1$. Then Hecke's integral formula is

$$w 2^{-r_1} n^{-1} R^{-1} \Lambda_K(s) = \sum_{C \in I_K} \int_{x \in [0,1]^r} \Lambda(Y(C, x)^0, ns/2) \, dx.$$

Here w is the number of roots of unity in K and R is the regulator of K.

PROOF.

Step 1. *The switch from a sum over ideals to a sum over elements of ideals.* Recall that I_K is the ideal class group. Let C be an ideal in I_K. Choose an ideal $\mathfrak{A}$ in the inverse class C^{-1}. Then $\mathfrak{A}\mathfrak{B} = \alpha O_K = (\alpha)$ for all ideals $\mathfrak{B}$ in C. Thus

$$\zeta_K(s) = \sum_{C \in I_K} \mathrm{N}\mathfrak{A}^s \sum_{\alpha \in (\mathfrak{A} - 0)/U_K} \mathrm{N}(\alpha)^{-s},$$

where U_K is the group of units in O_K and the quotient $(\mathfrak{A} - O)/U_K$ means take representatives for the equivalence relation $\alpha \sim \beta$ if and only if $\alpha = \beta u$ for some u in U_K. Here we have used the multiplicativity of the norm and the fact that $(\beta) = (\alpha)$ if and only if $\beta = \alpha u$, for some u in O_K.

Step 2. *Products of* Γ*-functions as products of Mellin transforms.* The Mellin transform formula for the gamma function is

$$a^{-s}\Gamma(s) = \int_{\mathbb{R}^+} y^{-s-1} \exp(-ay)\, dy, \qquad \text{for } a > 0,\ \mathrm{Re}\, s > 0.$$

Taking products of such integrals, we obtain

$$(2^{-r_2}|d_K|^{1/2}\pi^{-n/2})^s \Gamma(s/2)^{r_1} \Gamma(s)^{r_2} (N\mathfrak{A}/N(\alpha))^s$$
$$= \int_{y \in (\mathbb{R}^+)^{r_1+r_2}} \mathrm{N}y^{s/2} \exp\left(-\pi(|d_K| N\mathfrak{A}^2)^{-1/n}\, \mathrm{Tr}\,(\bar{\alpha}y\alpha)\right)\frac{dy}{y},$$

where

$$\mathrm{N}y = \prod_{j=1}^{r_1+r_2} y^{e_j}, \qquad \mathrm{Tr}\, y = \sum_{j=1}^{r_1+r_2} e_j y_j, \qquad \frac{dy}{y} = \prod_{j=1}^{r_1+r_2} \frac{dy_j}{y_j},$$

and

$$e_j = \begin{cases} 1 & \text{for } j = 1, \ldots, r_1, \\ 2 & \text{for } j = r_1 + 1, \ldots, r_1 + r_2. \end{cases}$$

The definitions are set up to make

$$\mathrm{N}y = \prod_{j=1}^{r_1+r_2} |\alpha^{(j)}|^{e_j} = \mathrm{N}(\alpha), \qquad \text{if } y = (|\alpha^{(1)}|, \ldots, |\alpha^{(r_1+r_2)}|),$$

for example. When we write $\bar{\alpha}y\alpha$ we mean the element of $(\mathbb{R}^+)^{(r_1+r_2)}$ given by the vector with jth component $|\alpha^{(j)}|^2 y_j$.

Step 3. *Hecke's change of variables.* Hecke decided to send $(y_1, \ldots, y_{r_1+r_2})$ to $(u, x_1, \ldots, x_r)$ via the equations

$$y_j = u\tau_j, \qquad j = 1, \ldots, r_1 + r_2.$$

Here the τ_j are defined in the statement of the theorem as products of powers of conjugates of fundamental units. If the fundamental units did not exist, then we would be able to deduce a contradiction to the convergence of Dedekind's zeta function. The following exercise is needed to perform the indicated substitution in the integral of step 2.

Exercise 14. Show that the Jacobian of the change of variables above is

$$\left|\frac{\partial y}{\partial(x,u)}\right| = y_1 \cdots y_{r+1} u^{-1} 2^{r_1 - 1} nR.$$

Thus we obtain

$$(2^{-r_2}|d_K|^{1/2}\pi^{-n/2})^s \Gamma(s/2)^{r_1}\Gamma(s)^{r_s}(\mathrm{N}\mathfrak{A}/\mathrm{N}(\alpha))^s$$

$$= n2^{r_1-1}R \int_{x\in\mathbb{R}^r}\int_{u>0} u^{-1+ns/2}\exp\left(-\pi u(|d_K|\mathrm{N}\mathfrak{A}^2)^{-1/n}\operatorname{Tr}(\bar{\alpha}\tau\alpha)\right) du\, dx.$$

Here $\tau = (\tau_1, \ldots, \tau_n)$.

Step 4. *Perform the integral over u and switch the quotient modulo units to the domain of the integral over x.* You obtain

$$(2^{-r_2}|d_K|^{1/2}\pi^{-n/2})^s \Gamma(s/2)^{r_1}\Gamma(s)^{r_2}\zeta_K(s)$$

$$= n2^{r_1}Rw^{-1}\pi^{-ns/2}\Gamma(ns/2)|d_K|^{s/2}\sum_{C\in I_K}\int_{x\in[0,1]^r} Z(Y(C,x), ns/2)\, dx.$$

To see this, note that the Dirichlet unit theorem says that if $u \in U_K$ then $u = \kappa\varepsilon_1^{a_1}\ldots\varepsilon_r^{a_r}$ for some $a_i \in \mathbb{Z}$ and κ is a root of unity in K. This decomposition is unique. Thus

$$|(\alpha u)^{(j)}|^2\tau_j = |\alpha^{(j)}|^2\prod_{i=1}^{r}|\varepsilon_i^{(j)}|^{2(x_i+a_i)}. \qquad \square$$

Exercise 15. Verify that $Y(C,x)$ is actually a positive definite symmetric matrix.

Hint. Suppose that α in $\mathfrak{A}$ has the form $\alpha = \Sigma\, a_j\omega_j$ for $a = {}^t(a_1, \ldots, a_n)$ in $\mathbb{Z}^n$. Then $Y(C,x) = \operatorname{Tr}(\bar{\alpha}\tau\alpha)$. This gives a positive real number even if you plug in a from $\mathbb{R}^n$. To see that the matrix $Y(C,x)$ is real and symmetric you need to look at the permutation

$$\sigma = \begin{pmatrix} 1\cdots r_1 & r_1+1 & \cdots & r_1+r_2 & r_1+r_2+1 & \cdots & n \\ 1\cdots r_1 & r_1+r_2+1 & \cdots & n & r_1+1 & \cdots & r_1+r_2 \end{pmatrix}$$

induced by complex conjugation on the conjugates of K. Then $\tau_{\sigma(j)} = \tau_j$.

Note that the value of the Epstein zeta function is independent of the choice of ideal in C and integral basis of that ideal. It is also independent of the particular set of fundamental units that we choose. Note that in particular, $|Y(C,x)| = \mathrm{N}\mathfrak{A}^2|d_K|$, since the integer matrix relating a $\mathbb{Z}$-basis of $\mathfrak{A}$ and a $\mathbb{Z}$-basis of O_K can be diagonalized by elementary divisor theory using elementary

1.4. Mellin Transforms, Epstein and Dedekind Zeta Functions

row and column operations. Changing the ideal in C which is used replaces the matrix $Y(C, x)$ by $Y(C, x)[A]$, for some A in $GL(n, \mathbb{Q})$. Since we are taking determinants to be one in the integral formula, we are transforming by $GL(n, \mathbb{Z})$, which leaves the Epstein zeta function invariant. A similar argument applies to changing the $\mathbb{Z}$-basis of the ideal.

If the field K is imaginary quadratic, the formula in Theorem 2 is much simpler, since then there are no integrals. Then it is in fact a very old formula, expressing the connection between ideals in imaginary quadratic fields and binary quadratic forms. We will see in Chapter 3 that this leads to an algorithm for the computation of class numbers of imaginary quadratic fields. This explains how Gauss could make class number conjectures before class numbers were invented. He was talking about class numbers of binary quadratic forms and conjectured that in the positive definite case there are only finitely many discriminants with a given class number.

Corollary (Properties of the Dedekind Zeta Function).

(1) $\zeta_K(s)$ can be continued to a meromorphic function of s with a simple pole at $s = 1$ having residue

$$\operatorname*{Res}_{s=1} \zeta_K(s) = 2^{r_1}(2\pi)^{r_2} hRw^{-1}|d|^{-1/2}$$

And $\zeta_K(s)$ satisfies the functional equation

$$\Lambda_K(s) = (2^{-r_2}|d|^{1/2}\pi^{-n/2})^s \Gamma(s/2)^{r_1} \Gamma(s)^{r_2} \zeta_K(s) = \Lambda_K(1-s).$$

(2) The Dedekind zeta function has the incomplete gamma expansion:

$$\Lambda_K(s) = \frac{\lambda}{s(s-1)} + n 2^{r_1-1} R w^{-1} \sum_{\substack{C \in I_K \\ a \in \mathbb{Z}^n_{-0}}} \int_{x \in [0,1]^r} \left(G\left(\frac{ns}{2}, Y(C,x)^\circ[a]\right) \right.$$
$$\left. + G\left(\frac{n(1-s)}{2}, Y(C,x)^\circ[a]\right) \right) dx$$

Here $\lambda = 2^{r_1} h R w^{-1}$.

PROOF. The Corollary follows from Theorems 1 and 2, using the fact that $Y(C, x)^{-1} = Y(C', -x)$, where C' denotes the ideal class containing the ideal $\mathfrak{A}'$ dual to the ideal $\mathfrak{A}$ in C that we had selected. If $\mathfrak{A}$ is an ideal the dual ideal $\mathfrak{A}'$ is defined to be

$$\mathfrak{A}' = \{\beta \in K \,|\, \operatorname{Tr}(\alpha\beta) \in \mathbb{Z} \text{ for all } \alpha \in \mathfrak{A}\}.$$

See Lang [3, pp. 57–58] where it is shown that $\mathfrak{A}'$ is a fractional ideal in general. You need to see that if ω' is a dual basis for the $\mathbb{Z}$-basis ω of the ideal $\mathfrak{A}$; i.e., if

$$\operatorname{Tr}(\omega_i \omega_j') = \delta_{ij} = \text{ the Kronecker delta},$$

then $(\omega_i^{(j)})^{-1} = {}^t(\omega_i'^{(j)})$ and ω' is a $\mathbb{Z}$-basis for the dual ideal to that generated by ω. The most important dual ideal is that for O_K itself. The *different* is defined to be $\delta_K = (O'_K)^{-1}$. And $N\delta_K = |d_K|$. The concept of dual ideal is really the same as the concept of dual lattice or dual subgroup of a locally compact abelian group. Poisson summation will always relate the sum over a lattice and that over the dual lattice, as we will show in detail in the next example from crystallography. See also Weil [2, Chapter 2]. When one replaces lattices by positive definite matrices, the dual lattice corresponds to the inverse matrix.

□

If the class number were infinite or if the r fundamental units did not exist, then one could show that $\zeta_K(s)$ is infinite, contradicting the bound that we found when $\operatorname{Re} s > 1$ in formula (1.24). This point of view is developed in Siegel [3, Vol. II, pp. 93–94], where it is noted that Dirichlet discovered his unit theorem "als er im Jahre 1846 in Rom der Ostermusik in der Sixtinischen Kapelle zuhörte."

We see from the Corollary that $\zeta_K(s)$ is negative when s is less than but sufficiently close to 1. It might be expected that the Dedekind zeta function would behave like Riemann's and thus its only possible zero in $(0, 1)$ would be at the point $\frac{1}{2}$, but no one has been able to prove this, even for quadratic fields (cf. Stark [1], Purdy et al [1]). In fact, one expects that any proof that works for the Riemann zeta function should be generalizable to the Dedekind zeta function. For example, Siegel says (see [3, Vol. II, p. 72]), "Sollte einmal ein Beweis für die Richtigkeit der Riemannschen Vermutung gelingen, so wird dieser vermutlich für alle, auch die von Hecke verallgemeinerten ζ-Funktionen gültig sein." However, although this statement appears true for the case of Landau's prime ideal theorem, it does not appear to work for real zeros of the Dedekind zeta function. For it is easy to show that the Riemann zeta function has no zeros in $(0, 1)$ (cf. Exercise 10). But no one knows how to generalize this to Dedekind's zeta function (i.e., there may exist "Siegel zeros").

Of course one still does not know whether the Riemann hypothesis holds for the Dedekind (or Riemann) zeta functions. Very little computer work has been done on any but Riemann's zeta function, but see Lagarias and Odlyzko [1].

We are interested in the sign of $\zeta_K(s)$ in $(0, 1)$ thanks to the Brauer-Siegel theorem. For the incomplete gamma expansion in the Corollary allows one to say that $\zeta_K(s) \leq 0$ implies

$$\frac{2^{r_1} w^{-1} hR}{s(1-s)} \geq \text{a sum of integrals of incomplete gamma functions.}$$

The best way to analyze the terms on the right side of this inequality seems to involve rewriting the terms as higher-dimensional incomplete gamma functions. If we can get good lower bounds on these integrals then we get good lower bound on the product of the class number and the regulator. Upper bounds come from (1.24), for example.

It is tempting to try to use Hecke's integral formula (Thm. 2) to deduce the behavior of $\zeta_K(s)$ for s in $(0, 1)$ from the behavior of $Z(Y, s)$ for s in $(0, n/2)$. But

1.4. Mellin Transforms, Epstein and Dedekind Zeta Functions

we know from Exercises 11 and 13 that the behavior of the Epstein zeta function is very different from the expected behavior of Dedekind's zeta function.

One can also interpret Hecke's integral formula (Theorem 2) as the computation of the 0th Fourier coefficient of the function $\Lambda_n(Y(C, x), ns/2)$, considered as a periodic function of x. Siegel computes all the Fourier coefficients for K a real quadratic field in [4]. One obtains Hecke L-functions with grossencharacters χ:

$$L(s, \chi) = \sum \chi(\mathfrak{A}) N \mathfrak{A}^{-s}.$$

For a real quadratic field of class number one and fundamental unit ε, an example of a Hecke grossencharacter is

$$\chi((\alpha)) = (\alpha/\alpha')^{\pi i / \log \varepsilon}, \quad \text{for } \alpha \text{ in } O_K.$$

Here α' means the conjugate of α in K. For more information on Hecke grossencharacters, see Hecke [1, pp. 215–234, 249–287].

If K is imaginary quadratic, then there is no integral in Hecke's integral formula. Then the result analogous to that of Siegel gives a connection between nonanalytic Eisenstein series and Hecke L-functions with grossencharacters. An example of grossencharacter χ for an imaginary quadratic field of class number one is:

$$\chi((\alpha)) = \alpha^e \bar{\alpha}^f, \quad \text{with } e + f \text{ even.}$$

In 1970 Damerell found $L(n, \chi)$, for large n in $\mathbb{Z}^+$, to be an algebraic number times a certain explicit transcendental number (see Weil [3]).

You might ask why one cares about values of L-functions at positive integers. The earliest result of this type was that of Euler, who showed that for Riemann's zeta function (see Exercise 7 of §3.5),

$$\zeta(2m) = (-1)^{m-1} \tfrac{1}{2} (2\pi)^{2m} B_{2m}/(2m)!, \quad m = 1, 2, 3, \ldots.$$

Here $B_m \in \mathbb{Q}$ is the nth *Bernoulli number* defined by

$$\frac{x}{e^x - 1} = \sum_{n \geq 0} B_n \frac{x^n}{n!}.$$

Analogous results, using similar methods, have recently been obtained by Shintani [1, 3]. Interpretations involving p-adic interpolation of these values have been much discussed (see Coates [1], Iwasawa [1]). There are also some very general conjectures of Lichtenbaum giving K-theoretic meaning to values of L-functions at negative integers (see Borel [2], Lichtenbaum [1], Serre [3]).

Values of L-functions are of interest also because of their connection with Hilbert's 12th problem which asks for an explicit construction of class fields or abelian extensions of number fields K. When $K = \mathbb{Q}$, a theorem of Kronecker and Weber says that a finite extension E of $\mathbb{Q}$ with abelian Galois group must lie in a cyclotomic field $E \subset \mathbb{Q}(\exp(2\pi i/m))$. When the base field is imaginary quadratic rather than $\mathbb{Q}$, one adjoins values of elliptic modular functions (see Shimura [1], Chowla et al. [1]). When K is arbitrary, Shimura has obtained results of this sort using algebraic geometry. Another point of view is expressed in Stark [3]. Stark's idea is that values of L-functions can be used to find units which generate class fields and express the reciprocity laws of class field theory. He has proved this in classical cases and has convincing computer examples in other cases (e.g., the base field real quadratic or cubic).

Application 2. Crystallography

A three-dimensional lattice is the set of all linear combinations with integer coefficients of three linearly independent vectors v_1, v_2, v_3 in $\mathbb{R}^3$. If you choose the standard basis vectors of $\mathbb{R}^3$, your lattice is $\mathbb{Z}^3$. If you think of atoms as points with legs, then by connecting atoms you can build up a crystal lattice such as that of salt (NaCl) pictured along with various other lattices in Figure 1.13 from Born and Huang [1, pp. 383–384].

A *crystallographic point group* G is a subgroup of the group $O(3)$ of rotations of 3-space which leaves a lattice L invariant; i.e., for each v in L the vector gv is in L, for any g in G. Note that you can represent the elements of G by matrices with integer entries or by rotation matrices. Thus G is compact and discrete which means that G must be a finite group. There are 18 abstract point groups. If instead of isomorphism classes, you look at conjugacy classes in the group $GL(3, \mathbb{R})$ of all nonsingular 3×3 real matrices, then you get 32 groups.

Abstract Point Groups

(1) Cyclic Groups
 C_n = cyclic group with n elements, $n = 1, 2, 3, 4, 6$.
(2) Dihedral Groups
 D_n = group of proper rotations (i.e., det = 1) of a regular n – gon, $n = 2, 3, 4, 6$.
(3) Tetrahedral Group
 T = group of proper rotations of the regular tetrahedron.
(4) Octahedral Group
 O = group of proper rotations of the cube.
(5) Point Groups of the 2nd Kind
 Add $-I = -$the identity matrix to the group. Write $C_2 = \{I, -I\}$.
 $C_4 \times C_2, C_6 \times C_2, D_2 \times C_2, D_4 \times C_2, D_6 \times C_2, T \times C_2, O \times C_2$

The groups C_n, D_n, for $n = 1, 2, 4$ and T, 0 leave invariant the lattice $\mathbb{Z}^3$. The groups C_n, D_n, for $n = 3, 6$ leave invariant a lattice with basis $(1, 0, 0)$, $(1/2, \sqrt{3}/2, 0)$, $(0, 0, 1)$. The matrix $-I$ leaves all lattices invariant. More information on these things can be found in Birman [1], Hilbert and Cohn-Vossen [1], Janssen [1], Loeb [1], Lomont [1], Nussbaum [1], and Schwarzenberger [1].

1.4. Mellin Transforms, Epstein and Dedekind Zeta Functions 77

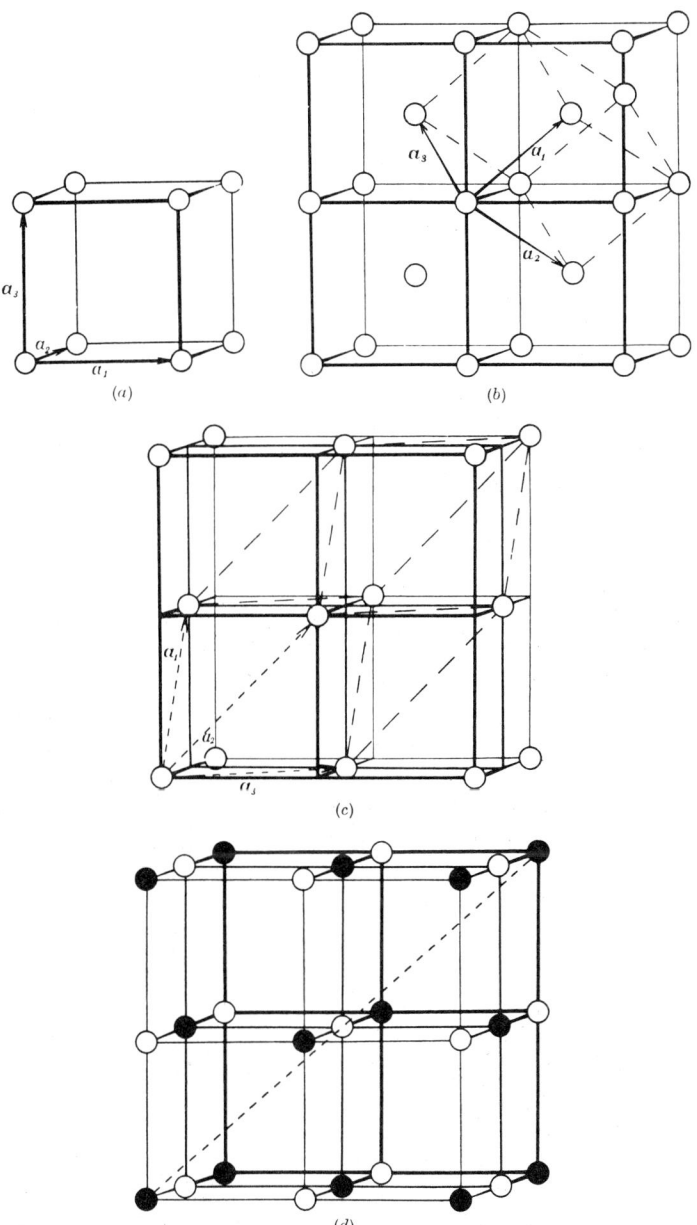

Figure 1.13. Various crystal lattices: (a) Simple cubic; (b) Body-centered cubic; (c) Face-centered cubic; (d) NaCl; (e) CsCl; (f) Diamond; (g) ZnS. (From Born and Huang, *Dynamical Theory of Crystal Lattices*, © 1956, pp. 383–384. Reprinted by permission of Oxford University Press.)

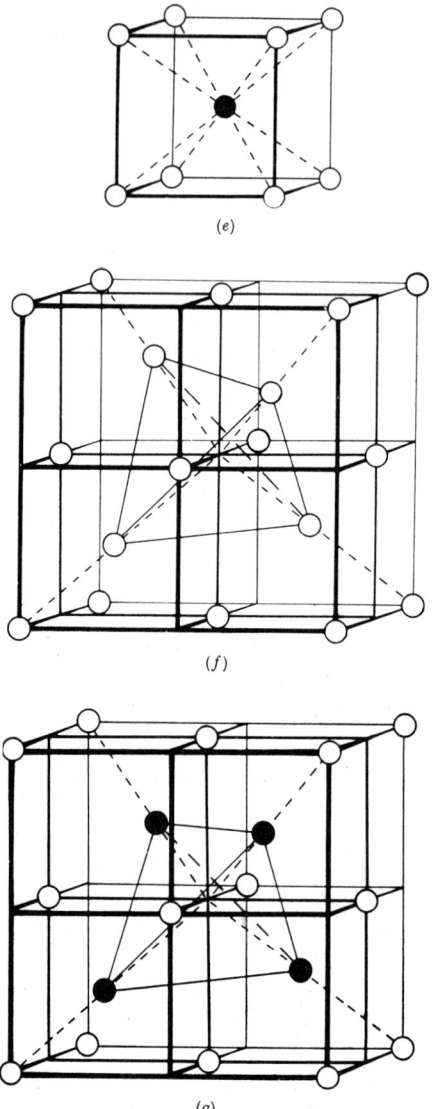

Figure 1.13. (continued)

1.4. Mellin Transforms, Epstein and Dedekind Zeta Functions

The *crystallographic space groups* are discrete subgroups of the group $E(3)$ of Euclidean motions of $\mathbb{R}^3$; i.e., the group generated by rotations and translations. The subgroup U of translations in the space group G is isomorphic to $\mathbb{Z}^3$ and G is an extension of U by the finite point group K. There are 219 nonisomorphic space groups. And there are 230 nonconjugate space groups. Schoenfliess, Federov, and Barlow worked this out around 1890, after Sohnke had listed the orientation-preserving space groups. See Schwarzenberger [1, pp. 132–133] for the fascinating history of this episode. It follows that there are only 230 ways to form crystals. And crystallography is based on the study of such groups and their representations. For the quantum mechanics of such a crystal will be ruled by the symmetry group of the crystal, and not the full group of rotations. This will become clearer after we discuss the Schrödinger equation in §2.2.

But we do not want to discuss the representations of finite groups here. Nor do we want to discuss the quantum mechanics that can be derived from knowledge of the character tables of these finite groups leaving crystals invariant. The interested reader could consult any of the references above for these things.

Instead we want to show how to use Theorem 1 to deal with some lattice sums that arise in physics. Given a crystal lattice $\mathbb{Z}v_1 \oplus \mathbb{Z}v_2 \oplus \mathbb{Z}v_3$, we can form a positive definite 3×3 matrix

$$Y = \begin{pmatrix} {}^tv_1 v_1 & {}^tv_1 v_2 & {}^tv_1 v_3 \\ {}^tv_2 v_1 & {}^tv_2 v_2 & {}^tv_2 v_3 \\ {}^tv_3 v_1 & {}^tv_3 v_2 & {}^tv_3 v_3 \end{pmatrix} = I[v_1 v_2 v_3].$$

For any sum over the crystal involving powers of the distance of a vector from the origin we are really computing Epstein zeta functions, since

$$w = a_1 v_1 + a_2 v_2 + a_3 v_3 = (v_1 v_2 v_3) \begin{pmatrix} a_1 \\ a_2 \\ a_3 \end{pmatrix}$$

and $\|w\|^2 = Y[a]$, if $a = {}^t(a_1, a_2, a_3)$ is in $\mathbb{Z}^3$. Many sums of this sort are considered in Born and Huang [1, pp. 248, 385–390, 413] and Ziman [1, pp. 37–41], for example.

Next let us consider the simplest example—the computation of the electrostatic energy of salt (NaCl). The size of this energy has an effect on the solubility of a salt in various solvents, as well as on the hardness and melting point of the crystal (see Højendal [1]). The potential energy of one ion with respect to all others in the lattice is

$$\phi = \frac{v v_1 e^2}{L_0} \sum (\pm) \frac{L_0}{L} = \frac{v v_1 e^2}{L_0} \alpha,$$

where ve, $v_1 e$ are the charges of the two kinds of ions, L_0 is the length of an edge of the unit cell, L is the distance between two ions, and α is *Madelung's constant*. Then sign ± 1 is determined according to charges of the two ions.

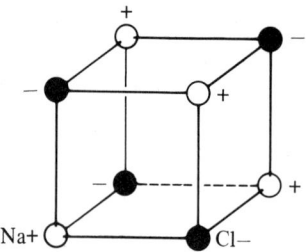

Figure 1.14. NaCl lattice.

A slight generalization of Theorem 1 leads to a simple way to compute Madelung's constant (see Exercise 17). This method and variants, some of which use the Fourier expansion of Epstein's zeta function in Chapter 3, have led to many papers in the crystallography literature (cf. the preceding references to Born and Huang or Ziman, plus Emersleben [1, 2], Ewald [1], Sakamoto [1], and Zucker [1], or the multitudes of papers cited by these authors). The series involved in the computation of Madelung's constant for NaCl (salt) is a difference of two Epstein zeta functions evaluated at $s = \frac{1}{2}$. Since Epstein's zeta function of a 3 × 3 matrix has a pole at $s = \frac{3}{2}$, the Dirichlet series does not converge at $s = \frac{1}{2}$. Thus the series expressing Madelung's constant does not converge absolutely, and a certain amount of care must be taken.

Højendal [1] invented a method that just sums the terms of the original series in a straightforward way. He says that he finds Ewald's method (see Exercise 17) "still less understandable to persons not acquainted with the highest of mathematics" than Madelung's which involves the Fourier series for Epstein's zeta function (see Chapter 3).

Exercise 16 (Højendal's Method for Evaluating Madelung's Constant for Salt (NaCl). Compute α defined above for the NaCl crystal pictured at the beginning of our discussion of crystallography. To do this begin by computing $\Sigma(\pm)L_0/L$ for the eight ions in the smallest possible crystal pictured in Fig. 1.14. You should obtain

$$\frac{3}{1} - \frac{3}{\sqrt{2}} + \frac{1}{\sqrt{3}}.$$

Then compute the sum over the next layer of six cubes surrounding this one completely. And continue in this way proceeding by layers similar to "a system of Chinese boxes." We have included Højendal's result so that you can check your program. He had no computer. The nth level has $(2n)^3 - 1$ ions to deal with. So his ninth division involves summing 8000 terms. Hopefully the reader will use a computer.

1.4. Mellin Transforms, Epstein and Dedekind Zeta Functions

$$1 \text{ division: } \frac{3}{1} - \frac{3}{\sqrt{2}} + \frac{1}{\sqrt{3}} \qquad = \quad 1.456030$$

$$2 \text{ division: } \frac{3}{1} - \frac{9}{\sqrt{2}} + \frac{7}{\sqrt{3}} - \frac{3}{\sqrt{4}} + \frac{12}{\sqrt{5}} -$$

$$-\frac{12}{\sqrt{6}} - \frac{3}{\sqrt{8}} + \frac{6}{\sqrt{9}} - \frac{1}{\sqrt{12}} \quad = \quad 0.295739$$

3 division: $\qquad = -0.004729$
4 division: $\qquad = 0.000679$
5 division: $\qquad = -0.000221$

Total $\qquad 1.747498 \pm 0.0005$

6 division: $\qquad = 0.000100$
7 division: $\qquad = -0.000059$
8 division: $\qquad = 0.000040$
9 division: $\qquad = -0.000028$
estimated for further divisions $\qquad = 0.000007$

Total $\qquad 0.000060$
added to the above $\qquad 1.747498$
we obtain as our most accurate value $\qquad 1.747558$

This is very close to the values found by Madelung and by Emersleben who found: 1.747557.

Exercise 17 (Ewald's Method or the Method of Theta Functions).

(a) Show that Madelung's constant for NaCl is

$$\sum_{n \in \mathbb{Z}^3 - 0} \exp[\pi i(n_1 + n_2 + n_3)] I[n]^{-1/2}, \qquad I = \text{identity matrix}.$$

This is a special case of Epstein's most general zeta function considered in Epstein [1, 2], for example. The *general Epstein zeta function* is

$$Z(Y, g, h, s) = \sum_{a \in \mathbb{Z}^n - 0} Y[a + g]^{-s} \exp(2\pi i^t h a),$$

if $\operatorname{Re} s > n/2$, $Y \in \mathscr{P}_n$, $g, h \in \mathbb{R}^n$.

(b) Define the *general theta function* by

$$\theta(Y, g, h, t) = \sum_{a \in \mathbb{Z}^n - 0} \exp\{-\pi t Y[a + g] + 2\pi i^t h a\}$$

for $Y \in \mathscr{P}_n$, $t > 0$, $g, h \in \mathbb{R}^n$. Show that theta satisfies the transformation formula

$$\theta(Y, g, h, t) = t^{-n/2} |Y|^{-1/2} \exp(-2\pi i^t g h) \theta(Y^{-1}, h, -g, t^{-1}).$$

(c) Imitate the proof of Theorem 1 to show that Epstein's general zeta function satisfies the functional equation

$$\Lambda(Y, g, h, s) = \pi^{-s}\Gamma(s)Z(Y, g, h, s)$$
$$= |Y|^{-1/2} \exp(-2\pi i\, {}^t gh) \Lambda(Y^{-1}, h, -g, n/2 - s).$$

Also obtain an expansion of $Z(Y, g, h, s)$ in incomplete gamma functions analogous to the expansion obtained in Theorem 1.

(d) Use the result of (c) to write a computer program to compute Madclung's constant for NaCl. Compare the results with those of Problem 16.

There are other applications of Epstein zeta functions. For example, this function is the simplest case of the zeta functions studied in Minakshisundaram and Pleijel [1] in order to obtain information about the eigenvalues of the Laplacian on compact differentiable manifolds (see also Singer [1]). And minima of Epstein zeta functions for fixed s have been investigated in order to find the best lattice to use in numerical integration (see Delone and Ryskov [1] and Fields [1]). Applications of Epstein zeta functions to quantum statistical mechanics are considered by Hurt and Hermann [1, Ch. 8].

CHAPTER II
A Compact Symmetric Space—The Sphere

It is important that you establish a steady, rhythmic tempo while doing the exercises and that you rest as little as possible between them. Also remember not to strain yourself.

> From *Jane Fonda's Year of Fitness and Health/1984, Desk Diary*, Simon and Schuster, NY, 1983. Reprinted by permission.

2.1. Spherical Harmonics

Whenever there is a large earthquake the Earth vibrates for days afterwards. The vibrations consist of the superposition of the elastic-gravitational normal modes of the Earth that are excited by the earthquake.

> From F. Gilbert [1, p. 107].

A *surface* (or Laplace) *spherical harmonic* is an eigenfunction of the Laplacian on the sphere. These are the analogues of exponentials for Fourier analysis on the sphere. Laplace and Legendre introduced these functions in order to study gravitational theory in the 1780s. Spherical harmonics are necessary for the analysis of any phenomena with spherical symmetry; e.g., earthquakes, the hydrogen atom, and the solar corona. Some of these topics will be discussed later in this section.

Since our treatment of harmonic analysis on the sphere is rather condensed, the reader may want to consult some of the following references for more

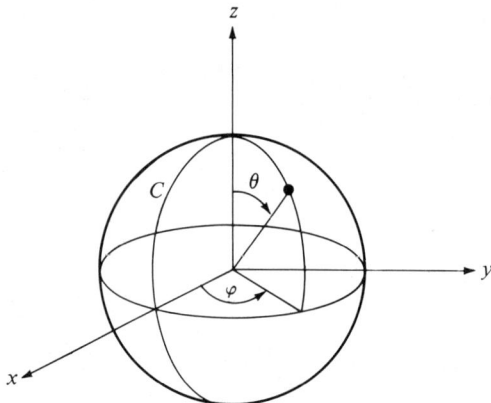

Figure. 2.1. Angular coordinates on the unit sphere.

information: Coifman and Weiss [1], Courant and Hilbert [1], Dym and McKean [1], Erdélyi et al. [1], Lebedev [1], Müller [1], Sugiura [1], Talman [1], and Vilenkin [1]. For the history of the subject, see Wangerin [1].

Before discussing spherical harmonics, we need to understand something about the geometry of the sphere. This symmetric space is closely related to the *orthogonal group* $O(n)$ of real $n \times n$ matrices U such that ${}^t U U = I$, where ${}^t U$ denotes the transpose of U and I denotes the $n \times n$ identity matrix. And $SO(n)$, the *special orthogonal group*, is the subgroup of matrices U in $O(n)$ such that the determinant of U is one. Note that you can regard U in $SO(n-1)$ as an element of $SO(n)$ by forming

$$\begin{pmatrix} U & 0 \\ 0 & 1 \end{pmatrix} \text{ in } SO(n).$$

Exercise 1 (The Sphere as a Quotient or Homogeneous Space). Consider the sphere $S^{n-1} = \{x \in \mathbb{R}^n \mid \|x\| = 1\}$. Show that S^{n-1} can be identified with the quotient space $SO(n)/SO(n-1)$, using the preceding identification of $SO(n-1)$ as a subgroup of $SO(n)$.

Hint. Map the coset $gSO(n-1)$ to the vector ge_n for g in $SO(n-1)$ and $e_n = {}^t(0,\ldots,0,1)$. You might also want to read the discussion of the topology of the spheres and orthogonal groups in Chevalley [1, pp. 52–67]. In particular, the fundamental (or Poincaré) group of $SO(2)$ is isomorphic to $\mathbb{Z}$, while that of $SO(n)$, $n \geq 3$, has order 2.

From now on we shall consider only the sphere S^2. See the references for the general case.

Now the sphere S^2 is a *differentiable manifold*. This means that locally it looks like two-dimensional Euclidean space. To make this precise, we use the usual angular coordinates (θ, φ), $0 < \varphi < 2\pi$, $0 < \theta < \pi$, pictured in Fig. 2.1, to

2.1. Spherical Harmonics

parameterize S^2 except for the semicircle C through the poles and $(1,0,0)$. A similar coordinate patch can be constructed to cover the rest of the sphere. The equations for the rectangular coordinates (x, y, z) of a point on S^2 in terms of the angular coordinates are

$$x = \sin\theta \cos\varphi$$
$$y = \sin\theta \sin\varphi \qquad (2.1)$$
$$z = \cos\theta$$

More information on differentiable manifolds can be found in Choquet-Bruhat et al. [1], Helgason [1, 2, 3], Loomis and Sternberg [1], Singer and Thorpe [1], or Spivak [1], for example.

Before proceeding further with our discussion of spherical geometry, let us review how to change variables in the Laplacian, using the method of Courant and Hilbert [1, Vol. 1, pp. 224–225]. Assume that the substitution mapping (x, y, z) to (u_1, u_2, u_3) is differentiable with differentiable inverse. Let the Jacobian matrix of the change of variables be

$$A = \frac{\partial(x, y, z)}{\partial(u_1, u_2, u_3)}. \qquad (2.2)$$

Then the *volume element* is

$$dx\, dy\, dz = \|A\|\, du_1\, du_2\, du_3, \qquad \|A\| = \text{absolute value of determinant of } A. \qquad (2.3)$$

And the *Euclidean arc length element* is

$$ds^2 = (dx\, dy\, dz)\begin{pmatrix} dx \\ dy \\ dz \end{pmatrix} = (du_1\, du_2\, du_3)^t A A \begin{pmatrix} du_1 \\ du_2 \\ du_3 \end{pmatrix}.$$

Thus we obtain

$$ds^2 = \sum_{i,j=1}^{3} g_{ij}\, du_i\, du_j \qquad \text{where } G = {}^t A A = (g_{ij})_{1 \le i,j \le 3}. \qquad (2.4)$$

Similarly, one uses the fact that

$$\left(\frac{\partial f}{\partial x}, \frac{\partial f}{\partial y}, \frac{\partial f}{\partial z}\right) = \left(\frac{\partial f}{\partial u_1}, \frac{\partial f}{\partial u_2}, \frac{\partial f}{\partial u_3}\right) A^{-1},$$

to see that

$$\left(\frac{\partial f}{\partial x}\right)^2 + \left(\frac{\partial f}{\partial y}\right)^2 + \left(\frac{\partial f}{\partial z}\right)^2 = \sum_{i,j=1}^{3} g^{ij} \frac{\partial f}{\partial u_i} \frac{\partial f}{\partial u_j} \qquad \text{where } G^{-1} = (g^{ij})_{1 \le i,j \le 3}. \qquad (2.5)$$

To see what happens to the Laplacian in the new coordinate system, one can use the calculus of variations (see Courant and Hilbert [1, Vol. 1, Ch. 4]).

Suppose that f minimizes the integral

$$J(f) = \int (f_x^2 + f_y^2 + f_z^2)\,dx\,dy\,dz$$

subject to the constraint

$$K(f) = \int f^2\,dx\,dy\,dz = \text{constant}.$$

Then f must satisfy the *Euler-Lagrange equation*

$$f_{xx} + f_{yy} + f_{zz} = \Delta f = \lambda f$$

(see Courant and Hilbert [1, Vol. 1, p. 192]). That is, f must be an eigenfunction of the Euclidean Laplacian. Suppose now that we change variables in the integrals J and K. The new constrained minimization problem should lead to a differential equation involving the transformed version of the Laplacian. And this is indeed the case. From (2.3) and (2.5), one obtains

$$J(f) = \int \sum_{i,j=1}^{3} g^{i,j} f_{u_i u_j} \sqrt{|G|}\,du_1\,du_2\,du_3 \qquad \text{with } |G| = \det G,$$

and

$$K(f) = \int f^2 \sqrt{|G|}\,du_1\,du_2\,du_3.$$

The Euler-Lagrange equation for the transformed problem is

$$\sum_i \frac{\partial}{\partial u_i} \sqrt{|G|} \sum_k g^{i,k} \frac{\partial f}{\partial u_k} = \lambda f \sqrt{|G|}.$$

Thus the *Laplacian* in the new coordinate system is

$$\Delta f = u_{xx} + u_{yy} + u_{zz} = |G|^{-1/2} \sum_{i=1}^{3} \frac{\partial}{\partial u_i} |G|^{1/2} \sum_{k=1}^{3} g^{i,k} \frac{\partial f}{\partial u_k}. \tag{2.6}$$

There are other ways to obtain these formulas. See, for example, the following references: Arfken [1, Ch. 2], Choquet-Bruhat et al. [1, p. 307], Churchill and Brown [1, pp. 16–19], or Helgason [2, pp. 386–387]).

Now we can show that the sphere is a *Riemannian manifold*, meaning that there is a notion of arc length which is defined from an inner product on the tangent space to the surface at each point. In order to obtain the arc length element on S^2, we shall use the preceding discussion to write the Euclidean arc length element in spherical polar coordinates. This is done in the next exercise.

Exercise 2 (Spherical Polar Coordinates). Spherical polar coordinates for a point (x, y, z) in $\mathbb{R}^3$ are defined by

2.1. Spherical Harmonics

$$x = r \sin \theta \cos \varphi$$
$$y = r \sin \theta \sin \varphi$$
$$z = r \cos \theta$$

where $0 \leq \varphi \leq 2\pi$, $0 \leq \theta \leq \pi$, $0 \leq r$. Use formulas (2.3), (2.4), and (2.6) above to transform the Euclidean volume, arc length, and Laplacian to spherical polar coordinates.

Answer.
$$d\mu = dx \, dy \, dz = r^2 \sin \theta \, dr \, d\theta \, d\varphi,$$
$$ds^2 = dx^2 + dy^2 + dz^2 = dr^2 + r^2 \, d\theta^2 + r^2 \sin^2 \theta \, d\varphi^2,$$
$$\Delta f = (r^2 \sin \theta)^{-1} \left(\frac{\partial}{\partial r}\left(r^2 \sin \theta \frac{\partial f}{\partial r}\right) + \frac{\partial}{\partial \theta}\left(\sin \theta \frac{\partial f}{\partial \theta}\right) + \frac{\partial}{\partial \varphi}\left(\frac{1}{\sin \theta}\frac{\partial f}{\partial \varphi}\right)\right).$$

It follows from Exercise 2 that the *element of arc length* on the unit sphere S^2 is
$$ds^2 = d\theta^2 + \sin^2 \theta \, d\varphi^2. \qquad (2.7)$$

Note that this element of arc length is invariant under $0(3)$ as is the corresponding *volume element*:
$$d\mu = \sin \theta \, d\theta \, d\varphi. \qquad (2.8)$$

Finally the Laplacian on S^2 is
$$\Delta^* = (\sin \theta)^{-1}\left(\frac{\partial}{\partial \theta}\left(\sin \theta \frac{\partial}{\partial \theta}\right) + (\sin \theta)^{-1}\frac{\partial^2}{\partial \varphi^2}\right). \qquad (2.9)$$

Exercise 3. Show that the Euclidean Laplacian $\Delta f = f_{xx} + f_{yy} + f_{zz}$ is invariant under $0(3)$ (i.e., that Δ commutes with rotation). Deduce that the spherical Laplacian (2.9) is also invariant under $0(3)$.

Hint. Dym and McKean [1, p. 243] has a proof using the Fourier transform.

Because the sphere is a Riemannian manifold, we can consider *geodesics* on the sphere. The geodesic through two points on the sphere is the curve through these two points that minimizes distance. Airplane pilots know that geodesics in S^2 are great circles; i.e., the intersection of S^2 with a plane through the origin and the two given points on S^2. To see this, suppose that you are given points p and q on S^2. You can rotate S^2 so that both points p and q have φ-coordinate in (2.1) equal to zero. See Exercise 11 if you do not believe this. Then the distance between p and q on S^2 is
$$\int_p^q \sqrt{d\theta^2 + \sin^2 \theta \, d\varphi^2} \geq \int_p^q d\theta = \theta(q) - \theta(p).$$

Thus the great circle route minimizes distance on the sphere (assuming that you go in the proper direction around the circle).

Exercise 4 (The Sphere is a Symmetric Space). Show that S^2 is a symmetric space. To do this, it only remains to show that for each point p in S^2 there is a diffeomorphism $f_p: S^2 \to S^2$ which preserves arc length (i.e., f_p is an isometry of the Riemannian manifold) and reverses geodesics.

Hint. At $p = (0, 0, 1)$, use $(\theta, \varphi) \mapsto (-\theta, -\varphi)$.

Spherical geometry is non-Euclidean because the geodesics cannot be extended indefinitely, which violates Euclid's second postulate. In fact, Euclid's fifth postulate fails as well. And Klein noticed that if you identify antipodal points on the sphere, then any two geodesic lines have a unique point in common. The geometry obtained by identifying antipodal points on the sphere is called *elliptic geometry* and the resulting compact surface is also called the *projective plane*. The name "elliptic" is due to Klein (from a Greek work *elleipein*, to fall short). It is motivated by the following picture. Suppose that two geodesic rays R_1 and R_2 emanate from the ends of a geodesic segment perpendicular to them both. The distance between the rays R_1 and R_2 will decrease or fall short.

In elliptic or spherical geometry, when A, B, C are the three angles of a geodesic triangle, then the area of the triangle is $A + B + C - \pi$, measuring the angles in radians, of course. This result goes back to Albert Girard (1595–1632) for the sphere. References for non-Euclidean geometry are Coxeter [1] and Hilbert and Cohn-Vossen [1].

Exercise 5. Prove Girard's formula mentioned above.

Note. This could be proved using the Gauss-Bonnet formula (see Singer and Thorpe [1]) and, in fact, it was the first known case of the Gauss-Bonnet theorem.

It is now possible to make sense of the definition of surface spherical harmonic which was given at the beginning of this section. A *surface spherical harmonic* Y is a function $Y: S^2 \to \mathbb{C}$ such that

$$\Delta^* Y = (\sin \theta)^{-1} \left((\sin \theta \, Y_\theta)_\theta + (\sin \theta)^{-1} Y_{\varphi \varphi} \right) = \lambda Y \tag{2.10}$$

for some eigenvalue $\lambda \in \mathbb{C}$. The theorem that follows characterizes these spherical harmonics very explicitly in terms of (associated) Legendre polynomials. Legendre had discussed these polynomials before Laplace's work, but Legendre did not manage to publish his results until after Laplace did (see Legendre [1, 2] and Laplace [1]). Spherical harmonics also played a role in Laplace's long treatise on celestial mechanics, as well as in Gauss's work on terrestrial

magnetism. The next Exercises give some of the properties of *Legendre polynomials* $P_n(x)$ defined by

$$2^n n! P_n(x) = \frac{d^n}{dx^n}[(x^2-1)^n] \qquad \text{for } n = 0, 1, 2, \ldots, \tag{2.11}$$

and *associated Legendre functions* $P_n^m(x)$ defined by

$$P_n^m(x) = (1-x^2)^{m/2} \frac{d^m P_n(x)}{dx^m}. \tag{2.12}$$

Note that some authors replace $(1-x^2)$ by (x^2-1) in formula (2.12). We follow Courant and Hilbert [1] and Arfken [1] rather than Lebedev [1] in our choice of notation.

Exercise 6 (Legendre Polynomials and Associated Legendre Functions).

(a) Show that the associated Legendre functions are solutions of the Sturm-Liouville equation:

$$(1-x^2)u'' - 2xu' + \{n(n+1) - (1-x^2)^{-1}m^2\}u = 0.$$

(b) If $n = 0, 1, 2, \ldots$, show that $P_n^m(x) = 0$ for all x, unless $m = 0, +1, -1, +2, -2, \ldots, +n, -n$. Note that we can allow m to be negative by writing $(d/dx)^{-1}(d/dx) = \text{identity}$. Show also that

$$P_n^{-m}(x) = (-1)^m \frac{(n-m)!}{(n+m)!} P_n^m(x).$$

(c) Show that for fixed m in $\{0, \pm 1, \pm 2, \ldots\}$, the set $\{P_n^m | n = |m|, |m|+1, |m|+2, \ldots\}$ is a complete orthonormal set for $L^2[-1, +1]$.

Hint. The orthogonality is easy. For completeness when $m = 0$, see Courant and Hilbert [1, Vol. 1, pp. 82–83]. In the general case, write $P_{m+k}^m(x) = (1-x^2)^{m/2} f_k^m(x)$. Then $\{f_k^m, k = 0, 1, 2, \ldots\}$ is a complete set of orthogonal polynomials for $L^2([-1, +1], (1-x^2)^m)$. This is the weighted L^2-space with inner product given by

$$(f, g) = \int_{-1}^{+1} f(x)\overline{g(x)}\,(1-x^2)^m\,dx.$$

This means that you must show that f_k^m is orthogonal to x^r when $r = 0, 1, \ldots, k-1$. Integration by parts, using the case $m = 0$, will do the trick.

Exercise 7 (Integral Formula for Associated Legendre Functions). Show that the associated Legendre P_n^m is represented by the following integral:

$$P_n^m(x) = i^m \frac{\Gamma(n+m+1)}{\pi \Gamma(n+1)} \int_0^\pi (x + \sqrt{x^2-1}\cos\varphi)^n \cos(m\varphi)\,d\varphi$$

for $\operatorname{Re} x > 0$, $m = 0, 1, 2, \ldots$.

Hint. See Courant and Hilbert [1, Vol. 1, p. 505].

Exercise 7 has been generalized to symmetric spaces by Harish-Chandra. This will be discussed in later chapters.

Theorem 1 (Spherical Harmonics).

(a) *The only eigenvalues λ of the spherical Laplacian Δ^* defined in (2.10) are $\lambda = -n(n+1)$, $n = 0, 1, 2, \ldots$. The vector space of eigenfunctions of Δ^* corresponding to the eigenvalue $\lambda = -n(n+1)$ has dimension $2n+1$. A complete orthogonal set of eigenfunctions of Δ^* is*

$$\exp(im\varphi) P_n^m(\cos\theta) \qquad \text{for } n = 0, 1, 2, \ldots, |m| \le n.$$

Here P_n^m is the associated Legendre function (2.12) and the coordinates (θ, φ) are defined by (2.1).

(b) *Let $f(x, y, z) = r^n Y(\theta, \varphi)$, using spherical polar coordinates (r, θ, φ) corresponding to the rectangular coordinates (x, y, z) as in Exercise 2. Then Y is a surface spherical harmonic satisfying $\Delta^* Y = -n(n+1)Y$ if and only if $f(x, y, z)$ is a homogeneous harmonic polynomial of degree n. When we say that f is harmonic, we mean that f satisfies Laplace's equation*

$$\Delta f = f_{xx} + f_{yy} + f_{zz} = 0.$$

Exercise 8. Prove part (a) of Theorem 1, using separation of variables on $\Delta^* Y = \lambda Y$ and Exercise 6.

Note. There are other proofs that $\lambda = -n(n+1)$, $n = 0, 1, 2, \ldots$ are the only possible eigenvalues of Δ^* on S^2. For example, see Dym and McKean [1, pp. 252–253], Kirillov [1, pp. 271–274], or Van der Waerden [1, p. 21].

Exercise 9. Use separation of variables on $\Delta f = 0$ to prove part (b) of Theorem 1. Note that Exercise 2 says that if $f(x, y, z) = R(r) Y(\theta, \varphi)$, then

$$\Delta f = r^{-2}(r^2 R'(r))' Y + r^{-2} R \Delta^* Y \qquad \text{where } ' = d/dr.$$

If $\Delta f = 0$, then $\Delta^* Y = \lambda Y$ and $(r^2 R'(r))' = -\lambda R$, for some constant λ.

Exercise 10 (Why the Eigenvalues of the Laplacian on the Sphere Are Negative). Suppose that f and $\Delta^* f$ are in $L^2(S^2)$. Show that $(f, \Delta^* f) \le 0$. Use Green's theorem. Then show that the eigenvalues of Δ^* are all negative or zero.

According to part (a) of Theorem 1, separation of variables in $\Delta^* Y = \lambda Y$ leads from functions $Y(\theta, \varphi)$ of the two angle variables in (2.1) to functions $\exp(im\varphi) P_n^m(\cos\theta)$. If we assume that Y is constant with respect to the angle φ, then $Y = Y(\theta) = P_n(\cos\theta)$. Such a spherical function $Y = Y(\theta)$ is called a *zonal spherical function*. The name results from the fact that the zeros of Y cut the sphere up into zones, as in Fig. 2.2. In general, $P_n(\cos\theta)$ has n distinct zeros in

2.1. Spherical Harmonics

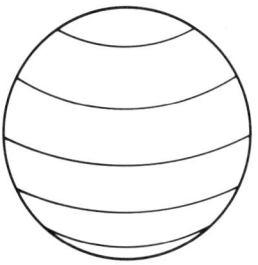

Figure 2.2. Zero set of $P_n(\cos\theta)$.

$0 \leq \theta < \pi$ which are positioned symmetrically about $\theta = \pi/2$ (see Fig. 2.2). There is a more group theoretical definition of zonal spherical function, which is developed in the following exercise.

Exercise 11 (Zonal Spherical Functions on $K\backslash G/K$, $G = SO(3)$, $K = SO(2)$). Let $G = SO(3)$ and $K = SO(2)$ be considered as a subgroup of G as in Exercise 1. We know from Exercise 1 that the sphere can be identified with G/K. Show that the space $K\backslash G/K$ of double cosets KgK, $g \in G$, can be represented by the cosets

$$Kg_\theta K, \quad \text{with} \quad g_\theta = \begin{pmatrix} \cos\theta & 0 & \sin\theta \\ 0 & 1 & 0 \\ -\sin\theta & 0 & \cos\theta \end{pmatrix}, \quad 0 \leq \theta < \pi.$$

When do two of these cosets coincide? The corresponding points to gK on the sphere are ge_3, where $e_3 = {}^t(0,0,1)$. So if

$$k_\varphi = \begin{pmatrix} \cos\varphi & \sin\varphi & 0 \\ -\sin\varphi & \cos\varphi & 0 \\ 0 & 0 & 1 \end{pmatrix},$$

we find that $k_{-\varphi} g_\theta e_3 = {}^t(\cos\varphi \sin\theta, \sin\varphi \sin\theta, \cos\theta)$. Thus a function on $K\backslash G/K$ is a function only of the angle θ.

Note. Let $A = \{g_\theta | 0 \leq \theta \leq \pi\}$. Then $SO(3) = KAK$ is the *Euler angle decomposition* of $SO(3)$. See Chapter 5 for a generalization of this decomposition and its application to harmonic analysis on general symmetric spaces.

Exercise 12. Check Table 2.1—surface spherical harmonics. The table uses the Condon-Shortley [1] convention.

Corollary (Harmonic Analysis on the Sphere). *Every function $f : S^2 \to \mathbb{C}$ with continuous second-order derivatives can be expanded in an absolutely and uniformly convergent series of spherical harmonics. If we take Y_n^m, $|m| \leq n$, to be an*

Table 2.1. Surface Spherical Harmonics

$$Y_n^m(\theta, \varphi) = (-1)^m \left[\frac{(2n+1)(n-m)!}{4\pi(n+m)!} \right]^{1/2} P_n^m(\cos\theta) \exp(im\varphi)$$

s $\quad Y_0^0 = (4\pi)^{-1/2}$

p $\begin{cases} Y_1^{\pm 1} = \mp (3/8\pi)^{1/2} \sin\theta \exp(\pm i\varphi) \\ Y_1^0 = (3/4\pi)^{1/2} \cos\theta \end{cases}$

d $\begin{cases} Y_2^{\pm 2} = (15/32\pi)^{1/2} \sin^2\theta \exp(\pm 2i\varphi) \\ Y_2^{\pm 1} = \mp (15/8\pi)^{1/2} \sin\theta \cos\theta \exp(\pm i\varphi) \\ Y_2^0 = (5/4\pi)^{1/2} [\frac{3}{2} \cos^2\theta - \frac{1}{2}] \end{cases}$

f $\begin{cases} Y_3^{\pm 3} = \mp (35/64\pi)^{1/2} \sin^3\theta \exp(\pm 3i\varphi) \\ Y_3^{\pm 2} = (105/32\pi)^{1/2} \sin^2\theta \cos\theta \exp(\pm 2i\varphi) \\ Y_3^{\pm 1} = \mp (21/64\pi)^{1/2} \sin\theta (5\cos^2\theta - 1) \exp(\pm i\varphi) \\ Y_3^0 = (7/4\pi)^{1/2} [\frac{5}{2} \cos^3\theta - (\frac{3}{2}) \cos\theta] \end{cases}$

orthonormal basis for $L^2(S^2, d\mu)$, with $d\mu$ as in Exercise 2, then

$$f(\theta, \varphi) = \sum_{n \geq 0} \sum_{|m| \leq n} \hat{f}(n, m) Y_n^m(\theta, \varphi),$$

where

$$\hat{f}(n, m) = \int_{S^2} f \overline{Y_n^m} \, d\mu.$$

Exercise 13. Prove the preceding corollary of Theorem 1.

Hint. The Corollary can be proved from the existence of Green's function for the problem (see Courant and Hilbert [1, Vol. 1, p. 369]). Or you might try to relate this problem to Theorem 1 of §1.3.

Exercise 14 (The Gibbs Phenomenon for the Sphere). Find an analogue of Exercises 1 and 2 of §1.3 for the sphere.

Hint. There is a discussion of the Gibbs phenomenon as well as summability procedures in Weyl [1, Vol. I, pp. 305–353, 376–389 and Vol. IV, pp. 432–456].

There are many facts about Fourier series that would be worthwhile to extend to the Laplace series of spherical harmonics. For example, it is possible to obtain the analogue of the results of Slepian and Pollak (see Grünbaum et al. [1]). And it is possible to obtain a central limit theorem for the sphere or even $SO(3)$ (see Clerc and Roynette [1]).

Application to Quantum Mechanics. The Hydrogen Atom

According to the most simplistic version of quantum mechanics (i.e., neglecting spin and relativistic effects), the wave function ψ for the hydrogen atom satisfies the *Schrödinger equation*

$$-\left(\frac{\hbar^2}{2m}\Delta + \frac{e^2}{r}\right)\psi = E\psi, \qquad r = (x^2 + y^2 + z^2)^{1/2}. \qquad (2.13)$$

Here e is the charge of the electron, $\hbar$ – Planck's constant divided by 2π, Δ is the Laplacian ($\Delta f = f_{xx} + f_{yy} + f_{zz}$), $m = m_e m_p/(m_e + m_p)$ is the reduced mass of the system if m_e is the mass of the electron, and m_p the mass of the proton. The eigenvalue E is the *energy level* of the system. Some references for these matters are Biedenharn and Louck [1], Castellan [1], Condon and Odabaşi [1], Courant and Hilbert [1, Vol. I, pp. 341–343], Eyring, Walter, and Kimball [1], Mackey [1, pp. 159–271], Messiah [1, especially pp. 362 and 412], Sommerfeld [1, pp. 200–206], Van der Waerden [1, Ch.1], Weyl [2, pp. 41–70], and Wigner [1].

The eigenvalues E which lie in the discrete spectrum of the Schrödinger operator in equation (2.13) will be the only ones of interest for our discussion. Such eigenvalues are negative (in fact, there is also a continuous spectrum of positive real numbers). Physicists interpret the values of E which lie in the discrete spectrum as *energy levels of bound states of hydrogen*. Spectroscopists going back to Balmer in the 1880s have found various series of lines in the spectrum of hydrogen corresponding to these energy levels. We can obtain some understanding of these spectral lines by finding the discrete spectrum of the Schrödinger operator very explicitly.

We shall use spherical harmonics to separate variables via $\psi(x,y,z) = w(r)Y(\theta,\varphi)$ in equation (2.13) for the purpose of obtaining the discrete spectrum eigenvalues E. Then Exercise 2 shows that (2.13) becomes

$$\frac{(r^2 w')' + (2mr^2/\hbar^2)(E + e^2/r)w}{w} = -\frac{\Delta^* Y}{Y}, \qquad \text{where } ' = d/dr.$$

Both sides of this equation must be constant. Thus, by Theorem 1,

$$\Delta^* Y = -n(n+1)Y, \, n = 0, 1, 2, \ldots,$$

$$(r^2 w')' + \frac{2mr^2}{\hbar^2}(E + e^2/r)w = n(n+1)w.$$

Using Exercise 15 on Laguerre polynomials we find that

$$w(r) = r^n \exp(-cr/(2k)) L_{k+n}^{(2n+1)}(cr/k), \qquad k \in \mathbb{Z}, \, k > n, \, c = 2me^2/(\hbar^2).$$

Furthermore, the corresponding *eigenvalues* are

$$E_k = -\frac{me^4}{2\hbar^2 k^2}. \qquad (2.14)$$

Physicists call k the *principal quantum number*, n the *azimuthal quantum number*. If $Y = Y_n^p$, $|p| \leq n$, from Theorem 1, then p is called the *magnetic quantum number*. For each k, there are

$$k^2 = \sum_{n=0}^{k-1} (2n+1) \tag{2.15}$$

linearly independent eigenfunctions of (2.13) with eigenvalue E_k. Physicists call this k^2-*fold degeneracy*. For example, spectroscopists use the letters $(s, p, d, f, q, \ldots)$ to indicate the value of n corresponding to the state of hydrogen. They would say the ground state corresponds to $k = 1$ and is a 1s state. The first excited state corresponds to $k = 2$. It is said to be fourfold degenerate because it contains one 2s state and three 2p states.

Thus for each value k, n of the principal and azimuthal quantum numbers there are $(2n + 1)$ experimentally indistinguishable states of the hydrogen atom. This mathematically explains the *Zeeman effect* in which spectral lines of hydrogen split into an odd number of lines when a magnetic field is switched on. The effect was first observed by Zeeman in 1896.

Note that E_k given by formula (2.14) must equal $h\nu$, where ν is the frequency of a spectral line. And we obtain

$$\nu = Rk^{-2},$$

where R is the Rydberg constant, $R = 2\pi^2 m e^4 h^{-3}$. When an atom moves from initial state 1 with principal quantum number k_1 to final state 2 with principal quantum number k_2, the frequency of the associated spectral line is

$$\nu = R(k_2^{-2} - k_1^{-2}).$$

The series of spectral lines of hydrogen observed by Balmer in 1885 has $k_2 = 2$. Thus the Balmer series lines have frequencies

$$\nu = R(2^{-2} - k^{-2}), \quad k = 3, 4, 5, \ldots.$$

These lines are in the visible range. Lyman found a series of spectral lines in the ultraviolet range in 1909 with $n_2 = 1$. The series with $n_2 = 3, 4$, are in the infrared range. The Balmer series is not obtained in the laboratory because it requires the temperature of stellar atmospheres.

The theory just described does not account for the fine structure of the spectrum. Relativistic effects and spin must be considered. Even then, agreement between theory and experiment is not perfect. See Messiah [1, Vol. II, pp. 930–933] for a discussion of the Lamb shift.

There is an additional degeneracy of $(n + 1)^2$ rather than $2n + 1$, which was explained by Fock, who showed that if one formulates the Schrödinger equation for the hydrogen atom as an integral equation, then it is also invariant under $SO(4)$. See Louck and Metropolis [1] for a discussion of a related problem which was motivated by Fock's result.

Spectroscopists daily analyze all sorts of materials—atoms, molecules, crystals, solids, gases. But, of course the theory becomes much more com-

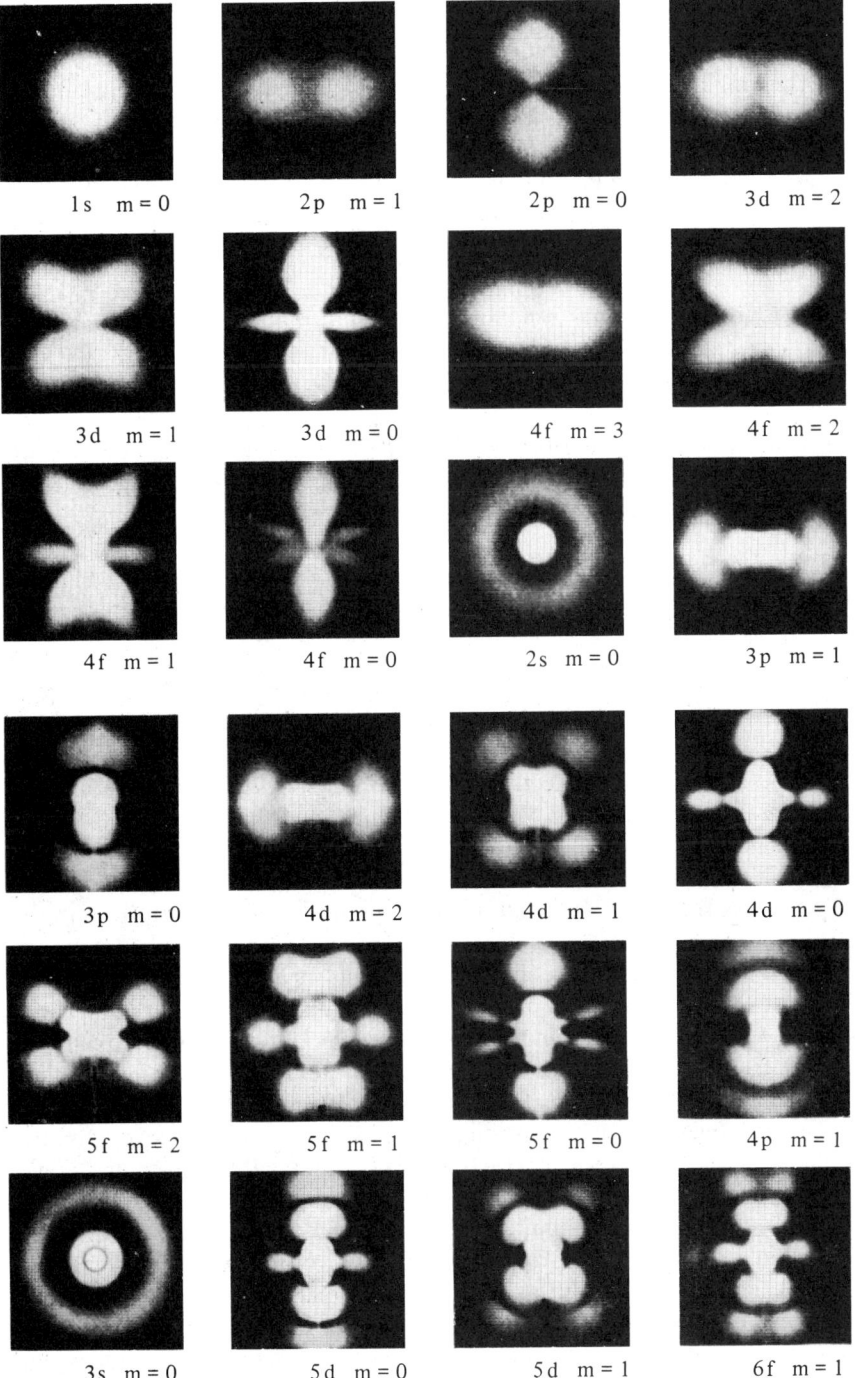

Figure 2.3. Photographs of the electron cloud for various states of the hydrogen atom as made from a spinning mechanical model. The probability-density distribution $\psi\psi^*$ is symmetrical about the φ-axis, which is vertical and in the plane of the paper. (From H.E. White, *Introduction to Atomic Spectra*, © 1934, p. 71. Reprinted by permission of McGraw-Hill Book Co.)

plicated for the many-body problem. Group theory helps, as many of the references mentioned at the beginning of this discussion show.

Exercise 15 (Laguerre Polynomials). Set

$$L_p = e^z \frac{d^p}{dz^p}(e^{-z}z^p) = \text{the } p\text{th Laguerre polynomial}.$$

Show that $w(r) = r^n \exp(-cr/(2k)) L_{k+n}^{(2n+1)}(cr/k)$ satisfies

$$(r^2 w')' + \frac{2mr^2}{\hbar^2}\left(E + \frac{e^2}{r}\right)w = n(n+1)w,$$

for E given by equation (2.14), $k \in \mathbb{Z}$, $k > n$, $c = 2me^2/(\hbar^2)$. Then show that we have found the only continuous bounded eigenfunctions $w(r)$ for the differential operator involved here.

Hint. See Arfken [1, Ch. 13] or Courant and Hilbert [1, Vol. 1, p. 330].

We have found a wave function ψ solving equation (2.13) of the form

$$\psi(r, \theta, \varphi) = Y_n^m(\theta, \varphi) r^n \exp(-cr/2k) L_{k+n}^{(2n+1)}(cr/k),$$

where Y_n^m denotes a spherical harmonic as in Theorem 1 and $L_a^{(b)}$ is the bth derivative of the Laguerre polynomial L_a. Then $|\psi|^2$ can be interpreted as the probability density of the electron for the various states of the hydrogen atom. Figure 2.3 from White [1, p. 71] illustrates the first few states.

Application to the Reconstruction of the Sun's Magnetic Field

Using the measured magnetic field of the sun's surface, the problem is to determine the magnetic field of the corona, assuming that the latter is current-free. The corona itself was not recognized until the last half of the 19th century. A new green spectral line was found to appear only in the corona and just above the thin layer next to the solar surface which emits the bright red Balmer line of hydrogen. By 1930 eighteen spectral lines had been discovered in the solar corona. And by the early 1940s it was realized that the green line comes from forbidden transitions in highly ionized metal atoms. The lack of spectral lines of hydrogen and helium in the corona imples that these elements are completely ionized. The corona is mostly hydrogen plasma at one to two million degrees K. There are also cooler regions and warmer regions. Since the corona is too hot to be in thermal equilibrium, there is a solar wind. And observed coronal holes are sources of streams of plasma, which ultimately give rise to terrestrial phenomena such as aurorae and disturbances in radio transmission. A more complete discussion of these basic facts can be found in Altschuler's article in Herman [1, pp. 105–145]. A fascinating account of solar theory can be found in Zirin [1].

2.1. Spherical Harmonics

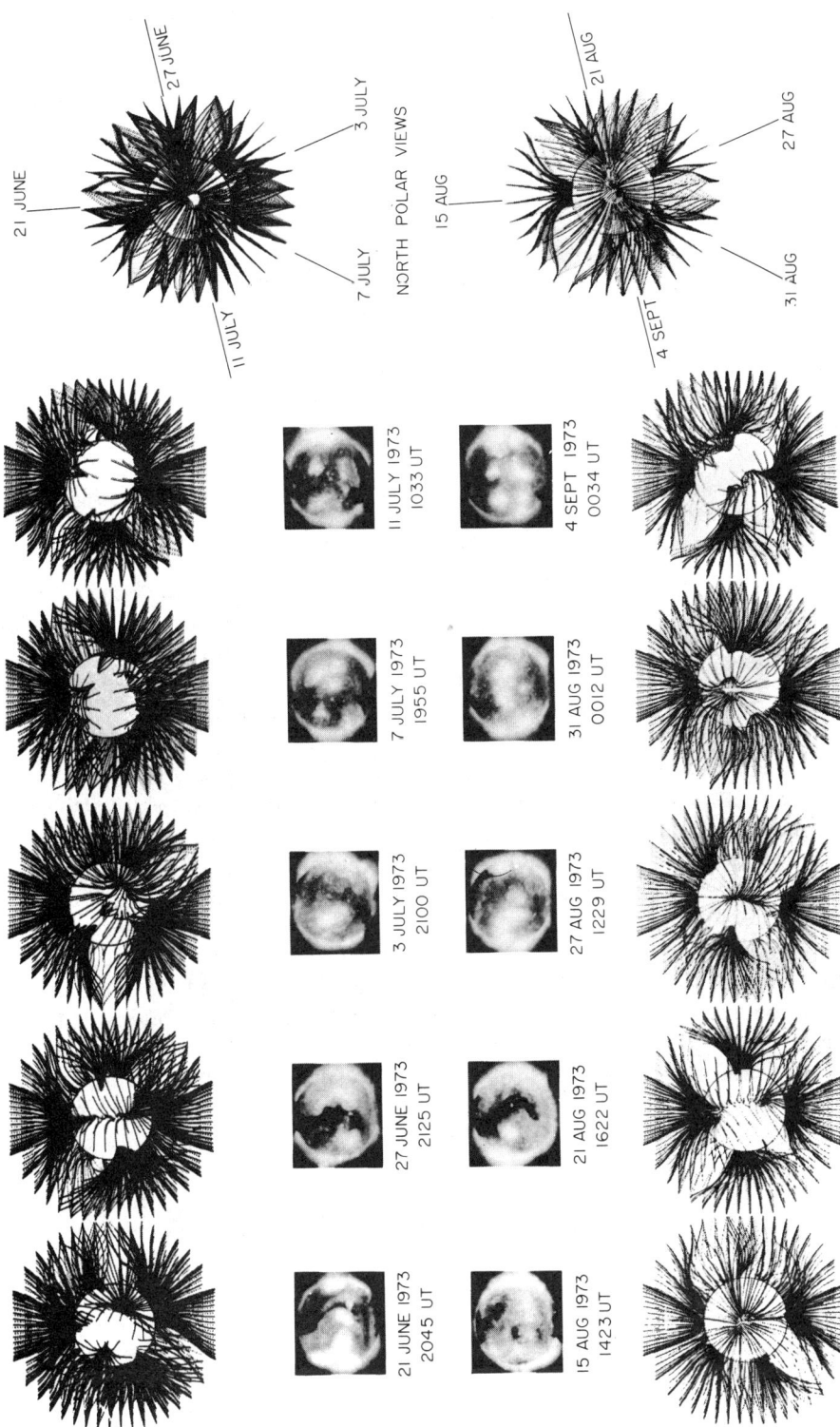

Figure 2.4. Open magnetic structures on the sun. (From Altschuler's article in Herman [1, p. 125]. Reprinted by permission of Springer-Verlag.)

The idealized mathematical problem involved in determining the magnetic field of the solar corona using measurements from the solar surface can be posed as follows:

Find ψ, where

$$\Delta\psi = 0, \qquad R < \|x\| < R_w,$$
$$\psi(x) = f(x), \qquad \text{if } \|x\| = R, \text{ the sun's surface,}$$
$$\psi(x) = 0, \qquad \text{if } \|x\| = R_w, \text{ sufficiently far from the sun.}$$

The function $f(x)$ is obtained from the measured magnetic field of the sun's surface. The relation $B = -\operatorname{grad} \psi$ gives the magnetic field itself. The assumption is that there are no magnetic monopoles and thus ψ must be harmonic. So we can use Exercise 9 and the Corollary to Theorem 1 to obtain

$$\psi(r,\theta,\varphi) = R \sum_{\substack{n\geq 0 \\ |m|\leq n}} (1-a^{2n+1})^{-1}\left((R/r)^{n+1} - a^{2n+1}(r/R)^n\right)\hat{f}(n,m)Y_n^m(\theta,\varphi),$$

where $a = R/R_w < 1$ and

$$\hat{f}(n,m) = \int_{S^2} f\,\overline{Y_n^m}\,d\mu.$$

This formula is satisfactory as a solution in a textbook, but the numerical problem remains to be solved. See Altschuler's article mentioned above for a description of the numerical work that led to the illustrations in Figure 2.4.

Another interesting application of spherical harmonics can be found in Chandrasekhar [1, Ch. VI]. The same sort of picture that gave us an idea of the shape of the probability distribution of the electron in a hydrogen atom should give an idea of the distribution of heat in a uniform sphere. Several applications of spherical harmonics to geophysics can be found in Gilbert [1], Kato [1], Stacey [1], and Trefil [1, pp. 106–111].

It is possible to generalize Theorem 1 to the unit sphere S^n in $\mathbb{R}^{n+1}$. In fact, spherical harmonics for such higher-dimensional spheres have been of use in quantum mechanics, as have the representations of many higher dimensional Lie groups. In the Lecture Note-Reprint collection Dyson [1] gives an interesting glimpse into the arguments concerning the choice of symmetry groups for nuclear and particle physics. The compact groups $SU(n)$, $n = 3, 6, 12$, seem to have attracted the most attention. Here $SU(n)$ is the *special unitary group* of $n \times n$ complex matrices U such that ${}^t\bar{U}U = I$ and $|U| = 1$. Some of the papers in the Dyson collection are devoted to showing that the representations of $SU(6)$ are somehow incompatible with those of the Lorentz group $O(3,1)$ which leaves the Maxwell equations invariant. See R. Hermann [1, pp. 144–149] and Talman [1, pp. 186–187] for discussions of Fock's work on applications of four-dimensional spherical harmonics to quantum mechanics. Hecke and others have made use of these higher-dimensional spherical harmonics in number theory and the kinetic theory of gases (see Hecke [1, pp, 361–373, 849–854] and Ogg [1, Ch. VI, p. 6]). We shall see in the next section that spherical harmonics also give a new description of harmonic analysis on $\mathbb{R}^n$.

Connections with Group Representations

It is not difficult to connect spherical harmonics with representations of $SO(3)$. Before doing so, let us sketch some of the *basic facts about group representations.* We will be sketchy because we do not intend to emphasize the representation-theoretic point of view. It would be useful for the reader to study these things however, and the following references will provide plenty of food for thought: Barut and Rączka [1], Boerner [1], Gelfand, Minlos, and Shapiro [1], Hamermesh [1], Mackey [1–3], Maurin [2], Sugiura [1], Talman [1], Van der Waerden [1], Varadarajan [1], Vergne [1], Vilenkin [1], Wallach [1, 2], Warner [1], Weyl [1–3], Wigner [1]. Mackey's introduction to Biedenharn and Louck [2] provides a translation of physicists' terminology for mathematicians (or vice versa).

Suppose that H is a *separable Hilbert space* and let $GL(H)$ be the group of invertible continuous linear maps from H to H. We want to consider representations of a topological group G by elements of $GL(H)$. By a *topological group* G we mean a group with a topology such that multiplication $(x, y) \mapsto xy$ and inversion $x \mapsto x^{-1}$ are continuous maps. A topological group is *locally Euclidean* if there is a neighborhood of the identity e in G which is homeomorphic to an open subset of $\mathbb{R}^n$. A *Lie group* is a locally Euclidean topological group whose group operations are infinitely differentiable maps. A *representation* of a topological group G is a pair (T, H) where H is a separable Hilbert space and $T: G \to GL(H)$ is a continuous group homomorphism, using the strong operator topology on $GL(H)$. Actually a stronger definition is made when G is not locally compact (see Kirillov [1, p. 111]).

Let (v, w), $v, w \in H$, denote the Hilbert space inner product on H. We say that the representation (T, H) is *unitary* if $(T(g)v, T(g)w) = (v, w)$ for all v, w in H and g in G. And (T, H) is said to be *finite dimensional* if H is finite dimensional and then $\dim H =$ *degree* of the representation.

Suppose that v, w are elements of H. The function $g \mapsto (T(g)v, w) = t_{v,w}(g)$, for $g \in G$, is called a *matrix entry* of the representation (T, H). Many of the special functions which are useful in applied mathematics are either matrix entries of representations of well-known Lie groups or are part of an orthogonal basis for the Hilbert space of such a representation.

Note that the theory of group representations is closely connected with the search for eigenfunctions of differential operators. For if a group G leaves a differential operator L invariant, then G takes eigenfunctions of L into eigenfunctions of L with the same eigenvalue. So G acts on the vector space H of eigenfunctions of L to give a representation of G.

We say that a representation (T, H) of G is *irreducible* if there is no closed subspace W of H with $\{0\} \subsetneqq W \subsetneqq H$ such that $T(g)W \subset W$ for all $g \in G$.

Two representations (T_1, H_1) and (T_2, H_2) of a Lie group G are said to be *equivalent* if there is a linear bicontinuous bijection $A: H_1 \to H_2$ such that $T_2(g)A = AT_1(g)$ for all g in G. Then A is called an *intertwining operator*. Or even more generally, if A is only continuous linear, it is still called an intertwining operator.

Exercise 16 (Schur's Lemma).

(a) Suppose that (T_1, H_1) and (T_2, H_2) are finite-dimensional irreducible representations of G which are inequivalent. If $A: H_1 \to H_2$ is a linear map such that $AT_1(g) = T_2(g)A$, for all g in G, then $A = 0$. Prove this.

(b) Suppose that (T, H) is an irreducible representation of G on a finite-dimensional complex vector space H. If $A: H \to H$ is a linear map such that $AT(g) = T(g)A$, for all g in G, show that there is a complex number c such that $A = cI$, where I is the identity operator; i.e., $Iv = v$ for all v in H.

Note. The spectral theorem for unitary operators allows you to extend this result to unitary representations on infinite-dimensional Hilbert spaces.

We say that a finite-dimensional representation (T, H) of G is *completely reducible* if $H = H_1 \oplus \cdots \oplus H_n$, with $T(g)H_i \subset H_i$, for all $g \in G$, provided that the representations (T_i, H_i) of G obtained by restricting $T(g)$ to H_i are irreducible for all $i = 1, \ldots, n$. We say $T = T_1 \oplus \cdots \oplus T_n =$ the *direct sum* of the T_i.

Exercise 17. Given a basis $e_1, \ldots, e_m$ of the representation space H of a finite-dimensional representation (T, H) of G, you can form the matrix corresponding to $T(g)$. If $e_1, \ldots, e_m$ is an orthonormal basis of H, then this matrix has (i,j)th entry $(T(g)e_i, e_j) = t_{j,i}(g)$, for $g \in G$.

(a) Show that two finite-dimensional representations (T_1, H) and (T_2, H) are equivalent if and only if the matrix of T_1 is obtained from the matrix of T_2 by changing the basis of H that is used.

(b) Show that a finite-dimensional representation (T, H) of G is completely reducible if and only if you can find a basis of H which puts the matrix of T in block diagonal form:

$$\begin{pmatrix} T_1 & & 0 \\ & \ddots & \\ 0 & & T_n \end{pmatrix}$$

so that the representation determined by the matrix T_i is irreducible for each $i = 1, \ldots, n$.

We will often need to be able to integrate functions on topological groups. The integral involved is an analogue of the Lebesgue integral on $\mathbb{R}^n$ and is supposed to come from a countably additive positive measure on the Borel

2.1. Spherical Harmonics

sets in the group. Such an integral on a topological group G is *right invariant* if

$$\int_G f(x)\,dx = \int_G f(xa)\,dx \qquad \text{for all } a \text{ in } G.$$

The invariant integral on a compact Lie group was used by Hurwitz, Schur, and Weyl beginning in the 1890s. It is not hard to construct this integral for any Lie group (see Helgason [2, p. 365]). Haar proved the existence of a right-invariant integral on any locally compact topological group in 1933 and thus the integral is called the *Haar integral*. Proofs of the existence of the Haar integral may be found in Lang [2], Pontryagin [1], or Weil [1], for example.

The invariant integral can be used to show that any irreducible unitary representation of a compact group must be finite dimensional (see Kirillov [1, p. 135]), for example. And you will need the invariant integral to do Exercise 19. In fact, integration on topological groups will be a necessary tool for the rest of this book.

The right-invariant Haar measure is unique up to a positive constant multiple. This means that for each g in G, there is a positive constant $\delta(g)$ defined by

$$\int f(gx)\,dx = \delta(g)\int f(x)\,dx.$$

Then $\delta: G \to \mathbb{R}^+$ is a continuous homomorphism, which is called the *modular function* of G. Clearly this function relates the right- and left-invariant Haar integrals. In particular, one has

$$\int f(x)\delta(x)\,dx = \int f(x^{-1})\,dx.$$

The group G is said to be *unimodular* when $\delta(g) = 1$ for all g in G. This means that right-invariant Haar integrals are also left invariant. Many groups that we shall consider are unimodular. In particular, all compact groups are unimodular. Most of the Lie groups that we shall consider in this book are what is called "semisimple"; e.g., the special linear group of all $n \times n$ real matrices of determinant one (see Chapter 5 for a definition). It can be shown that all semisimple Lie groups are unimodular (see Helgason [2, p. 366]). However, not all groups are unimodular. For example, the group of upper triangular matrices with positive diagonal entries is not unimodular if $n \geq 3$ (see Exercise 30 of §4.1).

Exercise 18. Show that any finite-dimensional unitary representation is completely reducible and that its decomposition into irreducible representations is unique up to equivalence.

Exercise 19 (The Haar Integral on a Compact Group).

(a) Show that any compact group is unimodular; i.e., right Haar measure = left Haar measure.

(b) Use Haar measure to show that any representation of a compact group is equivalent to a unitary representation.

Suppose that (T_1, H_1) and (T_2, H_2) are representations of G. Define the *tensor product representation* $(T_1 \otimes T_2, H_1 \otimes H_2)$ by

$$(T_1 \otimes T_2)(g)(v_1 \otimes v_2) = (T_1(g)v_1) \otimes (T_2(g)v_2) \quad \text{for } g \in G, v_i \in H_i, i = 1, 2.$$

See the references mentioned above for more information on tensor products. The decomposition of tensor products of representations into direct sums of irreducible representations has much importance for quantum mechanics. For example, if one ignores the interaction between two electrons in the field of a positive nucleus, the Schrödinger operator of the system has eigenfunctions which are products $\psi_1 \psi_2$ of eigenfunctions ψ_i corresponding to representations T_i of $O(3)$, $i = 1, 2$. And the product $\psi_1 \psi_2$ corresponds to the representation $T_1 \otimes T_2$. The Clebsch-Gordon series breaks $T_1 \otimes T_2$ up into its irreducible components. Thus one can conclude what sort of spectral lines should occur for such a situation. See the references mentioned in the application to quantum mechanics developed above for more details.

There are many other useful topics in representation theory such as induced representations, Frobenius reciprocity, Cartan's theorem on the highest weight, and Weyl's character formula. See the references on group representations for a discussion of these matters. The systematic study of group representations of finite groups began in the 1890s with work of Frobenius, Schur, and others. In the 1920s Weyl obtained the irreducible representations of the compact simple Lie groups such as $G = SO(3)$ (see Weyl [1, Vol. II, pp. 543–647]. To do this Weyl used his formula for the character of a representation T, which is the trace of $T(g)$, $g \in G$. The representations of compact Lie groups G are studied by restricting them to a maximal abelian subgroup A or torus in G. Any representation of A decomposes into a direct sum of one-dimensional representations called weights. Among the weights there is a highest one and this characterizes the original representation of G, up to equivalence. We have actually seen an example of the theorem on the highest weight in our study of spherical harmonics, but this will not be developed here. Another result called the Borel-Weil-Bott theorem realizes the representations of compact Lie groups on sheaf cohomology groups (see Warner [1]).

The *character* χ_T of a finite-dimensional representation T is defined by $\chi_T(g) = \text{Trace}(T(g))$ for $g \in G$. Define $\hat{G} =$ the *dual of* G to be the set of equivalence classes of irreducible unitary representations of G. If G is compact, describing $\hat{G}$ is equivalent to describing the characters of G.

When infinite-dimensional representations are required, the character is defined as a distribution when possible. If $f \in L^1(G)$, define

$$T(f) = \int_G f(g) T(g) \, dg,$$

which means that for $x, y \in H$, we have

$$(T(f)x, y) = \int_G f(g)(T(g)x, y)\,dg,$$

with (,) = the Hilbert space inner product on H. If H has an orthonormal basis e_i, then we can often define the character of T via

$$\operatorname{Tr} T(f) = \sum_i (T(f)e_i, e_i) \qquad \text{where Tr} = \text{Trace},$$

considering the map $f \mapsto \operatorname{Tr} T(f)$ as a distribution. When H is finite dimensional,

$$\operatorname{Tr} T(f) = \int_G (\operatorname{Tr} T(g)) f(g)\,dg.$$

When G is a "tame" unimodular Lie group the *abstract Plancherel theorem* (or *Fourier inversion theorem*) says there is a measure $d\mu$ on $\hat{G}$ called the Plancherel measure such that

$$f(e) = \int_{\hat{G}} \operatorname{Tr} T(f)\,d\mu(T),$$

for infinitely differentiable functions f on G with compact support. See Vergne [1] for a interesting discussion of this subject with many examples. Harish-Chandra obtained the Plancherel measure for real semisimple Lie groups, for which the Plancherel inversion formula does not always involve an integral over all of the dual $\hat{G}$.

One of the main goals of this set of notes is the explicit description of the Plancherel measure for the symmetric space $GL(n, \mathbb{R})/O(n)$. See Chapter 4 for this result and its history.

Formula (2.16) below gives the Plancherel theorem for the group $SO(3)$.

Exercise 20 (The Irreducible Unitary Representations of $SO(3)$). Let $\{Y_n^m, |m| \leq n, n = 0, 1, 2, \ldots\}$ denote a complete orthonormal set of spherical harmonics as in Theorem 1. Define a $(2n + 1) \times (2n + 1)$ matrix $A_n(g)$ for g in $SO(3)$ by

$$Y_n^m(gx) = \sum_{|k| \leq n} (A_n)_{m,k} Y_n^k(x).$$

(a) Justify this formula for $A_n(g)$ and then show that $A_n(g)$ defines a representation of $SO(3)$.
(b) Show that $A_n(g)$ is a unitary representation using

$$\int_{S^2} Y_n^m(gx)\overline{Y_n^k(gx)}\,dx = \int_{S^2} Y_n^m(x)\overline{Y_n^k(x)}\,dx.$$

(c) Show that the representation $A_n(g)$ is irreducible.

In fact, the representations $A_n(g)$ form a complete set of inequivalent irreducible unitary representations of $SO(3)$. This means that any function f in

$L^2(SO(3))$ with respect to Haar measure has a Fourier series expression, converging in the L^2 norm

$$f(g) = \sum_{n=0}^{\infty} (2n+1)\,\mathrm{Trace}\,[\hat{f}(n)A_n(g)],$$

where

$$\hat{f}(n) = \int_{SO(3)} f(g)\overline{{}^tA_n(g)}\,dg, \qquad dg = \text{Haar measure on } G. \tag{2.16}$$

This is proved, for example, in Dym and McKean [1, pp. 256–261]. Vilenkin [1, pp. 440–457] examines the representations of $SO(n)$ defined in an analogous way to that used for $n = 3$ in Exercise 20 and shows that if $n > 3$, then these representations do not exhaust all of the irreducible unitary representations of $SO(n)$. Only the class-one representations are obtained in this way. A *class-one* representation (A, H) of G has a vector v in H such that $A(k)v = v$ for all k in K. Here $G = SO(n)$ and $K = SO(n-1)$ (embedded in G as in Exercise 1).

Formula (2.16) giving Fourier analysis on the group $SO(3)$ is a special case of the *Peter-Weyl Theorem* for any compact group (see Weyl [1, Vol. III, pp. 58–75], Pontryagin [1], or Weil [1]). The theorem says that if $\{T^\alpha = (t_{ij}^\alpha), \alpha \in \mathrm{A}\}$ is a complete system of irreducible unitary representations of a compact topological group G, then $\sqrt{d_\alpha}\, t_{ij}^\alpha$ forms a complete orthonormal set in $L^2(G)$, where d_α = degree of T^α. The proof is not hard using the spectral theorem for compact self-adjoint operators (Theorem 7 of §1.3). The compact self-adjoint operator used in the proof of the Peter-Weyl theorem is the convolution operator $Tf = g * f$, where g is a fixed nonzero continuous function on G such that $g(x) = g(x^{-1})$ for all x in G. The convolution is taken with respect to Haar measure on G. The eigenfunctions for a fixed eigenvalue of T form a finite-dimensional vector space on which G operates by sending $f(x)$ to $f(xa)$ for $a \in G$. This gives a representation of G. The eigenfunctions for T form a complete orthonormal set in $L^2(G)$ and thus so do the matrix entries of the unitary irreducible representations of G. It follows that the analogue of formula (2.16) replaces A_n by the irreducible unitary representations of G and $(2n+1)$ by the degree of the irreducible representation.

The fact that special functions come from representations often leads to an easier understanding of the myriads of formulas lised in books such as Erdélyi et al. [1, 2].

Exercise 21 (Addition Formulas for Matrix Entries). Suppse that (T, H) is a finite-dimensional representation of G and that $e_1, \ldots, e_n$ is an orthonormal basis of H. Let the i,jth matrix entry of T be $t_{ij}(g) = (T(g)e_i, e_j)$ for g in G. Here (,) denotes the inner product on H. Show that $T(gh) = T(g)T(h)$ implies the following addition formula for the matrix entries:

$$t_{ij}(gh) = \sum_{k=1}^{n} t_{ik}(h)t_{kj}(g).$$

2.1. Spherical Harmonics

Exercise 22 (Addition Formula for Spherical Harmonics). For x, y in S^2, prove that

$$\sum_{|m| \leq n} Y_n^m(x) \overline{Y_n^m(y)} = \frac{2n+1}{4\pi} P_n({}^t xy).$$

Note that both sides are unchanged if you replace x by gx and y by gy for g in $SO(3)$. So the right and left sides of the equation can only differ by a constant. Find the constant by setting $x = y$ and integrating over S^2.

It is possible to characterize spherical harmonics by integral equations rather than differential equations. This method often simplifies the theory, for the same reason that Green's functions simplify the theory of differential operators (see §1.3 and the discussion surrounding equation (1.15)). Examples of the method appear in Weyl's work on spherical harmonics (see Weyl [1, Vol. III, pp. 386–399]), Selberg's work on the trace formula (see Selberg [1]), and Harish-Chandra's work on group representations (see Harish-Chandra [2, 3]). The method of integral equations also makes it easier to extend results to finite, p-adic and adelic groups (see Gelbart [1], Godement [1], Macdonald [1], Tamagawa [1]). This should not, however, cause us to forget the differential equations and the contact with applied mathematics.

The following theorem was proved in 1916 by Funk [1] and in 1918 by Hecke [1, pp. 208–214].

Theorem 2 (The Funk-Hecke Theorem). *Suppose that Y_n is a spherical harmonic of degree n. Let $f: [-1, +1] \to \mathbb{C}$ be continuous. Then*

$$\int_{S^2} f({}^t ux) Y_n(u) \, du = 2\pi Y_n(x) \int_{-1}^{+1} f(t) P_n(t) \, dt.$$

PROOF. Since every continuous function on $[-1, +1]$ can be uniformly approximated by zonal spherical functions $P_n(x)$, it suffices to do the following exercises. □

Exercise 23 (Properties of Spherical Harmonics).

(a) Show that

$$\int_{S^2} Y_m(x) P_n({}^t xy) \, dx = \frac{4\pi}{2n+1} Y_n(y) \delta_{mn}.$$

Here $\delta_{mn} = 0$ if $m \neq n$ and 1 if $m = n$.

Hint. Use the addition formula of Exercise 22.

(b) Prove that

$$\int_{-1}^{+1} P_n(x) P_m(x) \, dx = \frac{2\pi}{2n+1} \delta_{mn}.$$

The Funk-Hecke theorem actually characterizes spherical functions. This result is generalized in Helgason [2, p. 439. Cor. 7.4]. And we shall also use this approach in later chapters. In order to generalize the Funk-Hecke theorem, one should view the spherical harmonic Y as a function on $G = SO(3)$ and replace the formula in Theorem 2 by

$$Y(I)\int_K Y(xky)\,dk = Y(x)\int_K Y(ky)\,dk \qquad \text{for all } x, y \text{ in } G, \qquad (2.17)$$

where dk is Haar measure on $K = SO(2)$ embedded in G as in Exercise 1, and I = the identity matrix. Furthermore zonal spherical harmonics P are characterized by the integral equation

$$P(I)\int_K P(xky)\,dk = P(x)P(y) \qquad \text{for all } x, y \text{ in } G. \qquad (2.18)$$

This characterization of zonal spherical harmonics is due to Gelfand [1] in the general case.

Exercise 24 (The Group-Theoretical Version of the Funk-Hecke Theorem). Show that formula (2.17) implies the Funk-Hecke theorem by integrating (2.17) times $f(y)$ where $f: K\backslash G/K \to \mathbb{C}$.

Exercise 25 (The Convolution Theorem for $SO(3)$). Let f, g be integrable functions on $SO(3)$ with respect to Haar measure. Define the *convolution* of f and g by

$$f * g(x) = \int_{SO(3)} f(u)g(xu^{-1})\,du \qquad \text{for } x \text{ in } SO(3).$$

Define the Fourier transform $\hat{f}(n)$ for $n = 0, 1, \ldots$ as in formula (2.16). Show that

$$\widehat{(f * g)}(n) = \hat{f}(n)\hat{g}(n) \qquad \text{for all } n = 0, 1, 2, \ldots.$$

Hint. Use the fact that A_n is a representation.

Exercise 26 (The Fundamental Solution to the Heat Equation on the Sphere $S^2 \subset \mathbb{R}^3$). Given the initial heat distribution $f(\theta)$, solve the following initial value problem:

$$\Delta^* u(\theta, \phi, t) = u_t, \qquad u(\theta, \phi, 0) = f(\theta), \qquad t > 0.$$

Here Δ^* is the Laplacian on the sphere given by formula (2.10).

Answer. $u = G_t * f$, where

$$G_t(\theta, \phi) = G_t(\theta) = \sum_{n>0} c_n P_n(\cos\theta)\exp[-n(n+1)t].$$

The constants c_n are chosen so that $G_t \to \delta$ as $t \to 0^+$.

Note. G.S. Watson [1] considers various analogues of the Gaussian distribution for the sphere, in connection with various statistical problems such as that of studying the distribution of the unit normal vectors to the planes of all known comets.

2.2. $O(3)$ and $\mathbb{R}^3$. The Radon Transform

> It is interesting to consider the method by which algorithms have been 'justified' mathematically in this field [computerized tomography]. While this method consists of mathematical reasoning in a certain sense, the reasoning is far from rigorous. Approximations are introduced at many steps with only intuition as a guide to the error involved. We do not know of a single instance in which a tomographic algorithm has been justified in a truly rigorous sense. Thus, in contrast to some other workers in this field, we do not feel that one derivation is more rigorous than another, whether it is based on Radon's inversion formula, the Fourier inversion formula or any other foundation.
>
> From Shepp and Kruskal [1, p. 421].

It is easy to see (cf. Exercise 1) that the Fourier transform on $\mathbb{R}^n$ commutes with rotation. This leads to a new formulation of harmonic analysis on $\mathbb{R}^n$ in terms of spherical harmonics. For simplicity, we shall consider only the case $n = 3$ here. The generalization to $\mathbb{R}^n$, n arbitrary, can be found in Coifman and Weiss [1] and Stein and Weiss [1]. These results are due to Cauchy and Poisson for radial functions, and Bochner and Hecke in general (see Bochner [1, p. 235] and the article of Stein in J.M. Ash [1, pp. 104–105]).

Exercise 1 (The Fourier Transform on $\mathbb{R}^n$ Commutes with Rotation). If $g \in O(n)$ and $f\colon \mathbb{R}^n \to \mathbb{C}$, set $(L(g)f)(x) = f(gx)$ for $x \in \mathbb{R}^n$. Show that $L(g)\hat{f} = \widehat{L(g)f}$, when f is a sufficiently nice function, that the Fourier transform of f exists (as defined in §1.2 for various classes of functions).

Exercise 2 (J-Bessel Functions). Define the Bessel function of the first kind by the power series

$$J_\nu(z) = \sum_{k=0}^{\infty} \frac{(-1)^k (z/2)^{\nu+2k}}{\Gamma(k+1)\Gamma(k+\nu+1)} \qquad \text{for } |\arg z| < \pi.$$

(a) Show that $y(z) = J_\nu(z)$ satisfies Bessel's equation:

$$y'' + (1/z)y' + (1 - (\nu/z)^2)y = 0.$$

(b) Show that $J_\nu(z)$ is represented by the integral formula:

$$J_\nu(z) = \frac{(z/2)^\nu}{\Gamma(1/2)\Gamma(\nu+1/2)} \int_{-1}^{+1} (1-t^2)^{\nu-1/2} \exp(izt)\,dt,$$

when $\operatorname{Re}\nu > -\tfrac{1}{2}$ and $|\arg z| < \pi$.

(c) Show that $J_{\nu-1}(z) + J_{\nu+1}(z) = \dfrac{2\nu}{z} J_\nu(z)$.

(d) Prove that

$$J_{n+1/2}(z) = (-1)^n (2/\pi)^{1/2} z^{n+1/2} \left(\frac{1}{z}\frac{d}{dz}\right)^n \frac{\sin z}{z}, \qquad n = 0, 1, 2, \ldots.$$

These functions are often called spherical Bessel functions.

Exercise 3 (More Properties of Bessel Functions).

(a) *Asymptotics.* Show that

$$J_\nu(z) \sim \Gamma(1+\nu)^{-1}(z/2)^\nu, \qquad \text{as } z \to 0.$$

$$J_\nu(z) \sim \sqrt{\frac{2}{\pi z}} \cos\left(z - \frac{\nu\pi}{2} - \frac{\pi}{4}\right), \qquad \text{as } z \to \infty.$$

(b) Show that $J_n(z) = (-1)^n J_{-n}(z)$, when $n = 0, 1, 2, \ldots$. Then show that if ν is not an integer, J_ν and $J_{-\nu}$ are linearly independent.

(c) *Functional Equation.* Prove that $J_\nu(-z) = \exp(\nu\pi i) J_\nu(z)$.

(d) *Addition Formula.* Set $R = (r_1^2 + r_2^2 - 2r_1 r_2 \cos\theta)^{1/2}$ and prove that

$$J_{1/2}(R) = (2R)^{1/2}(r_1 r_2)^{-1/2}\Gamma(\tfrac{1}{2}) \sum_{m=0}^\infty (m+\tfrac{1}{2}) J_{m+1/2}(r_1) J_{m+1/2}(r_2) P_m(\cos\theta).$$

Theorem 1 (The Bochner-Hecke Formula). *Suppose that $Y(u)$, $u \in S^2$, is a surface spherical harmonic of degree n (as in Theorem 1 of §2.1) and let $f(ru) = g(r)Y(u)$ for $r \in \mathbb{R}^+$, $u \in S^2$, where $g: \mathbb{R}^+ \to \mathbb{C}$, is such that*

$$\int_0^\infty |g(r)| r^2 \, dr < \infty.$$

Then the Euclidean Fourier transform of f (as defined in §1.2) is given by

$$\hat{f}(ru) = 2\pi i^{-n} Y(u) r^{-1/2} \int_{s \in \mathbb{R}^+} g(s) s^{3/2} J_{n+1/2}(2\pi rs)\,ds.$$

PROOF. Let dv denote the element of surface area on S^2. Then, using Theorem 2 of §2.1, we have

$$\hat{f}(ru) = \int_{s \in \mathbb{R}^+} \int_{v \in S^2} \exp(-2\pi i r s^t uv) g(s) Y(v) s^2 \, ds\, dv$$

$$= 2\pi Y(u) \int_{s \in \mathbb{R}^+} g(s) s^2 \int_{-1}^{+1} \exp(-2\pi rst) P_n(t)\, dt\, ds.$$

The proof of this theorem is completed in Exercise 4. □

Exercise 4. Show that $P_n(x) = (2^n n!)^{-1} \dfrac{d^n}{dx^n}(x^2 - 1)^n$ implies that

$$\int_{-1}^{+1} \exp(-2\pi irst) P_n(t)\, dt = (-i)^n (rs)^{-1/2} J_{n+1/2}(2\pi rs).$$

Use integration by parts and Exercise 2. Then complete the proof of Theorem 1.

How does the formula of Bochner and Hecke fit into the general scheme of harmonic analysis on symmetric spaces? Why did the Bessel functions suddently appear out of the blue? To understand this, one must view $\mathbb{R}^3$ as a symmetric space rather than just as an additive group. This leads one to think of the full group of isometries of $\mathbb{R}^3$, namely, the Euclidean group of rigid motions of $\mathbb{R}^3$. If we consider only the orientation preserving motions G, then G is the semidirect product of the groups T and K, where T consists of translations and is isomorphic to $\mathbb{R}^3$, and $K = SO(3)$. Clearly one can identify $\mathbb{R}^3$ with G/K by mapping a coset gK, $g \in G$, to gO = the result of applying the motion g to the origin in $\mathbb{R}^3$.

The connection of Theorem 2 with the representations of the Euclidean group (as well as the generalization to $\mathbb{R}^n$) is made in Vilenkin [1, Ch. 11]. The case of $\mathbb{R}^2$ is discussed in Dym and McKean [1, pp. 263–273]. See also Helgason [2, pp. 402–403] and Talman [1, Ch. 12].

Exercise 5 (The Eigenfunctions of the Euclidean Lapacian in Spherical Polar Coordinates). Use Exercise 2 of §2.1 to show that if $\Delta f = f_{xx} + f_{yy} + f_{zz} = \lambda f$, and $f(ru) = g(r)Y(u)$ for $r \in \mathbb{R}^+$, $u \in S^2$, and Y is a surface spherical harmonic of degree n, then $g(r) = (rt)^{-1/2} J_{n+1/2}(2\pi rt)$, $\lambda = (2\pi t)^2$.

Theorem 2 (Harmonic Analysis on $\mathbb{R}^3$ in Spherical Polar Coordinates). *Let $e_{n,m,t}(ru) = 2\pi Y_n^m(u)(rt)^{-1/2} J_{n+1/2}(2\pi rt)$ for $r \in \mathbb{R}^+$, $u \in S^2$, where $\{Y_n^m \mid n = 0, 1, 2, \ldots, |m| \leq n\}$ denotes a complete orthonormal set of surface spherical harmonics of degree n (as in Theorem 1 of §2.1). Then any f in $L^2(\mathbb{R}^3)$ has a Fourier expansion (converging in the L^2-norm) of the form*

$$f(x) = \sum_{n \geq 0} \sum_{|m| \leq n} \int_{t>0} \hat{f}(n, m, t) e_{n,m,t}(x) t^2\, dt,$$

where

$$\hat{f}(n, m, t) = \int_{\mathbb{R}^3} f(y) \overline{e_{n,m,t}(y)}\, dy.$$

Proof. It suffices to assume that $f(ru) = g(r) Y_n^m(u)$ for $r \in \mathbb{R}^+$, $u \in S^2$. Then we must show that

$$g(r) = (2\pi)^2 \int_{s>0} \int_{t>0} g(t)(st)^{-1/2} J_{n+1/2}(2\pi st) t^2\, dt (rs)^{-1/2} J_{n+1/2}(2\pi rs) s^2\, ds. \tag{2.19}$$

This is done in the following exercise. □

Exercise 6.

(a) Prove formula (2.19) by writing down the inversion formula for the ordinary Fourier transform (from Theorem 1 of §1.2) in spherical polar coordinates. Then use Theorem 1 to evaluate the inner Fourier transform. Finally, Theorem 2 of §2.1 and Exercise 4 should be used to complete the proof.

(b) Show that formula (2.19) is equivalent by change of variables to a special case of *Hankel's inversion formula*:

$$f(r) = \int_0^\infty y J_v(yr) \int_0^\infty x J_v(xy) f(x) \, dx \, dy. \tag{2.20}$$

Hankel's inversion formula (2.20) gives the spectral resolution of the singular Sturm-Liouville operator

$$Lf = \frac{1}{x}(-(xf')' + v^2 f/x) \quad \text{for } x \in (0, \infty). \tag{2.21}$$

This spectral resolution can be derived from a formula of Stieltjes, Stone, Kodaira, and Titchmarsh, which is itself a corollary of the Von Neumann Spectral Theorem for unbounded operators. Let us summarize the theory briefly, following Lang [2], Reed and Simon [1, Ch. 7], Stakgold [1, 2], and some unpublished notes of J. Korevaar. More details can be found in these references as well as Dunford and Schwartz [1], Gelfand and Vilenkin [1], Levitan and Sargsjan [1], Maurin [2], Naimark [1], Titchmarsh [2], Weyl [1, Vol. I, pp. 195–297], and Yosida [1, 2]. In particular, Dunford and Schwartz [1, Vol. II, pp. 1333–1392, 1532–1533] provides an extremely careful discussion of spectral theory for singular differential operators, including the Bessel operator (2.21).

We want to consider linear, possibly unbounded, operators $T: D \to V$, where D is a dense subspace of a Hilbert space V. Define $D^* = \{y \in V \mid \text{there is a } z \text{ in } V \text{ with } (Tx, y) = (x, z), \text{ for all } x \text{ in } D\}$. Define $T^* =$ the adjoint of T by $(Tx, y) = (x, T^*y)$. We say that T is *self-adjoint* if $D = D^*$ and $T = T^*$. The *spectrum* $\sigma(T)$ of the operator T is defined to be $\sigma(T) = \{\lambda \in \mathbb{C} \mid (T - \lambda I)^{-1}$ does not exist as a bounded linear operator on $V\}$. The *point spectrum of* T (or the set of *eigenvalues* of T) consists of $\lambda \in \mathbb{C}$ such that $(T - \lambda I)$ is not one-to-one. The *continuous spectrum* of T is the set of $\lambda \in \mathbb{C}$ such that $(T - \lambda I)$ is 1 : 1 and $D = \text{range}(T - \lambda I)$ is dense in V, but $(T - \lambda I)^{-1}$ is an unbounded linear operator with domain D. The *von Neumann Spectral Theorem* says that if T is any self-adjoint operator (bounded or not), there is a Stieltjes integral representation

$$I = \int dE_\lambda, \quad T = \int \lambda \, dE_\lambda, \tag{2.22}$$

for some family of projection-valued measures dE_λ (cf. Reed and Simon [1,

2.2. $O(3)$ and $\mathbb{R}^3$. The Radon Transform

Ch. 7]). The integrals are over the spectrum of T, which is real, because T is self-adjoint. The spectral theorem implies that for polynomials $p(\lambda)$, $p(T) = \int p(\lambda) dE_\lambda$. It is possible to define $f(T)$ for continuous functions f on the spectrum of T and then this same formula holds upon replacing p by f. In order to obtain the formula of Stieltjes, Stone, Titchmarsh, and Kodaira, one must be able to compute the *Green's function* or resolvent kernel $G(\lambda; x, y)$ defined by

$$(T - \lambda I)^{-1} v(x) = \int G(\lambda; x, y) v(y) \, dy. \tag{2.23}$$

For if we have a convergent integral of the form

$$f(\mu) = \int_{-\infty}^{+\infty} (\lambda - \mu)^{-1} g(\lambda) \, d\lambda,$$

then at any point c of continuity of $g(\lambda)$, we have

$$\frac{1}{2\pi i} \lim_{\varepsilon \to 0+} (f(c + i\varepsilon) - f(c - i\varepsilon)) = \lim_{\varepsilon \to 0+} \left\{ \frac{1}{\pi} \int_{-\infty}^{+\infty} \frac{\varepsilon}{(\lambda - c)^2 + \varepsilon^2} g(\lambda) \, d\lambda \right\} = g(c).$$

This is Poisson's integral formula for a half-plane (see Ahlfors [1, p. 171]). It follows that the spectral measure for the operator T is given by the *formula of Stieltjes, Stone, Kodaira, and Titchmarsh*:

$$2\pi i \frac{dE_\lambda}{d\lambda} = (T - (\lambda + i0)I)^{-1} - (T - (\lambda - i0)I)^{-1}, \tag{2.24}$$

assuming that E_λ is well-behaved. This formula is proved in Dunford and Schwartz [1, Vol. II], Lang [2, pp. 412–413], and Reed and Simon [1, p. 237]. Stieltjes [1] found the result in 1894.

In order to apply formula (2.24), we need a way to compute the Green's function $G(\lambda; x, y)$. We use the method outlined in Stakgold [1, 2]. Consider a *Sturm-Liouville operator* which is singular at a and b:

$$Lf = \frac{1}{w}(-(pf')' + qf), \qquad a < x < b. \tag{2.25}$$

Here "singular" means that either the interval is infinite or the functions $w(x)$ or $q(x)$ blow up, or $p(x)$ vanishes at some point in $[a, b]$. The Hilbert space associated to (2.25) is $L^2([a, b], w)$ consisting of Lebesgue measurable functions f on $[a, b]$ such that

$$\int_a^b |f(x)|^2 w(x) \, dx < \infty.$$

The *inner product* for this weighted L^2-space with weight w is

$$(f, g) = \int_a^b f(x) \overline{g(x)} w(x) \, dx.$$

If one makes the correct assumptions about p, q, w and if one imposes the correct sort of boundary conditions at a and b, the operator L will be self-adjoint. In particular, p, q, w should be real and w should be positive.

The Green's function $G(\lambda; x, y)$ of formula (2.22) must satisfy

$$-\frac{\partial}{\partial x}\left(p\frac{\partial}{\partial x}G(\lambda; x, y)\right) + (q - \lambda w)G(\lambda; x, y) = \delta(x - y).$$

Pick c in (a, b). We want to choose two linearly independent solutions φ_λ and ψ_λ of $Lu = \lambda u$, as follows. Suppose that φ_λ solves $Lu = \lambda u$ and in addition φ_λ lies in $L^2([a, c], w)$. And let ψ_λ solve $Lu = \lambda u$ and $\psi_\lambda \in L^2([c, b], w)$. Then

$$G(\lambda; x, y) = \frac{1}{c_\lambda}\begin{cases}\varphi_\lambda(x)\psi_\lambda(y), & a < x < y < b\\ \varphi_\lambda(y)\psi_\lambda(x), & a < y < x < b\end{cases}$$

and (2.26)

$$c_\lambda = \begin{vmatrix}\varphi_\lambda(x) & -\psi_\lambda(x)\\ p\varphi_\lambda'(x) & -p\psi_\lambda'(x)\end{vmatrix}.$$

Example (Spectral Measure for the Hankel Transform Using the Titchmarsh-Kodaira Formula). Consider the Sturm-Liouvulle operator given by formula (2.21). Then formula (2.26) becomes

$$G(\lambda; x, y) = \frac{\pi i}{2}\begin{cases}J_\lambda(\sqrt{\lambda}x)H_\nu^{(1)}(\sqrt{\lambda}y), & 0 < x < y < \infty\\ J_\lambda(\sqrt{\lambda}y)H_\nu^{(1)}(\sqrt{\lambda}x), & 0 < y < x < \infty.\end{cases}$$

Here J_ν is the Bessel function of the first kind from Exercise 2 and $H_\nu^{(1)}$ is the Hankel function (or Bessel function of the third kind) defined by

$$H_\nu^{(1)}(z) = \frac{J_{-\nu}(z) - \exp(-\nu\pi i)J_\nu(z)}{i\sin(\nu\pi)}$$

when ν is not an integer. Take limits as $\nu \to n$ to define $H_n^{(1)}$ when n is an integer. It follows from Lebedev [1, pp. 112–113] that the determinant $c_\lambda = -2i/\pi$ in this case.

In order to check this formula for $G(\lambda; x, y)$, one needs the following asymptotic properties of the Bessel functions from Lebedev [1, Ch. 5] $(\nu > 0)$:

$$\begin{aligned}
J_\nu(x) &\sim (x/2)^\nu\Gamma(1 + \nu)^{-1}, & x &\to 0,\\
J_\nu(x) &\sim \sqrt{\frac{2}{\pi x}}\cos\left(x - \frac{\nu\pi}{2} - \frac{\pi}{4}\right), & x &\to \infty,\\
H_\nu^{(1)}(x) &\sim -i(2/x)^\nu\Gamma(\nu)/\pi, & x &\to 0,\\
H_\nu^{(1)}(x) &\sim \sqrt{\frac{2}{\pi x}}\exp\left[i\left(x - \frac{\nu\pi}{2} - \frac{\pi}{4}\right)\right], & x &\to \infty.
\end{aligned} \tag{2.27}$$

Note that if $0 < \nu < 1$, both $J_\nu(\sqrt{\lambda}x)$ and $H_\nu^{(1)}(\sqrt{\lambda}x)$ are in $L^2([0, 1], x)$. One says that 0 is then in the limit-circle case in Weyl's theory (cf. Stakgold [1, 2]). If

2.2. $O(3)$ and $\mathbb{R}^3$. The Radon Transform

$v \geq 1$, then $J_v(\sqrt{\lambda x})$ is in $L^2([0, 1], x)$ and $H_v^{(1)}(\sqrt{\lambda x}) \notin L^2([0, 1], x)$. One says that 0 is then in the limit-point case. Note that $J_v(\sqrt{\lambda x}) \notin L^2([1, \infty], x)$ and $H_v^{(1)}(\sqrt{\lambda x}) \in L^2([1, \infty], x)$ so that infinity is always in the limit-point case.

To compute the jump of $G(\lambda; x, y)$ required by formula (2.24), assume that $x < y$. Then for $v > 0$,

$$G(\lambda + i0; x, y) - G(\lambda - i0; x, y)$$
$$= \frac{i\pi}{2}(J_v(\lambda^{1/2}x)H_v^{(1)}(\lambda^{1/2}y) - J_v(-\lambda^{1/2}x)H_v^{(1)}(-\lambda^{1/2}y)).$$

So we need the functional equations

$$\begin{aligned} J_v(-x) &= \exp(v\pi i)J_v(x), \\ H_v^{(1)}(-x) &= \exp(-v\pi i)[H_v^{(1)}(x) - 2J_v(x)]. \end{aligned} \quad (2.28)$$

These formulas imply

$$G(\lambda + i0; x, y) - G(\lambda - i0; x, y) = i\pi J_v(\lambda^{1/2}x)J_v(\lambda^{1/2}y).$$

Thus we obtain from (2.22) and (2.24),

$$\frac{\delta(x-y)}{x} = \frac{1}{2}\int_0^\infty J_v(\lambda^{1/2}x)J_v(\lambda^{1/2}y)\,d\lambda = \int_0^\infty J_v(\rho x)J_v(\rho y)\rho\,d\rho.$$

This is Hankel's integral theorem.

For higher-rank symmetric spaces, the method just used to compute the spectral measure proves rather unwieldy because the Green's functions are much more complicated. Thus we will formulate another method in Chapter 3 and 4 (see Exercise 6 of §3.2, for example). For this method, one needs only the asymptotics and functional equations of the eigenfunctions appearing in the spectral expansion of the operator L, rather than the expression for the resolvent kernel, involving other eigenfunctions as well. The method is due to Harish-Chandra and it would be very interesting to derive Harish-Chandra's Plancherel measure form the spectral theorem in a similar way to that which gives formula (2.24).

One can also ask for the spectral decomposition of the operator in (2.21) on the finite interval $(0, a)$. The problem still has a singularity at 0. But on $(0, a)$, the spectral resolution of the operator L in (2.21) is given by a series rather than an integral, when $v > -1$:

$$f(x) = \sum_{m=1}^\infty c_m J_v(\alpha_{vm} x/a), \quad 0 \leq x \leq a, \quad v > -1, \quad (2.29)$$

where $\{\alpha_{vm}\}_{m \geq 1}$ is the set of all positive roots of $J_v(\alpha_{vm}) = 0$ and

$$c_m = 2a^{-2}[J_{v+1}(\alpha_{vm})]^{-2}\int_0^a f(r)J_v(\alpha_{vm}r/a)r\,dr.$$

The *Fourier-Bessel series* expansion (2.29) is derived in Titchmarsh [Vol. 1, pp. 81–86]. See also Stakgold [2, pp. 305–308, 313–315]. It is quite interesting to let a approach infinity in (2.29) and watch (2.29) approach (2.20) (cf. Morse and Feshbach [1, Vol. I, pp. 762–766]).

Exercise 7 (The Kontorovich-Lebedev Transform). Consider Bessel's equation with the role of the parameters interchanged:

$$-(xw')' + \mu xw - \lambda w/x = 0, \qquad 0 < x < \infty.$$

Here $\mu > 0$ and λ is the eigenvalue. Show that the Green's function is

$$G(\lambda; x, y) = \begin{cases} I_{-i\sqrt{\lambda}}(\sqrt{\mu}x)K_{-i\sqrt{\lambda}}(\sqrt{\mu}y), & 0 < x < y < \infty, \\ I_{-i\sqrt{\lambda}}(\sqrt{\mu}y)K_{-i\sqrt{\lambda}}(\sqrt{\mu}x), & 0 < y < x < \infty, \end{cases}$$

where I and K are the Bessel functions of imaginary argument. The definitions are

$$I_\nu(z) = \exp(-\nu\pi i/2)J_\nu(z \exp(\pi i/2)), \qquad -\pi < \arg z < \pi/2,$$

$$K_\nu(z) = \frac{\pi i}{2} \exp(\nu\pi i/2)H_\nu^{(1)}(z \exp(\pi i/2)), \qquad -\pi < \arg z < \pi/2.$$

See Chapter 3 and Lebedev [1] for more information on these functions. Show that $K_{-\nu}(x) = K_\nu(x)$ and $I_{-\nu}(x) - I_\nu(x) = 2\sin(\nu\pi)K_\nu(x)/\pi$. Then prove the Kontorovich-Lebedev inversion formula:

$$x\delta(x - y) = 2\pi^{-2} \int_0^\infty \nu \sinh \pi\nu K_{i\nu}(\sqrt{\mu}x) K_{i\nu}(\sqrt{\mu}y) \, d\nu.$$

This will be a central result in §3.1.

Application of Fourier Analysis on $\mathbb{R}^2$ to Medicine. CAT Scanners and the Radon Transform

Modern X-ray scanners can reproduce the tissue density function $f(x)$, $x \in \mathbb{R}^2$, for a plane slice of a person's head, for example. They operate by inverting the *Radon transform*, defined for $k \in \mathbb{R}$, $u \in S^1$, by

$$Rf(k, u) = \int_{{}^t xu = k} f(x) \, ds, \tag{2.30}$$

where the integral is over a line in the plane and ds is the element of arc length. The inversion formula for this transform goes back to Radon [1] in 1917. It says that

$$f(x) = -\pi^{-1} \int_{q > 0} q^{-1} d\left((2\pi)^{-1} \int_{u \in S^1} Rf(q + {}^t xu, u) \, du \right). \tag{2.31}$$

Here the integral over q in $\mathbb{R}^+$ can be viewed as a Stieltjes integral or as a Cauchy principal value integral, assuming that f is continuous with compact

support. References for the Radon transform include Dym and McKean [1], Gelfand, Graev, and Vilenkin [1, Vol. 5], Herman [1], Helgason [4], Louis [1, 2], Ludwig [1], and Shepp and Kruskal [1]. In fact, Funk [1] proved the analogue of (2.31) for S^2 rather than $\mathbb{R}^2$ one year earlier than Radon proved his result. Helgason [4] shows that a vast generalization of this theory is possible, viewing the Radon inversion formula as involving two dual integrations—one over the points in a hyperplane and the other over the hyperplanes through a point.

The next sequence of exercises presents a derivation of Radon's inversion formula.

Exercise 8.

(a) Suppose that $f: \mathbb{R}^2 \to \mathbb{C}$ is a Schwartz function. Show that the Fourier inversion formula can be written in the form

$$f(x) = \int_{u \in S^1} Ef({}^t ux, u)\, du,$$

where

$$Ef(t, u) = \frac{1}{2} \int_{r \in \mathbb{R}} \int_{k \in \mathbb{R}} |r| Rf(k, u) \exp[2\pi i r(t - k)]\, dk\, dr.$$

In the first formula du denotes the angle measure on the unit circle S^1. And $Rf(k, u)$ is defined by (2.30).

(b) Let $\mathrm{abs}(r) = |r|$, $r \in \mathbb{R}$. Show that $\mathrm{abs}(r)$ is not the Fourier transform of a Lebesgue integrable function. Hint: recall the Riemann-Lebesgue lemma. Note that if $\widehat{\mathrm{abs}(r)}$ were a function, then you could write:

$$Ef(t, u) = \frac{1}{2} \int_{k \in \mathbb{R}} \widehat{\mathrm{abs}(t + k)} Rf(k, u)\, dk,$$

using properties of the Fourier transform from Theorem 1 of §1.2.

Exercise 9 (Derivatives and Fourier Transforms of Some Distributions). Define the Cauchy principal value integral by

$$PV\left(\int f(x)\, dx\right) = \lim_{\varepsilon \to 0} \left(\int_{|x| > \varepsilon} f(x)\, dx\right).$$

Define

$$(x^{-1}, \varphi) = PV \int x^{-1} \varphi(x)\, dx$$

and

$$(x^{-2}, \varphi) = PV \int x^{-2} (\varphi(x) - \varphi(0))\, dx$$

for test functions φ as in §1.1. Prove that in the sense of distributions (see §1.1) the following formulas are valid:

(a) $x^{-1} = (\log|x|)'$.
(b) $x^{-2} = -(x^{-1})'$.
(c) $\widehat{x^{-1}} = -i\pi \operatorname{sgn}(x)$, where $\operatorname{sgn}(x) = \operatorname{abs}(x)/x$.
(d) $\widehat{\operatorname{abs}(x)} = -(2\pi^2)^{-1}x^{-2}$.

Hint. See Vladimirov [1, pp. 75, 86, 134] or Bracewell [1, p. 130].

Exercise 10. Derive Radon's inversion formula from Exercises 8 and 9, using properties of Fourier transforms of distributions.

Note. The *Hilbert transform* of a function is

$$Hf(x) = -\pi^{-1} \int_{-\infty}^{+\infty} f(t)(x-t)^{-1}\,dt = -\pi^{-1} f * (x^{-1}).$$

Thus part (c) of Exercise 9 implies that

$$\widehat{Hf(x)} = i \operatorname{sgn}(x) \hat{f}.$$

Because Radon's inversion formula (2.31) contains a derivative, it does not appear to be as useful for numerical calculation as the formula of Exercise 8. So those that have computed these transforms in practice have used approximations to $\widehat{\operatorname{abs}(x)}$ (cf. Herman [1, p. 19ff] and Shepp and Logan [1]).

Exercise 11. Compute the Fourier transform of the following approximation to $\operatorname{abs}(x)$, for $x \in \mathbb{R}^2$:

$$f(x) = \begin{cases} |x|, & \text{if } |x| < A, \\ 0, & \text{otherwise.} \end{cases}$$

Compare the decay at infinity with that of $|x|^{-2}$ from part (d) of Exercise 9.

Note. You can write the Fourier transform in the plane as a Bessel transform, must as we did in Theorem 1 for the Fourier transform in 3-space.

The four illustrations in Fig. 2.5 from Shepp and Kruskal [1] show a mathematical phantom [Fig. 2.5(a)] representing a slice of a human head and reconstructions of this phantom by three different algorithms. The Hounsfield algorithm used in the first commerical CAT scanner [see Fig. 2.5(b)] was not based on Radon inversion, but instead on an iterative relaxation method. Figure 2.5(c) was produced with a Fourier-based algorithm of Shepp. Figure 2.5(d) employs a more recent algorithm used by EMI, Ltd. Clearly progress has been made. A description of the progress is contained in the following quotation from Science *214* (1981), p. 1327:

2.2. $O(3)$ and $\mathbb{R}^3$. The Radon Transform

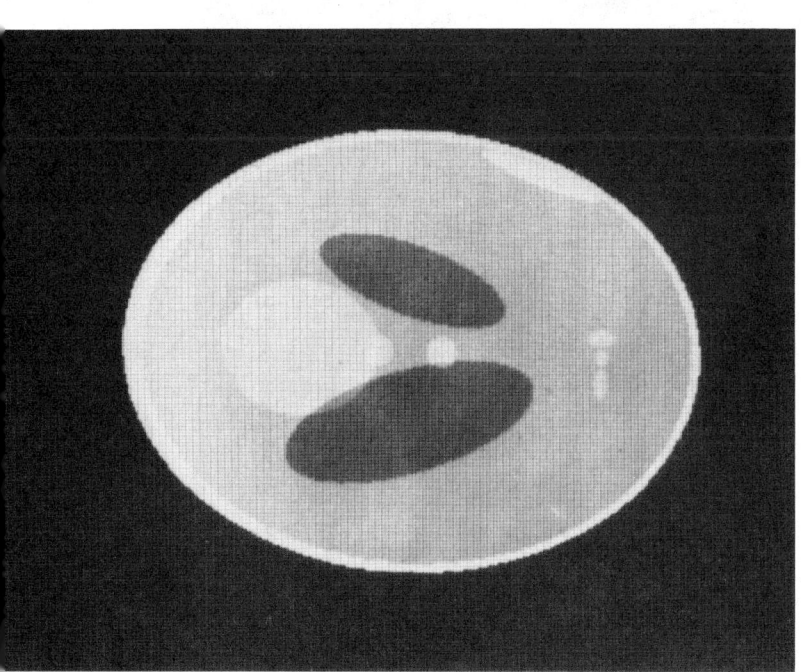

(a)

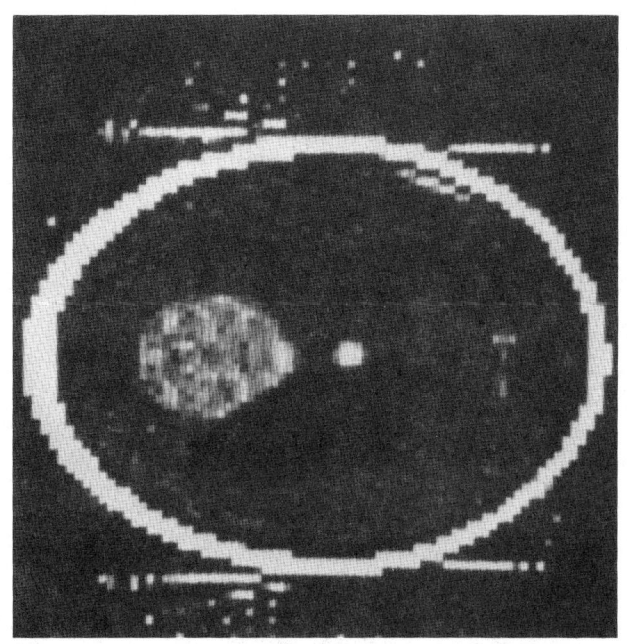

(b)

Figure 2.5. (a) Simulation of human head using 11 ellipses. The density of the skull is 2.0 and of the ventricles, tumors, etc. is 1.0–1.05; (b) Reconstruction using the algorithm embodied in the first commercial machine (EMI Ltd.) from 180×160 strip projection data obtained by exact calculation from (a); (c) Reconstruction from the same data using the Fourier based algorithm of Shepp; (d) Reconstruction using the algorithm now embodied in the EMI machine from 180×239 strip projection data obtained by exact calculation from (a). (From Shepp and Kruskal [1, pp. 422–424]. Reprinted by permission of American Mathematics Monthly.)

118   II. A Compact Symmetric Space—The Sphere

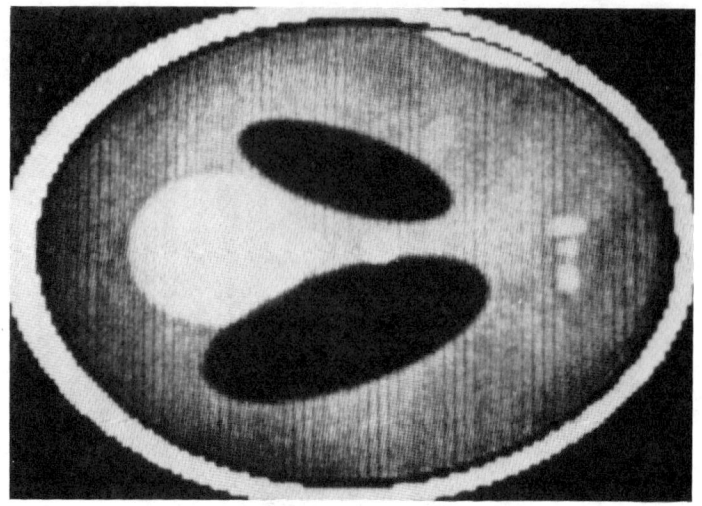

(d)

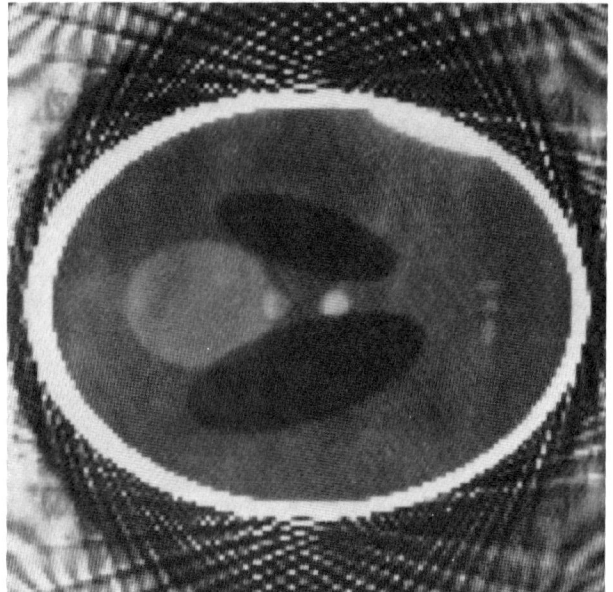

(c)

Figure 2.5. (continued)

2.2. $O(3)$ and $\mathbb{R}^3$. The Radon Transform

When CT scanners first became commercially available about 8 years ago it took 5 minutes to scan a patient's head and 5 minutes for each computerized reconstruction of an image from the x-ray data. Now, because of advances in the design of the scanners and in computer technology, the newest machines can scan a head in just 10 seconds and can reconstruct an image virtually instantaneously. According to Jay Thomas Payne of Abbott Northwestern Hospital in Minneapolis, the Mayo Clinic's first CT scanner, which is only 5 years old, has been relegated to the clinic's historical museum.

Nuclear magnetic resonance tomography is perhaps destined to be the tomography of the future. It uses magnetic fields rather than x-rays and thus is presumably less damaging to the body. Louis [1, 2] considers applications of the three-dimensional Radon transform to NMR tomography. Spherical harmonics are used to improve the algorithm by studying the kernel of the transform (ghosts).

There are many other applications of the Radon transform; for example, in radio astronomy (see Bracewell's article in Herman [1, pp. 81–104]). And there are applications to partial differential equations; e.g., to find solutions of the wave equation (see Dym and McKean [1, pp. 137–139] and Helgason [4]).

This concludes our discussion of harmonic analysis on the sphere. It would be possible to consider spherical analogues of many more results from Chapter 1. We shall leave this to the interested reader. For example, the group of isometries of the sphere is just $O(3)$. A discontinuous subgroup $\Gamma \subset O(3)$ has the property that any domain in the sphere can contain only finitely many points equivalent under Γ to any given point. There are very few such discontinuous Γ for the sphere or elliptic plane. They correspond to the regular polyhedra (see Hilbert and Cohn-Vossen [1, p. 242]). These are the analogues of the space groups considered in §1.4. It is also possible to discuss the analogue of the Poisson summation formula for $\Gamma\backslash S^2 \cong \Gamma\backslash G/K$, $G = SO(3)$, $K = SO(2)$. We shall not do this here, since our main interest it is the Selberg trace formula for noncompact fundamental domains (see later chapters). In fact, we never examined the Selberg trace formula when G is the Euclidean group either (but see Hejhal [5]). It would be useful to find fast Fourier transforms for the sphere.

CHAPTER III

The Poincaré Upper Half-Plane

But why then, this mystical set-up of putting the definition before the proof?

From Lakatos [1, p. 154].

3.1. Hyperbolic Geometry

...la géométrie non-Euclidienne...est la clef véritable du problème qui nous occupe.*

From a letter of Poincaré to Klein in 1881 appearing in *Acta Math.* 39 (1923), p. 100.

In Chapter 2, we considered a model for elliptic geometry, in which any two geodesics intersect, so that there are no parallels. Now we want to investigate a model for hyperbolic geometry, in which there are infinitely many parallels to a given geodesic through a given point. This sort of geometry was discovered by Bolyai, Gauss, and Lobatchevsky in the 1820s. However, Gauss never published his results, perhaps because the idea was controversial. In fact, Gauss embittered Bolyai by claiming precedence in a letter to Bolyai's father (see Gauss [1, Vol. 8, pp. 220–225]). The subject of non-Euclidean geometry was still controversial when Lewis Carroll "repudiated hyperbolic geometry in 1888 as being too fanciful" (see the article of Coxeter in *COSRIMS* [1, p. 55]). Models for hyperbolic geometry were obtained first by Liouville, then by

---

* ...Non-Euclidean geometry...is the real key to the problem that concerns us.

3.1. Hyperbolic Geometry

Beltrami in 1868, and by Klein in 1870. Poincaré rediscovered the Liouville-Beltrami upper half-plane model in 1882 and this space is usually called the Poincaré upper half-plane, though some call it the Lobatchevsky upper half-plane (but see Milnor [2]). Poincaré [1] also considered discontinuous groups of transformations of the hyperbolic upper half-plane as well as the functions left invariant by these groups and we intend to do the same. The geometric foundations for such work were laid by Gauss in 1827 (see Gauss [1, Vol. 4, pp. 217–258]) and by Riemann in 1854 (see Riemann [1, pp. 272–287]). These important papers of Gauss and Riemann are discussed from a modern perspective in Spivak [2, Vol. II]. Further progress was made from the 1880s to the 1930s with the investigation of Lie groups and symmetric spaces by Lie, Cartan, and others. Ultimately, non-Euclidean geometry has become about as controversial as $\sqrt{-1} = i$. Like complex analysis, analysis on Lie groups and symmetric spaces such as the Poincaré upper half-plane has become an indispensible tool for modern work in physics and engineering. We will see many examples, including work on the design of microwave transmission lines and the computation of the electrostatic field due to a thin charged conductor in the shape of the surface of two intersecting spheres. There are many applications to quantum physics in Hurt [1,2] and Gutzwiller [1–3]. Some references for this section are Auslander [1], Dym and McKean [1], Hilbert and Cohn-Vossen [1], Maass [1], and Siegel [1].

We shall see that the Poincaré upper half-plane is the symmetric space of the Lie group $SL(2, \mathbb{R})$, the *special linear group* of all 2×2 real matrices with determinant 1. This group has made many appearances in number-theoretic investigations, going back to work of Gauss and others on quadratic forms with integer coefficients. Now $SL(2, \mathbb{R})$ is isomorphic to the group $SU(1, 1)$ of 2×2 complex matrices of determinant 1 which preserve the hermitian form $-z_1 \bar{z}_1 + z_2 \bar{z}_2$. This means that the corresponding symmetric space can also be viewed as the unit disc (see the last part of this section concerning applications to electrical engineering). The group $SL(2, \mathbb{R})$ is locally isomorphic to the Lorentz-type group $SO(2, 1)$ of real 3×3 matrices of determinant 1 preserving the quadratic form $x_1^2 + x_2^2 - x_3^2$, a group whose analogue $SO(3, 1)$ is quite important in physics, since it leaves Maxwell's equations invariant (see Exercise 13).

The *Poincaré upper half-plane* H is defined by

$$H = \{x + iy | x, y \in \mathbb{R}, y > 0\}, \quad \text{where } i = \sqrt{-1}, \tag{3.1}$$

using a new notion of *arc length* given by

$$ds^2 = y^{-2}(dx^2 + dy^2). \tag{3.2}$$

This arc length is not "too fanciful," because it is invariant under the *action* of g in $SL(2, \mathbb{R})$ on z in H defined by fractional linear transformation:

$$gz = g(z) = (az + b)/(cz + d), \tag{3.3}$$

$$\text{if } g = \begin{pmatrix} a & b \\ c & d \end{pmatrix}, \text{ with } a, b, c, d \in \mathbb{R}, ad - bc = 1.$$

More information on fractional linear transformations can be found in Ahlfors [1], Siegel [1, Ch. 3], and almost any book on complex analysis.

Exercise 1. Show that the map $z \mapsto g(z)$ defined by (3) gives a conformal or angle-preserving mapping of H onto H.

Hint. Note that $\text{Im}(g(z)) = y|cz + d|^{-2}$, if $z = x + iy$, and $g'(z) = (cz + d)^{-2}$.

Exercise 2 (Group Invariance of the Riemannian Metric on H). Show that the Poincaré arc length element (3.2) is invariant under the action (3.3) of g in $SL(2, \mathbb{R})$.

Hint. If $w = f(z)$ is a holomorphic function and $w = u + iv$, $z = x + iy$, with u, v, x, $y \in \mathbb{R}$, then the Jacobian matrix of the change of variables from z to w is

$$J = \begin{pmatrix} \dfrac{\partial u}{\partial x} & \dfrac{\partial u}{\partial y} \\ \dfrac{\partial v}{\partial x} & \dfrac{\partial v}{\partial y} \end{pmatrix} = \begin{pmatrix} \dfrac{\partial u}{\partial x} & \dfrac{\partial u}{\partial y} \\ -\dfrac{\partial u}{\partial y} & \dfrac{\partial u}{\partial x} \end{pmatrix},$$

by the Cauchy-Riemann equations. Now

$$\begin{pmatrix} du \\ dv \end{pmatrix} = J \begin{pmatrix} dx \\ dy \end{pmatrix},$$

and the determinant of J is $|f'(z)|^2$. In our case $f(z) = (az + b)/(cz + d)$ with a, b, c, $d \in \mathbb{R}$ and $ad - bc = 1$. Thus, you easily compute that

$$v = y|cz + d|^{-2} \quad \text{and} \quad f'(z) = (cz + d)^{-2}.$$

The *geodesics*, or curves minimizing the Poincaré arc length in H, are straight lines or circles orthogonal to the real axis. To prove this, we imitate the argument that works for Euclidean space and the sphere. It suffices to show that the y-axis is the curve minimizing distance and passing through i and $y_0 i$, $y_0 > 0$. For we can find an element g of $SL(2, \mathbb{R})$ which moves any two given points in H to i and iy_0, for some $y_0 > 0$, by Exercise 3. Now let

$$z(t) = x(t) + iy(t), \qquad a \leq t \leq b,$$

denote any curve in H with $z(a) = i$ and $z(b) = iy_0$. Then the Poincaré length of this curve is (for $y_0 > 1$)

$$\int_a^b y^{-1}(x'^2 + y'^2)^{1/2}\, dt \geq \int_1^{y_0} y^{-1}\, dy = \log y_0.$$

And $|\log y_0|$ is the Poincaré length of the segment of the y-axis joining i and iy_0. We will see that a similar argument works in the higher-dimensional symmetric spaces to be considered in later chapters.

3.1. Hyperbolic Geometry

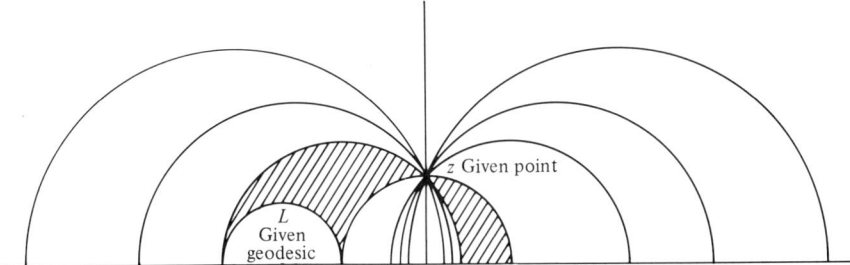

Figure 3.1. The failure of Euclid's fifth postulate. All geodesics through z outside the shaded angle fail to meet L.

Exercise 3. Show that given p, q in H, there is a matrix $g \in SL(2, \mathbb{R})$ such that $g(p) = i$ and $g(q) = iy_0$ for some $y_0 > 0$.

Hint. First move p to i. Then note that $k \in SO(2) \subset SL(2, \mathbb{R})$ implies $k(i) = i$.

We can use the Poincaré arc length to define the *non-Euclidean distance* between two points in H. This distance is the Poincaré length of the unique geodesic segment connecting the two points. The geometry obtained by considering geodesics to be the straight lines is called *hyperbolic geometry*. It is similar to Euclidean geometry in many respects. For example, two points in H determine a unique geodesic passing through them. And the non-Euclidean distance makes H a metric space. So, for example, the distance on H satisfies the triangle inequality, meaning that the length of one side of a triangle is less than or equal to the sum of the lengths of the other two sides.

Exercise 4. Show that two points in H determine a unique geodesic passing through them. Then show that the non-Euclidean distance satisfies the triangle inequality.

Exercise 5. We can use the Poincaré metric, which is really an inner product on the tangent space to H at a point, to define the angle between tangent vectors or curves in H. Show that this notion of angle is the same as the Euclidean angle measure.

But the hyperbolic geometry does not satisfy Euclid's fifth postulate. For given a point z in H not on a geodesic L, there are infinitely many geodesics through z which do not meet L (see Figure 3.1, in which all geodesics through z outside the shaded angle fail to meet L).

The group of all isometries of H is generated by the fractional linear transformations (3.3) and $z \mapsto -\bar{z}$. Here isometry means Poincaré-distance preserving. See Exercise 12 for a proof.

We can now use the standard formulas from Riemannian geometry to write down the $SL(2, \mathbb{R})$-invariant area element and Laplacian (see formulas (2.2)–

(2.6) and Auslander [1]). The $SL(2,\mathbb{R})$-invariant *area element* on H is
$$d\mu = y^{-2} \, dx \, dy. \tag{3.4}$$
The *Laplace operator* on H is
$$\Delta_H = \Delta = y^2 \left(\frac{\partial^2}{\partial x^2} + \frac{\partial^2}{\partial y^2} \right). \tag{3.5}$$

Exercise 6 ($SL(2,\mathbb{R})$-Invariance of the Area Element and the Laplacian on H). Show that $d\mu$ in (3.4) and Δ_H in (3.5) are $SL(2,\mathbb{R})$ invariant.

Hint. Recall Exercise 2 and formulas (2.2)–(2.6).

Exercise 7. Show that the non-Euclidean area of a hyperbolic triangle is π minus the sum of the angles.

Hint. You can prove this directly or note that this is a special case of the Gauss-Bonnet formula, since the Gaussian curvature of H is -1 (see Singer and Thorpe [1, pp. 175, 191–192] or Auslander [1, pp. 265, 268]).

It is now possible to explain the name "hyperbolic" in at least two ways. The name is due to Klein and comes from the Greek word *hyperballein* meaning to throw beyond. The first justification for the name "hyperbolic" is that if two geodesic rays R_1 and R_2 emanate from the ends of a geodesic segment perpendicular to them both, then the non-Euclidean distance between R_1 and R_2 will increase. The second explanation for the name "hyperbolic" comes from the fact that the Gaussian curvature of H is -1 and thus H resembles a hyperboloid of one sheet or a hyperbolic paraboloid (e.g. $z = x^2 - y^2$ at the origin). References for this are Auslander [1], Flanders [1], Guggenheimer [1], and Hilbert and Cohn-Vossen [1]. One could similarly justify the word "elliptic" in Chapter 2.

There are other useful realizations of H. The most abstract of them is given in the following exercise. Compare it with Exercise 1 of §2.1.

Exercise 8 (H as a Quotient or Homogeneous Space). Show that H can be identified with G/K where $G = SL(2,\mathbb{R})$ and $K = SO(2)$.

Hint. Map G/K one-to-one onto H by sending the coset gK, $g \in G$, to $g(i) =$ the image of $i = \sqrt{-1}$ under the fractional linear transformation (3.3) defined by g.

For connecting Chapters 3 and 4, the most useful realization of H is as *the space of positive definite binary quadratic forms of determinant 1* or what amounts to the same thing:

$$\mathscr{SP}_2 = \{ Y \in \mathbb{R}^{2 \times 2} \mid Y \text{ positive definite symmetric of determinant } 1 \}. \tag{3.6}$$

3.1. Hyperbolic Geometry

A 2×2 symmetric matrix Y with real coefficients is *positive* if $Y[x] = {}^t x Y x > 0$ for every nonzero column vector x in $\mathbb{R}^2$. See Chapter 4 for more information about positive matrices. Note that g in $SL(2, \mathbb{R})$ acts on Y in $\mathscr{SP}_2$ via $Y \to Y[g] = {}^t g Y g$. We shall use left G-actions in Chapter 3 and right G-actions in Chapter 4. Hopefully this will not cause too much confusion.

Exercise 9 (*H* as a Space of Positive Quadratic Forms of Determinant 1).

(a) Show that the following maps are identifications and preserve the group action of $SL(2, \mathbb{R})$ on the three homogeneous spaces:

$$K \backslash G \to \mathscr{SP}_2 \to H$$

$$Kg \mapsto {}^t gg = P \mapsto z \in H \text{ with } P[{}^z_1] = 0,$$

where $P[A] = {}^t APA$ for any suitable matrix of vector A. Here $K \backslash G$ is used to denote the homogeneous space of right cosets Kg, $g \in G = SL(2, \mathbb{R})$, $K = SO(2)$. Note that you can write P in $\mathscr{SP}_2$ as follows:

$$P = \begin{pmatrix} y^{-1} & 0 \\ 0 & y \end{pmatrix} \begin{bmatrix} 1 & -x \\ 0 & 1 \end{bmatrix} = \begin{pmatrix} y^{-1} & -x/y \\ -x/y & (x^2 + y^2)/y \end{pmatrix},$$

with $x, y \in \mathbb{R}$, $y > 0$.

This is clearly possible. Then $P[{}^z_1] = 0$ for $z \in H$ implies $z = x + iy$.

(b) Show that a geodesic-reversing isometry of $\mathscr{SP}_2$ at the identity I is $Y \mapsto Y^{-1}$. The corresponding mapping in H is $z \mapsto -z^{-1}$.

Given $P \in \mathscr{SP}_2$, the spectral theorem for self-adjoint operators on finite-dimensional Hilbert space says there is a matrix k in $SO(2)$ such that $P[k] = {}^t kPk$ is diagonal with first diagonal entry $t > 0$ and second entry $1/t$. We can think of k and t as *polar coordinates* for P. If you want to use geodesic polar coordinates, you would use k and $|\log t|$. This amounts to defining *geodesic polar coordinates* (r, u) of $z = x + iy$ in H by

$$z = \begin{pmatrix} \cos u & \sin u \\ -\sin u & \cos u \end{pmatrix} \begin{pmatrix} \exp(-r/2) & 0 \\ 0 & \exp(r/2) \end{pmatrix} (i), \tag{3.7}$$

$$x = y \sinh r \sin 2u, \quad y = (\cosh r + \cos 2u \sinh r)^{-1}.$$

As u runs from 0 to 2π and r from 0 to infinity, the upper half-plane is covered once. To see this, note that the eigenvalues t of the positive definite matrix $P \in \mathscr{SP}_2$ corresponding to z as in Exercise 9 are uniquely determined up to order. And the formula (3.7) fixes that order by requiring that the first eigenvalue be the smaller one. And the subgroup of $SO(2)$ preserving a diagonal matrix, $P \in \mathscr{SP}_2$, $P \neq I$, has order 2. But this subgroup does not affect elements of H at all.

Exercise 10. Show that in formula (3.7), r is the Poincaré distance between z and $i = \sqrt{-1}$. Then show that the orthogonal grid in H created by the curves

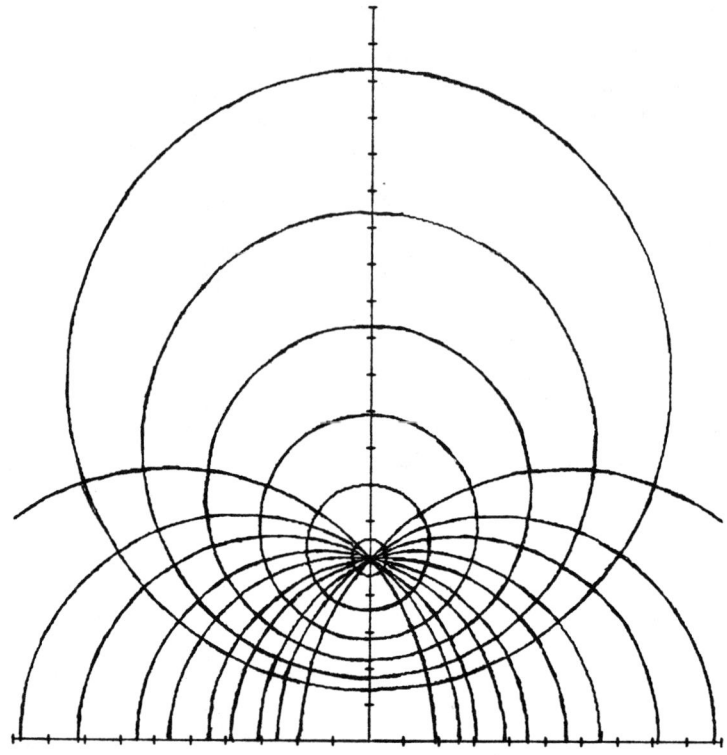

Figure 3.2. Coordinate grid for geodesic polar coordinates.

$u = $ constant and $r = $ constant consists of circles, such as those pictured in Fig. 3.2. In particular,

$\{k(\exp(-r)i)|r > 0\}$ = a circle through $i = \sqrt{-1}$ orthogonal to the x-axis,

$\{k_u(e^{-r}i)|k \in SO(2)\}$ = a circle with center $i\cosh r$ and radius $\sinh r$.

Exercise 11 (Changing Variables in Geodesic Polar Coordinates). Show that in geodesic polar coordinates (r, u) as in (3.7), we have

$$ds^2 = dr^2 + (\sinh r)^2 \, du^2,$$

$$d\mu = y^{-2} \, dx \, dy = \sinh r \, dr \, du,$$

$$\Delta = (\sinh r)^{-1} \frac{\partial}{\partial r}\left(\sinh r \frac{\partial}{\partial r}\right) + (\sinh r)^{-2} \frac{\partial^2}{\partial u^2}.$$

Use formulas (2.2)–(2.6).

Another realization of hyperbolic geometry is the *unit disc*:

$$U = \{z \in \mathbb{C} | |z| \leq 1\}. \tag{3.8}$$

The *Cayley transform* $z \mapsto w = (z - i)(z + i)^{-1}$ maps H conformally onto U. See Siegel [1, Ch. 3], Maass [1, Ch. 1], and Helgason [1, pp. 4–7] for more information on this model. The group of orientation-preserving isometries of U is $SU(1,1)$ which is the group of matrices $g \in \mathbb{C}^{2\times 2}$ such that

$$^t\bar{g}\begin{pmatrix}1 & 0\\ 0 & -1\end{pmatrix} g = \begin{pmatrix}1 & 0\\ 0 & -1\end{pmatrix}.$$

Such g have the form

$$g = \begin{pmatrix}a & b\\ \bar{b} & \bar{a}\end{pmatrix}, \qquad \text{with } a, b \in \mathbb{C} \text{ and } |a|^2 - |b|^2 = 1.$$

This version of hyperbolic geometry is much used by electrical engineers, as we shall see.

Exercise 12. Use Schwarz's lemma from complex analysis (see Ahlfors [1, p. 135]) to show that the set of all orientation-preserving isometries of the unit disc U is the group $SU(1,1)$. Show also that $SU(1,1)$, together with the isometry given by $z \mapsto \bar{z}$, generates all the isometries of U. Then use the Cayley transform to translate this result to H. Finally, characterize the geodesics in U, using the appropriate $SU(1,1)$-invariant arc length.

Application to Microwave Engineering: The Smith Chart

Microwave engineers deal with electromagnetic waves at high frequencies and short wavelengths from 30 cm to 0.3 mm (see the chart of the electromagnetic spectrum appearing in Fig. 1.7). Microwaves were first used in World War II to detect planes and ships via radar. Now there are numerous applications to such diverse areas as astronomy, communications, cooking, location of speeders on the highways, and nuclear physics. Non-Euclidean geometry comes into the design of microwave transmission lines because the basic quantitites of interest are related by fractional linear maps (or higher-dimensional analogues). Here we consider only a very simple example in which power transfer is increased on a lossless transmission line by connecting short-circuited stubs to the line. Further information on microwave theory can be found in the books of Altman [1], Baden Fuller [1], Collin [1], Helszajn [1], Hewlett-Packard [1], Lance [1], Magnusson [1], and Staniforth [1], for example. See Border [1] and Helton [1, 2, 3] for a general treatment of power transfer problems involving more complicated circuits with waves of varying frequency. Here we shall demonstrate that we are not engineers by using $i = \sqrt{-1}$ and not j. We shall also write $\bar{z}$ for the complex conjugate of z and not z^*, as the engineers do.

At short wavelengths, electronic circuits behave differently than they do at long wavelengths which are larger than the dimensions of the circuit under consideration. For example, the circuit parameters vary with position and

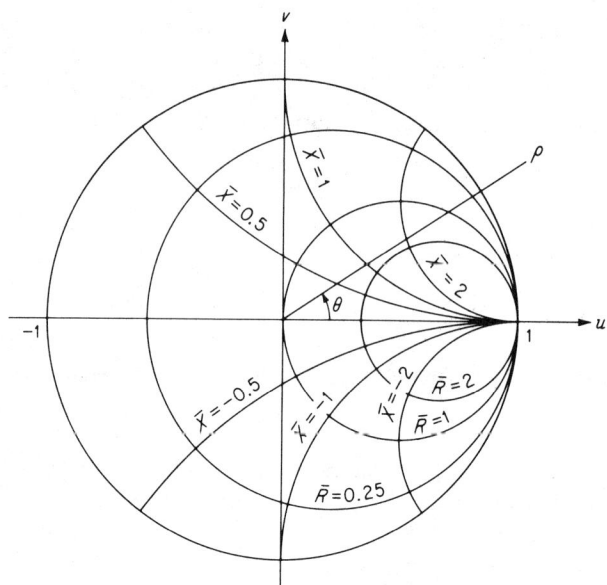

Figure 3.3. Constant $\bar{R}$ and $\bar{X}$ circles in the reflection-coefficient plane. (From R.E. Collin, *Foundations for Microware Engineering*, © 1966, p. 205. Reprinted by permission of McGraw-Hill Book Co.)

radiation becomes important. If a voltage wave $V^+ \exp(-i\beta d)$ with current $I^+ \exp(-i\beta d)$ arrives at the end of a lossless transmission line, a reflected voltage wave $V^- \exp(i\beta d)$ with current $-I^- \exp(i\beta d)$ is produced. Here the phase constant $\beta = 2\pi/\lambda$, if λ = wavelength and d measures distance along the line. If the load is at $d = 0$, $V_L = V^+ + V^-$, $I_L = I^+ - I^-$. The *load impedance* is $Z_L = V_L/I_L$ and the *characteristic impedance* is $Z_c = V^+/I^+ = V^-/I^-$. Define the *voltage reflection coefficient* $\Gamma = V^-/V^+$ and the *normalized load impedance* $z = Z_L/Z_c$. It follows that

$$\Gamma = (z-1)(z+1)^{-1} \quad \text{and} \quad z = (1+\Gamma)(1-\Gamma)^{-1}. \tag{3.9}$$

The load is said to be *matched* with the line when $\Gamma = 0$. This is the condition for maximizing power transfer because the *power flow* $P = V^+\bar{I}^+ - V^-\bar{I}^- = |V^+|^2\{1 - |\Gamma|^2\}/Z_c$. These definitions extend to give the reflection coefficient at position $-d$:

$$\Gamma(d) = [V^- \exp(-i\beta d)][V^+ \exp(i\beta d)]^{-1} = \Gamma_L \exp(-2i\beta d). \tag{3.10}$$

Then the normalized input impedance (seen looking toward the load) at position $-d$ is

$$z_{in} = (1 + \Gamma(d))(1 - \Gamma(d))^{-1}. \tag{3.11}$$

For lossy lines, $\Gamma(d) = \Gamma_L \exp(-2i\beta d - 2\alpha d)$.

Now $|\Gamma| \leq 1$ and the mapping (3.11) from z_{in} to Γ is a slight variation of the Cayley transform just discussed. Write $z_{in} = R + iX$, where R = input re-

3.1. Hyperbolic Geometry

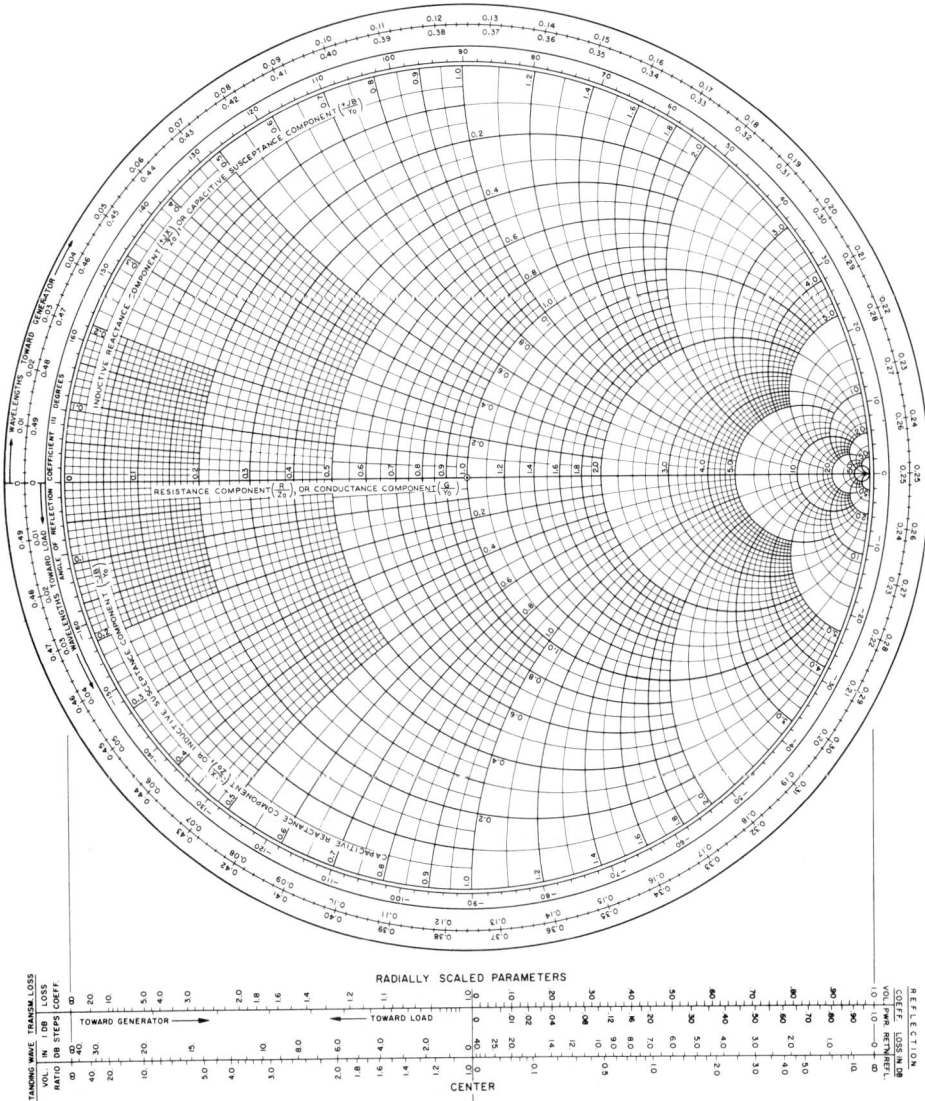

Figure 3.4. The Smith chart. (Reproduced by permission of Phillip H. Smith, Analog Instruments Co.)

sistance and X = input reactance (both normalized). Plotting the image of the lines R = constant and X = constant in the Γ-plane gives the picture in Fig. 3.3. This sort of graph is called a "Smith chart" because it was proposed by P.H. Smith in 1939. However, Matsumoto [1, p. 41] reports that Mitzuhashi had discussed such a chart in 1937. A copy of the actual graph paper available in the usual university book store is given in Fig. 3.4. Note that a movement a distance d along the transmission line changes Γ by $\exp(-2\beta id)$; i.e., the point on the Smith chart rotates by an angle $2\beta d$.

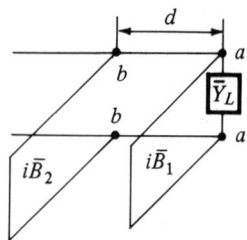

Figure 3.5. Double-stub tuner. (From R.E. Collin, *Foundations for Microwave Engineering*, © 1966, p. 212. Reproduced by permission of McGraw-Hill Book Co.)

Next let us consider an example of the use of the Smith Chart from Collin [1, pp. 212–215]. The general idea is to see what various network elements do to a given point Γ. The goal is to finish with $\Gamma = 0$ so that the load is matched. First a few more definitions. Let the normalized input admittance $y_{in} = z_{in}^{-1} = (1 - \Gamma)(1 + \Gamma)^{-1} = G + iB$, where G = conductance and B = susceptance. A stub is a short length of transmission line with a short circuit at its end. Consider the double-stub tuner pictured in Fig. 3.5. Let P_1 on the Smith Chart correspond to $y_L = G_L + iB_L$ = the normalized load admittance. The first stub at the plane aa adds a susceptance iB_1 which moves P_1 along the G = constant circle to P_2. Then we get to the second stub at the plane bb by moving in the Γ-plane from P_2 to P_3, through the angle $2\beta d = 4\pi d/\lambda$, clock-

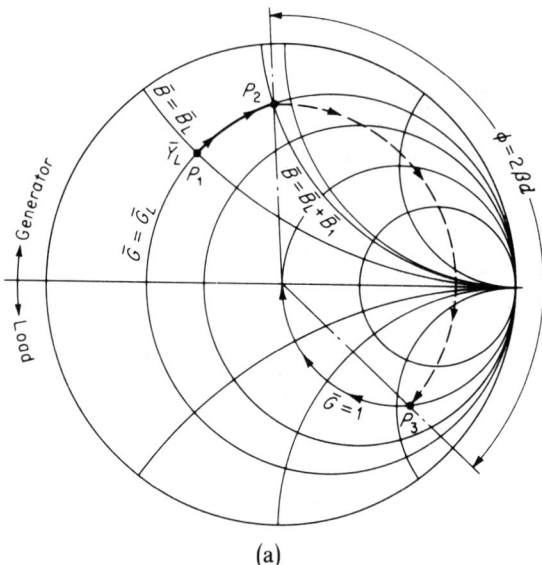

(a)

Figure 3.6. (a) Representation of the operation of a double-stub tuner; (b) Determination of required susceptance for the first stub in a double-stub tuner; (c) Range of load impedance which cannot be matched when $d = \lambda/8$. (From R.E. Collin, *Foundations for Microware Engineering*, © 1966, pp. 213–214. Reproduced by permission of McGraw-Hill Book Co.)

3.1. Hyperbolic Geometry

wise. The point P_4 is obtained by moving along G = constant. In order to make $P_4 = 0$ and achieve matching, the point P_3 must lie on the circle $G = 1$. You can figure out graphically the value B_1 must have to do this. In fact, there may be two solutions or no solution, depending on the value of G_L and d. See Fig. 3.6 from Collin [1, pp. 213–214]. It turns out that one can always obtain matching with three stubs at fixed positions (see Collin [1, pp. 214–217]).

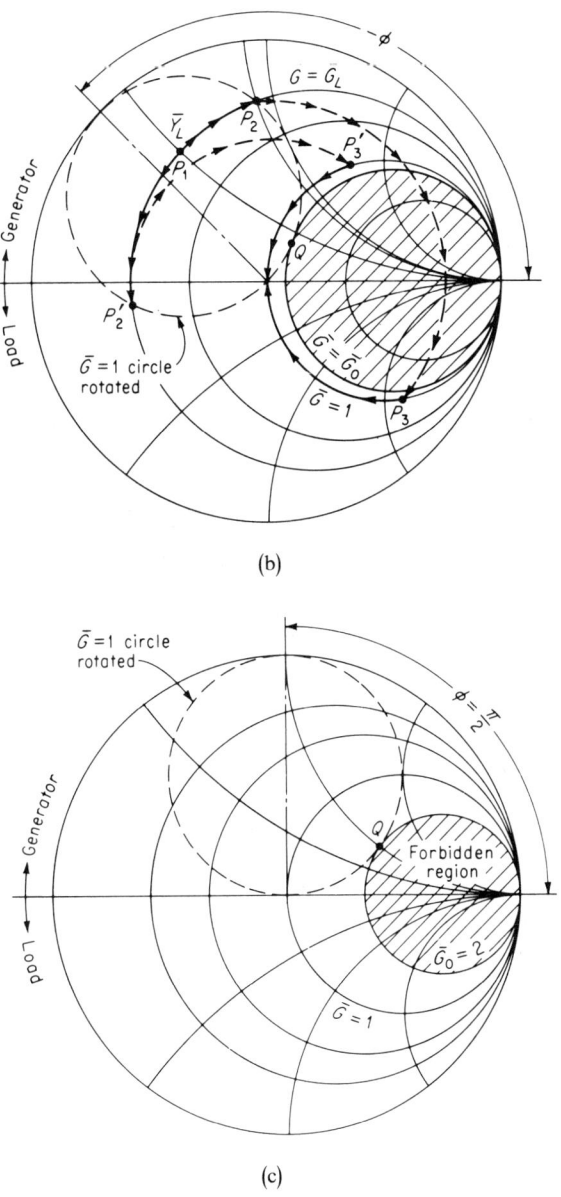

Figure 3.6. (continued)

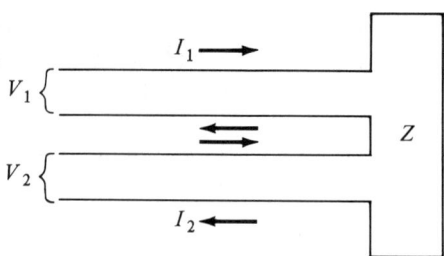

Figure 3.7. A Two-Port circuit.

Note that z or y lie in the right half-plane, rather than in the Poincaré upper half-plane H. To put y in H multiply it by i:

$$y = G + iB \mapsto iY = -B + iG.$$

Then the first stub in the double-stub tuner transforms iY to $iY + B_1$. This means that the stub corresponds to the following matrix in $SL(2, \mathbb{R})$:

$$\begin{pmatrix} 1 & B_1 \\ 0 & 1 \end{pmatrix}.$$

In general, two-port circuits correspond to 2×2 matrices. The terminology does not appear to be totally standardized. But we shall describe the correspondence given in Helton [1]. The two-port circuit has the diagram shown in Fig. 3.7. For a linear two-port circuit, the *impedance matrix* $Z \in \mathbb{C}^{2 \times 2}$ satisfies $ZI = V$, where

$$I = \begin{pmatrix} I_1 \\ I_2 \end{pmatrix} \quad \text{and} \quad V = \begin{pmatrix} V_1 \\ V_2 \end{pmatrix}.$$

In this section we shall use $\mathbb{L}$ rather than I as the 2×2 identity matrix. The *scattering matrix* for the circuit is $S = (\mathbb{L} - Z)(\mathbb{L} + Z)^{-1}$. And the *chain matrix* Ч (the Russian letter cha) is defined by

$$\text{Ч} \begin{pmatrix} V_2 \\ -I_2 \end{pmatrix} = \begin{pmatrix} V_1 \\ I_1 \end{pmatrix}.$$

The *power consumption* is $P = {}^t\bar{V}I$ = the hermitian product of the two vectors. The *admittance matrix* is $Y = Z^{-1}$. In Fig. 3.8 we list the chain matrices for some common two-ports.

There is yet one more realization of the group $SL(2, \mathbb{R})$ which we should mention. This connects the group with Lorentz-type groups which are of great interest in physical applications. The connection is developed in the following exercise.

Exercise 13. Let

$$W = \left\{ X(x_0, x_1, x_2) = \begin{pmatrix} x_0 + x_1 & x_2 \\ x_2 & x_0 - x_1 \end{pmatrix} \middle| x_i \in \mathbb{R}, 0 \le i \le 2 \right\},$$

3.1. Hyperbolic Geometry

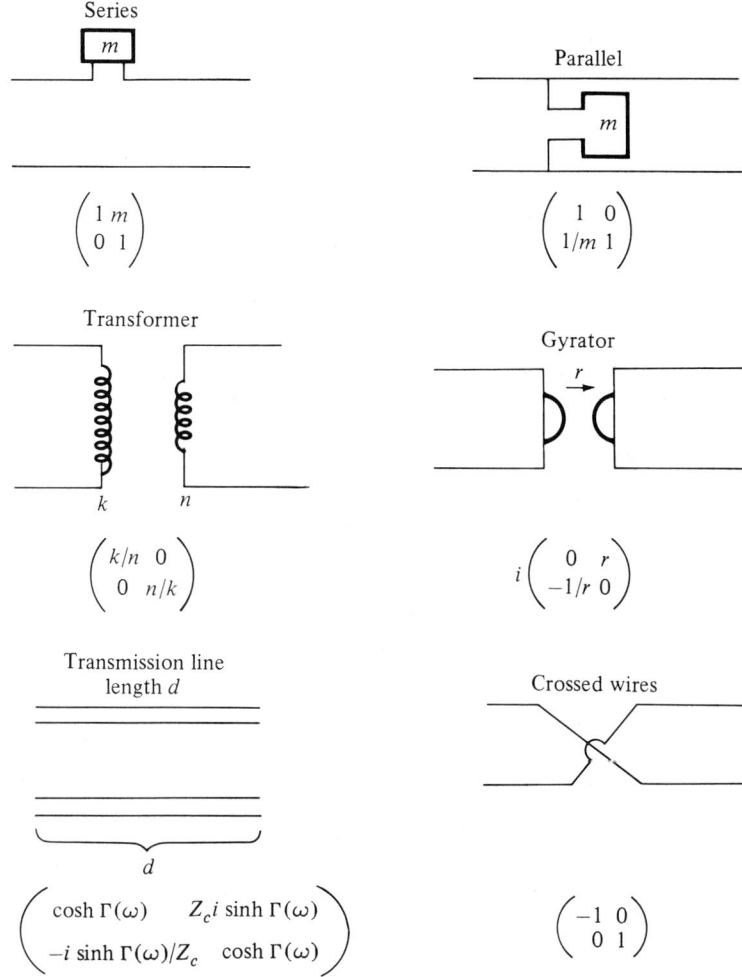

Figure 3.8. Chain matrices for some common two-ports. (Based on an illustration in *Bulletin of the American Mathematical Society*, "Noneuclidean Functional Analysis and Electronics", J. William Helton, (1982) Volume 7, pp. 1–64, by permission of the American Mathematical Society.)

and $e_0 = X(1, 0, 0)$, $e_1 = X(0, 1, 0)$, $e_2 = X(0, 0, 1)$. For each g in $SL(2, \mathbb{R})$, let $s(g)$ denote the matrix of the linear transformation of W given by sending x in W to $gx^t g$, using the basis $\{e_0, e_1, e_2\}$. Show that $s: SL(2, \mathbb{R}) \to G_+(2)$ is a continuous homomorphism onto $G_+(2) =$ the connected component of the identity in the three-dimensional Lorentz group $O(1, 2)$. Prove that the kernel of s is $\{\pm 1\}$. The Lorentz group $O(1, 2)$ is defined to consist of all real 3×3 matrices which preserve the quadratic form $x_0^2 - x_1^2 - x_2^2$.

Hint. Note that $\det(X(x_0, x_1, x_2)) = x_0^2 - x_1^2 - x_2^2$ is invariant under $s(g)$. You can find more details in Sugiura [1, pp. 206–207].

Alternative Discussion. Consider a 2×2 matrix A. If e_1 and e_2 are basis vectors and $Ae_i = a_{1i}e_1 + a_{2i}e_2$, then we obtain the second symmetric power, denoted $\text{Sym}^2(A)$, by considering symmetric products (analogous to the alternating products appearing in the theory of differential forms):

$$Ae_1 \cdot Ae_1 = a_{11}^2 e_1^2 + 2a_{11}a_{21}e_1 e_2 + a_{21}^2 e_2^2,$$

$$Ae_2 \cdot Ae_2 = a_{12}^2 e_1^2 + 2a_{12}a_{22}e_1 e_2 + a_{22}^2 e_2^2,$$

$$Ae_1 \cdot Ae_2 = a_{11}a_{12}e_1^2 + (a_{11}a_{22} + a_{21}a_{12})e_1 e_2 + a_{21}a_{22}e_2^2.$$

So the matrix of $\text{Sym}^2(A)$, using the basis $e_1^2, e_1 e_2, e_2^2$, is:

$$\begin{pmatrix} a_{11}^2 & a_{11}a_{12} & a_{12}^2 \\ 2a_{11}a_{21} & a_{11}a_{22} + a_{21}a_{12} & 2a_{12}a_{22} \\ a_{21}^2 & a_{21}a_{22} & a_{22}^2 \end{pmatrix}.$$

This is a representation; i.e., $\text{Sym}^2(gh) = \text{Sym}^2(g)\text{Sym}^2(h)$.

Next note that we can view $SL(2, \mathbb{R})$ as the group $Sp(1, \mathbb{R})$, the symplectic group of 2×2 real matrices preserving the alternating form represented by the equation

$${}^t g \begin{pmatrix} 0 & 1 \\ -1 & 0 \end{pmatrix} g = \begin{pmatrix} 0 & 1 \\ -1 & 0 \end{pmatrix} = J.$$

Now

$$\text{Sym}^2(J) = \begin{pmatrix} 0 & 0 & 1 \\ 0 & -1 & 0 \\ 1 & 0 & 0 \end{pmatrix} \quad \text{which is equivalent to} \quad \begin{pmatrix} 1 & 0 & 0 \\ 0 & 1 & 0 \\ 0 & 0 & -1 \end{pmatrix}.^*$$

So now we see that the representation $Sp(1, \mathbb{R}) = SL(2, \mathbb{R}) \xrightarrow{\text{Sym}^2} SO(2, 1)$, provided that $\text{Sym}^2({}^t g) = {}^t\text{Sym}^2(g)$. To make this true, replace $\text{Sym}^2(A)$ by the equivalent matrix obtained by multiplying the 2nd row of $\text{Sym}^2(A)$ by $1/\sqrt{2}$ and the 2nd column by $\sqrt{2}$,

3.2. Harmonic Analysis on H

Wie man u.a. in den Werken von Euler und Gauss feststellen kann, haben in früheren Jahrhunderten die Mathematiker nicht nur allgemeine Theoreme veröffentlicht, sondern dazu auch Beispiele, die ihnen wohl Vergnügen machten und den Leser weiter belehrten.

...

Mir selber erscheint es allerdings nach den entscheidenden Ergebnissen von Gödel und Cohen geraten, mit dem ungehemmten axiomatischen Verfahren weiterhin vorsichtig zu sein. So hat mir schon immer die Anwendung des Auswahlaxioms widerstrebt, und ich fühle mich sogar jedesmal auf meinen

---

* Here equivalence means by change of basis.

3.2. Harmonic Analysis on H

Füssen sicherer, wenn ich auf zwei verschiedenen Wegen einsehen kann, dass zweimal zwei gleich vier ist.*

From Siegel [5, Vol. IV, p. 33].

Complete sets of eigenfunctions of the non-Euclidean Laplacian $\Delta = y^2(\partial^2/\partial x^2 + \partial^2/\partial y^2)$ can be found by separation of variables in either rectangular or geodesic polar coordinates. For rectangular coordinates, the spectral resolution of Δ reduces to the inversion formula for the Kontorovich-Lebedev transform which was discussed in Exercise 7 of §2.2. This result was used by Kontorovich and Lebedev [1, 2, 3] in the late 1930s to solve various boundary value problems in mathematical physics, such as Dirichlet's problem for a wedge (see Exercise 8). For geodesic polar coordinates, the spectral decomposition of Δ boils down to the inversion formula for the Mehler-Fock transform. This is due to Mehler [1] in 1881 and was proved by Fock [1] in 1943. One motivation for studying the Mehler-Fock transform was the need to solve such physical problems as the computation of the electrostatic field due to a thin charged conductor in the shape of the surface of the region bounded by two intersecting spheres (see Exercise 23). After considering these two different reductions of the spectral decomposition of Δ to that of certain one-variable singular Sturm-Liouville operators, we shall consider another approach due to Helgason [1, 5]. This result will then be used to find the fundamental solution of the heat equation on H. Then we shall discuss the central limit theorem on H, which was first proved by Karpelevich, Tutubalin, and Shur [1]. This can be used to study the power reflected by random inhomogeneities in a long transmission line, as was first noted by Gertsenshtein and Vasil'ev [1]. Such work is related to investigations of Clerc and Roynette [1], Faraut [1], Heyer [1], Keller and Papanicolaou [1], Papanicolaou [1], and Trimèche [1]. There are many applications of such central limit theorems; e.g., in demography, learning theory, and atomic physics (see Cohen [1], LePage [1], and Hantsch and von Waldenfels[1]). Thus, by the end of this section, we shall have discussed three versions of harmonic analysis on H, along with a few applications. In 1947, V. Bargmann essentially obtained the analogue of such Fourier decompositions for the group $SL(2, \mathbb{R})$ itself (see Bargmann [1]). We will do little more than mention this work, along with a few of the implications for quantum mechanics, at the end of the section.

---

* As one can see from the works of Euler and Gauss, mathematicians of previous centuries did not only publish general theorems but also examples which they might have enjoyed and which might have educated the reader.

...

I believe that after the decisive results of Gödel and Cohen one should be cautious with unrestrained axiomatic techniques. I was always reluctant to use the axiom of choice and am always more comfortable when I have two different techniques to see that $2 \times 2 = 4$.

References for this section include Dym and McKean [1], Gangolli [1–4], Helgason [1–9], Lang [4], Lebedev [1, 2], Rühl [1, 2], and Sugiura [1].

Harmonic analysis on H in rectangular coordinates x, y will be discussed first. These coordinates are the natural ones to use in the study of automorphic forms for $SL(2, \mathbb{Z})$, as we shall see in later sections of this chapter. In order to carry out harmonic analysis on H in rectangular coordinates, one must know about the eigenfunctions of the non-Euclidean Laplacian that transform nicely under translation. More precisely, we are seeking functions $f(z)$ which we call *k-Bessel functions* associated to $a \in \mathbb{R}$ and having the following properties:

(i) $f(z + u) = \exp(2\pi i a u) f(z)$, for all $u \in \mathbb{R}, z \in H$,

(ii) $\Delta f = \lambda f$, (3.12)

(iii) $|f(z)| \leq C y^p$ for some positive constants C, p.

We call these functions "k-Bessel functions" because they are closely related to the classical *K-Bessel functions* (also called Bessel functions of imaginary argument, Macdonald's functions, and modified Bessel functions of the third kind).

In order to build up such functions, we make use of the simpler function $f(z) = (\operatorname{Im} z)^s = y^s$, which we call the *power function*. It is easily seen that this power function is an eigenfunction of the non-Euclidean Laplacian; more precisely:

$$\Delta y^s = s(s-1) y^s. \tag{3.13}$$

So now we define the *k-Bessel function* at $s \in \mathbb{C}, z \in H, a \in \mathbb{R}$ by

$$k(s|z, a) = \int_{u=-\infty}^{+\infty} \{\operatorname{Im}[(-z + u)^{-1}]\}^s \exp(2\pi i a u) \, du. \tag{3.14}$$

Here we assume that $\operatorname{Re} s > 0$.

Exercise 1 (*K*-Bessel Functions).

(a) The K-Bessel function may be defined (see Lebedev [1] or Watson [1]) by

$$K_s(z) = \frac{1}{2} \int_0^\infty \exp\left[-\frac{z}{2}\left(t + \frac{1}{t}\right)\right] t^{s-1} \, dt,$$

for $\operatorname{Re} z > 0$. Show that if $\operatorname{Re} s > 0$,

$$\Gamma(s) \int_{-\infty}^{+\infty} (q^2 + x^2)^{-s} \exp(2irx) \, dx$$

$$= \begin{cases} 2\pi^{1/2} \left|\frac{r}{q}\right|^{s-1/2} K_{s-1/2}(2|rq|), & \text{if } r, q \neq 0 \\ |q|^{1-2s} \Gamma(\tfrac{1}{2}) \Gamma(s - \tfrac{1}{2}), & \text{if } r = 0, q \neq 0. \end{cases}$$

Hint. Use $\Gamma(s) = \int_0^\infty t^{s-1} e^{-t} \, dt$, for $\operatorname{Re} s > 0$, to see that

$$\Gamma(s)\int_{-\infty}^{+\infty}(q^2+x^2)^{-s}\exp(2irx)\,dx$$
$$=\int_{-\infty}^{+\infty}\exp(2irx)\int_0^{\infty}t^{s-1}(q^2+x^2)^{-s}e^{-t}\,dt\,dx$$
$$=\int_{-\infty}^{+\infty}\exp(2irx)\int_0^{\infty}u^{s-1}\exp[-u(q^2+x^2)]\,du\,dx.$$

Then complete the square in $-ux^2+2irx-uq^2=-u[(x+ir/u)^2-q^2+(r/u)^2]$. Now change variables via $w=x+ir/u$ and use the fact that

$$\int_{-\infty}^{+\infty}\exp(-w^2)\,dw=\sqrt{\pi}.$$

(b) Deduce from (a) that

$$k(s|z,a/\pi)=\exp(2iax)2\pi^{1/2}\Gamma(s)^{-1}|a|^{s-1/2}y^{1/2}K_{s-1/2}(2|ay|),$$

when $a\neq 0$, $z=x+iy\in H$, $\operatorname{Re}s>0$. Show also that $k(s|z,0)=y^{1-s}\Gamma(\frac{1}{2})\Gamma(s-\frac{1}{2})/\Gamma(s)$.

(c) Show that $\Delta_z k(s|z,a)=s(s-1)k(s|z,a)$ and $k(s|z+x,a)=\exp(2\pi iax)k(s|z,a)$.

Exercise 2 (Asymptotics and Functional Equations of K-Bessel Functions).

(a) *Functional Equation.* Show that $K_s(z)=K_{-s}(z)$, if $\operatorname{Re}z>0$.

(b) *Asymptotics.* Show that if s with $\operatorname{Re}s>0$ is fixed and $\operatorname{Re}z>0$, then

$$K_s(z)\sim 2^{s-1}\Gamma(s)z^{-s}\qquad\text{as } z\to 0,$$
$$K_s(z)\sim(\pi/(2z))^{1/2}e^{-z}\qquad\text{as } z\to\infty.$$

(c) Show that if $x>0$ is fixed, then

$$K_{it}(x)\sim(2\pi/t)^{1/2}\exp(-\pi t/2)\sin\left(\frac{\pi}{4}+t\log t-t-t\log\frac{x}{2}\right)\qquad\text{as } t\to\infty.$$

Hint. For the first formula in part (b), you can use the first integral formula in Exercise 1(a), after making the change of variables $v=tz$. For the second formula in part (b), see Lebedev [1, p. 123] or Olver [1]. For hints on part (c), see Exercise 9 of §3.7.

Exercise 3 (The Negativity of Δ on H).

(a) If Δf and f are both in $L^2(H)$, show that $(\Delta f, f)\leq 0$, using Green's theorem. Here $(f,g)=\int f(z)\overline{g(z)}y^{-2}\,dx\,dy$.

(b) Show that, in fact, $(\Delta f,f)\leq -\frac{1}{4}(f,f)$.
Hint. This is done by Dym and McKean [pp. 277–278].

(c) Show that if $s(s-1)\leq 0$, then $s=\frac{1}{2}+it$, $t\in\mathbb{R}$, or $s\in[0,1]$. Prove that if $s(s-1)<-\frac{1}{4}$, then $s\notin[0,1]$.

Exercise 4 (Separation of Variables in $\Delta f = \lambda f$, using Rectangular Coordinates). Seek solutions to $\Delta f = \lambda f$ of the form $f(x, y) = v(x)w(y)$. This leads to two ordinary differential equations:

$$\frac{w''}{w}(y) - \lambda y^{-2} = k = -\frac{v''}{v}(x),$$

where k is the separation constant. Then $v(x) = \exp(2\pi i a x)$ with $k = 4\pi^2 a^2$. If we set $w(y) = y^{1/2} u(y)$, show that u satisfies

$$y^2 u'' + y u' - (4\pi^2 a^2 y^2 + (\lambda + 1/4))u = 0.$$

Then show that if $a = 0$, a solution is $w(y) = y^s$, $\lambda = s(s-1)$, and if $a \neq 0$, a solution is $u(y) = K_{s-1/2}(2\pi|a|y)$, $\lambda = s(s-1)$. See Lebedev [1, pp. 109–110] if you need hints.

Theorem 1 (Fourier Inversion on H in Rectangular Coordinates). *Suppose that $f \in C_c^\infty(H)$; i.e., that f is infinitely differentiable with compact support in H. Then, setting $s = \frac{1}{2} + it$,*

$$f(z) = \pi^{-2} \int_{a \in \mathbb{R}} \int_{\text{Re}\,s = 1/2} \hat{f}(a, s) e_{a,s}(z)\, t \sinh \pi t\, dt\, da$$

where

$$e_{a,s} = \begin{cases} \exp(2\pi i a x) y^{1/2} K_{s-1/2}(2\pi|a|y), & \text{if } a \neq 0, \\ y^s, & \text{if } a = 0. \end{cases}$$

and

$$\hat{f}(a, s) = \int_H f(z) \overline{e_{a,s}(z)}\, y^{-2}\, dx\, dy.$$

PROOF. By the Fourier inversion formula of §1.2, this result is easily reduced to the inversion formula of Kontorovich and Lebedev [1–3] for functions $h(y)$, $y > 0$:

$$h(y) = 2\pi^{-2} \int_0^\infty t \sinh \pi t\, y^{-1/2} K_{it}(y) \int_0^\infty h(u) u^{-1/2} K_{it}(u)\, du\, dt$$

$$f(t) = 2\pi^{-2} t \sinh \pi t \int_0^\infty K_{it}(y) y^{-1} \int_0^\infty f(u) K_{iu}(y)\, du\, dy.$$

(3.15)

For it suffices to prove the theorem when $f(z) = v(x)w(y)$, if $z = x + iy$.

There are many ways to find the spectral measure $2\pi^{-2} t \sinh \pi t$ in equation (3.15). Three ways are given in Exercise 7 of §2.2, Exercise 16 and Exercise 21 of this section. What we call the *asymptotics/functional equations principle* gives a simpler method. We shall not prove this principle rigorously here, but instead give a plausible argument for its truth. The method comes from work of Harish-Chandra and Helgason. There is a rigorous proof using Paley-Wiener theory in Helgason [5], for the special case under consideration. The gen-

3.2. Harmonic Analysis on H

eralization to symmetric spaces of semisimple Lie groups has been obtained. But no one seems to have developed the principle as a general method in spectral theory. See Exercise 6, the discussion of the Mehler-Fock inversion formula (3.22), and §4.2 for more examples of the asymptotics/functional equations principle. From Exercise 2 above, we believe that the asymptotic behavior of K_{it} as y approaches 0 is given by

$$K_{it}(y) \sim 2^{it-1}\Gamma(it)y^{-it} + 2^{-it-1}\Gamma(-it)y^{it} \qquad \text{as } y \to 0^+. \qquad (3.16)$$

This can also be proved using the relation of K to the I-Bessel function and the power series for I. Note that the line of integration in the Fourier inversion integral is the line fixed by the functional equation of $K_s = K_{-s}$. And the two terms in (3.16) come from the functional equation. To prove (3.15), one wants to see that the kernel

$$W_R(x, y) = \pi^{-2} \int_{-R}^{+R} t \sinh \pi t (xy)^{-1/2} K_{it}(x) K_{it}(y) \, dt$$

approaches $\delta(x - y)$ as $R \to \infty$. Since our problem is invariant under $SL(2, \mathbb{R})$, it is reasonable to expect that it suffices to consider only $x, y \sim 0$. But then formula (3.16) and

$$\Gamma(it)\Gamma(-it) = \pi(t \sinh \pi t)^{-1} \qquad (3.17)$$

(see Lebedev [1, p. 3]) imply that

$$W_R(x, y) \sim (2\pi)^{-1} \int_{-R}^{+R} y^{-1/2 - it} x^{-1/2 + it} \, dt, \qquad x, y \to 0^+. \qquad (3.18)$$

And the right-hand side of (3.18) is a Dirac delta family by the Mellin inversion formula (see Exercise 1 of §1.4). So the spectral measure is chosen just to cancel out the gamma factors in (3.16). There are also discussions of the Kontorovich-Lebedev transform in Lebedev [2] and Sneddon [1, Ch. 6]. Note that one must show that the other terms in the asymptotic expansion (3.16) do not contribute to the inversion formula. One might also ask how one knows that there is no discrete spectrum. □

Exercise 5. Show that it does not matter which of the two inversion formulas in (3.15) you prove.

Exercise 6. Can the asymptotics/functional equations principle be used to compute the spectral measure for the Hankel integral formula of Exercise 6(b) in §2.2? You want to show that:

$$\int_0^R r J_v(rx) J_v(ry) r \, dr \sim \delta(x - y) \qquad \text{as } R \to \infty,$$

using

$$J_v(x) \sim (2/(\pi x))^{1/2} \cos\left(x - \frac{v\pi}{2} - \frac{\pi}{4}\right), \qquad \text{as } x \to \infty, v > -\tfrac{1}{2},$$

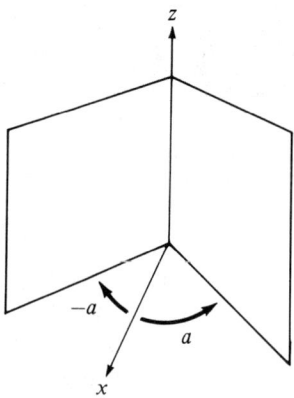

Figure 3.9. A wedge.

and
$$J_v(-x) = \exp(v\pi i) J_v(x).$$

Exercise 7 (Some Kontorovich-Lebedev Transforms).

(a) Show that the integral formula in Exercise 1(a) implies
$$K_{it}(z) = \int_0^\infty \exp[-z\cosh u] \cos(tu)\, du, \qquad \mathrm{Re}\, z > 0,\ t\text{ arbitrary}.$$

Use this to deduce that
$$\frac{\pi}{2} \exp(-z\cosh u) = \int_0^\infty K_{it}(z) \cos(tu)\, dt, \qquad \mathrm{Re}\, z > 0.$$

(b) Show that when $\mathrm{Re}(r) > |\mathrm{Re}(s)|$,
$$\int_0^\infty y^{r-1} K_s(y)\, dy = 2^{r-2} \Gamma\!\left(\frac{r+s}{2}\right) \Gamma\!\left(\frac{r-s}{2}\right).$$

Again you can use the integral formula in Exercise 1(a). For more hints, see Exercise 3 of §3.6 (the section in which we apply this result).

Note. You can find more examples in Erdélyi [2, Vol. II, pp. 125–153, 175–177] and Oberhettinger [2].

Exercise 8 (An Application of the Kontorovich-Lebedev Transform). Find the harmonic function in the wedge in Fig. 3.9 taking on given values on the boundary of the wedge; i.e., solve
$$\begin{cases} \Delta u(r,\phi,z) = 0 & \text{in } 0 < r < \infty,\ |\phi| \le a,\ z \in \mathbb{R}, \\ u(r,-a,z) = f(r,z), \\ u(r,a,z) = g(r,z). \end{cases}$$

3.2. Harmonic Analysis on H

Here (r, ϕ, z) denote cylindrical coordinates in $\mathbb{R}^3$ and f, g are given functions. Obtain solutions of

$$\Delta u = \frac{\partial^2 u}{\partial r^2} + \frac{1}{r}\frac{\partial}{\partial r} + \frac{1}{r^2}\frac{\partial^2 u}{\partial \phi^2} + \frac{\partial^2 u}{\partial z^2} = 0$$

of the form $u(r, \phi, z) = \exp(\pm \phi t \pm icz) R(r)$. Then $R(r)$ satisfies

$$R'' + r^{-1} R' + (t^2 r^{-2} - c^2) R = 0.$$

One solution is $K_{it}(r|c|)$. Use the Kontorovich-Lebedev transform to construct u from this. You can find the answer in Sneddon [1, pp. 363–366] or Lebedev [1, pp. 150–153].

Next we consider harmonic analysis on H in *geodesic polar coordinates* (r, u) given by

$$z = x + iy = k_u e^{-r} i, \quad \text{with} \quad k_u = \begin{pmatrix} \cos u & \sin u \\ -\sin u & \cos u \end{pmatrix}. \quad (3.19)$$

In §3.1, after formula (3.7), we noted that as u runs from 0 to 2π, and r runs from 0 to ∞, the upper half-plane is covered once. Recall also that we computed ds^2, $d\mu$, and Δ in geodesic polar coordinates in Exercise 11 of §3.1.

In order to do harmonic analysis on H in geodesic polar coordinates, one must have some knowledge of the associated Legendre functions $P^a_{-1/2+it}$. These are essentially the eigenfunctions f of Δ on H which transform under k_u defined in (3.19) according to the law

$$f(k_u(z)) = \exp(2iau) f(z), \quad z \in H, u \in \mathbb{R}. \quad (3.20)$$

Using the power function y^s to construct functions satisfying (3.20), as we did in formula (3.14), we define the *associated spherical function*

$$h(s|z, a) = (2\pi)^{-1} \int_0^{2\pi} \operatorname{Im}(k_{-t}(z))^s \exp(2iat) \, dt. \quad (3.21)$$

Exercise 9 (Associated Legendre Functions). The associated Legendre function $P^a_s(z)$ can be defined for $\operatorname{Re} z > 0$ (see Lebedev [1, p. 199]) by

$$P^a_s(z) = \frac{\Gamma(s + a + 1)}{2\pi \Gamma(s + 1)} \int_0^{2\pi} [z + \sqrt{z^2 - 1} \cos u]^s \exp(iau) \, du.$$

Show that

$$h(s|z, a) = \exp(2iau) \frac{\Gamma(1 - s)}{\Gamma(1 - s + 2a)} P^{2a}_{-s}(\cosh r), \quad \text{if } z = k_u e^{-r} i.$$

Hint. Note that $\operatorname{Im}(k_u e^{-r} i) = [\cosh r + \cos(2u) \sinh r]^{-1}$.

Exercise 10 (Asymptotics and Functional Equation). Define $P_s = P^0_s$.

(a) *Functional Equation.*
Show that if $\operatorname{Re} z > 0$, then $P_{-s}(z) = P_{s-1}(z)$.

(b) *Asymptotics.*

Show that if $\operatorname{Re} z > 0$, $\operatorname{Re} s > -\frac{1}{2}$, then
$$P_s(z) \sim \pi^{-1/2}\Gamma(s + \tfrac{1}{2})\Gamma(s+1)^{-1}(2z)^s, \qquad \text{as } z \to \infty.$$

Hint. Use the integral formula in Exercise 9 and make the change of variables $x = \tan(u/2)$. That leads to the formula
$$P_s(z) = \pi^{-1} \int_{-\infty}^{\infty} \{z + \sqrt{z^2 - 1}(2/(x^2+1) - 1)\}^s (x^2+1)^{-1}\, dx.$$

This integral approaches
$$\pi^{-1}(2z)^s \int_{x \in \mathbb{R}} (x^2+1)^{-s-1}\, dx = \pi^{-1}(2z)^s B(\tfrac{1}{2}, s + \tfrac{1}{2}), \qquad \text{as } z \to \infty,$$

where $B(a,b) = $ the beta function $= \Gamma(a)\Gamma(b)/\Gamma(a+b)$ (see Lebedev [1, Ch. 1]).

Exercise 11 (Separation of Variables in $\Delta f = \lambda f$ in Geodesic Polar Coordinates). Write $f(k_u e^{-r} i) = v(r)w(u)$ with $(u,r) = $ geodesic polar coordinates in H and solve $\Delta f = \lambda f$. You will obtain
$$\sinh r (v(r))^{-1}[\sinh r v'(r)]' - \lambda (\sinh r)^2 = k = -w''(u)/w(u),$$

where k is the separation constant. Set $w(u) = \exp(iau)$, $a \in \mathbb{Z}$. Then $k = a^2$ and v must be a solution of
$$\sinh r (\sinh r v'(r))' - (\lambda \sinh^2 r + a^2) v(r) = 0.$$

Set $x = \cosh r$, $V(x) = v(r)$. Show that V must satisfy the ODE
$$(1 - x^2) V'' - 2x V' + (\lambda + a^2(x^2 - 1)^{-1}) V = 0.$$

Set $\lambda = s(s-1)$ and obtain a solution $V(x) = P_{-s}^a(x)$ (see Lebedev [1, p. 214]).

The functions $P_{-1/2 + it}^a(x)$ were first considered by Mehler [1]. They came to be called conical functions because they arise in physical problems involving cones. We call them spherical functions because they are a special case of the (associated) spherical functions attached to any symmetric space, examples of which will be considered in later chapters. The Legendre functions can also be viewed as Gauss hypergeometric functions $_2F_1$ (see Lebedev [1, p. 165]). This gives the power series in the next exercise by definition.

Exercise 12 (A Few More Properties of Associated Legendre Functions).

(a) Show that if $|z - 1| < 2$ we have
$$P_s^0(z) = P_s(z) = \sum_{k=0}^{\infty} \frac{(-s)_k (s+1)_k}{(k!)^2} \left(\frac{1-z}{2}\right)^2 = {}_2F_1\left(-s, s+1; 1; \frac{1-z}{2}\right),$$

where

$$(s)_k = \begin{cases} 1 & \text{if } k = 0 \\ s(s+1)\cdots(s+k-1) & \text{if } k = 1, 2, \ldots. \end{cases}$$

and $_2F_1 = $ the Gauss hypergeometric function (see Lebedev [1]).

(b) Show that if $a = 0, 1, 2, \ldots$, then $P_s^a(z) = (z^2-1)^{a/2} D^a P_s(z)$, if $D = d/dz$.
(c) Show that $P_s^a(z) = P_s^{-a}(z)\Gamma(s+a+1)/\Gamma(s-a+1)$ for $a = 0, 1, 2, \ldots$.
(d) Show that if

$$I = \int_1^z$$

then

$$P_s^{-a}(a) = (z^2-1)^{-a/2} I^a P_s(z).$$

Note that $P_v^a(x)$ give polynomials when both a and v are nonnegative integers; in fact, they are the functions which arose in the solution of the analogue of Exercise 11 for the sphere in Chapter 2. We will not have anything to say about Legendre functions of the second kind until §3.7 (but see R. Hermann [2, Ch. 3]). Another reference for conical and spherical functions is Robin [1].

Exercise 13 (Laplace Transforms Relating K and P). Prove that

(a) $$\int_0^\infty \exp(-pt) t^{u-1/2} K_{v+1/2}(t)\, dt$$
$$= (\pi/2)^{1/2}\Gamma(u-v)\Gamma(u+v+1)(p^2-1)^{-u/2} P_v^{-u}(p),$$

if $\operatorname{Re}(u+v) > -1$, $\operatorname{Re}(u-v) > 0$, $\operatorname{Re} p > -1$.

(b) $$\int_1^\infty P_{v-1/2}(u)\exp(-ut)\, du = \sqrt{2/(\pi t)}\, K_v(t) \quad \text{if } \operatorname{Re} t > 0.$$

Theorem 2 (Harmonic Analysis on H in Geodesic Polar Coordinates). *Let (r, u) denote geodesic polar coordinates as in formula (3.19). And define*

$$\varepsilon_{a,t}(k_u e^{-r} i) = \exp(iau) P_{-1/2+it}^a(\cosh r).$$

Then, if $f \in C_c^\infty(H) = $ the infinitely differentiable functions with compact support,

$$f(z) = (2\pi)^{-1} \sum_{a \in \mathbb{Z}} (-1)^a \int_{t \in \mathbb{R}^+} \tilde{f}(a,t)\varepsilon_{-a,t}(z) t \tanh \pi t\, dt.$$

where

$$\tilde{f}(a,t) = \int_H f(z)\varepsilon_{a,t}(z)\, d\mu.$$

PROOF. We show in Exercise 14 that it suffices to prove the *inversion formula for the Mehler-Fock transform* (see Mehler [1] and Fock [1]):

$$g(v) = \int_0^\infty t \tanh \pi t P_{-1/2+it}(v) \int_1^\infty g(w) P_{-1/2+it}(w) \, dw \, dt,$$

$$f(w) = w \tanh \pi w \int_1^\infty P_{-1/2+iw}(u) \int_0^\infty P_{-1/2+it}(w) f(t) \, dt \, du. \tag{3.22}$$

There are discussions of this result in Exercises 15 and 17 below. We can also obtain (3.22) from the asymptotics/functional equations principle which we used to discuss Theorem 1 and the Kontorovich-Lebedev inversion formula. For by Exercise 10, if $|\arg x| < \pi$, then

$$P_{-1/2+it}(x) \sim \frac{\Gamma(it)}{\sqrt{\pi} \Gamma(\tfrac{1}{2}+it)} (2x)^{-1/2+it} + \frac{\Gamma(-it)}{\sqrt{\pi} \Gamma(\tfrac{1}{2}-it)} (2x)^{-1/2-it} \qquad \text{as } x \to \infty$$

for fixed real t. This can also be proved by noting that the Legendre function is a Gauss hypergeometric function (see Lebedev [1]). Helgason [5, pp. 62–82] gives a discussion of the complete asymptotic expansion of the spherical function and its application to Fourier inversion on the symmetric space via Paley-Wiener theory. Our goal is to show that the kernel

$$V_R(x, y) = \int_0^R t \tanh \pi t P_{-1/2+it}(x) P_{-1/2+it}(y) \, dt$$

approaches $\delta(x - y)$, as $R \to \infty$. As for the kernel (3.18), one can argue that the $SL(2, \mathbb{R})$-invariance of the problem means that it suffices to consider $x, y \sim \infty$. Now by a standard property of the gamma function (see Lebedev [1, p. 3]) we have

$$\Gamma(it)\Gamma(-it)\pi^{-1}\Gamma(\tfrac{1}{2}+it)^{-1}\Gamma(\tfrac{1}{2}-it)^{-1} = (\pi t \tanh \pi t)^{-1}. \tag{3.23}$$

It follows that

$$V_R(x, y) \sim \pi^{-1} \int_0^R x^{-1/2+it} y^{-1/2-it} \, dt \qquad \text{for } x, y \sim \infty.$$

The right-hand side of the asymptotic relation is a Dirac delta family by Mellin inversion (see Exercise 1 of §1.4). Thus the spectral measure was chosen so as to cancel out the gamma factors in the asymptotic formula for the Legendre function, as was the case for the Kontorovich-Lebedev transform.

The rest of the discussion of Theorem 2 is outlined in the next exercise. □

Exercise 14.

(a) Use Exercise 12 and integration by parts to show that if $f(k_u e^{-r} i) = g(x) h(u)$, with $x = \cosh r$, then

$$\tilde{f}(a, t) = \int_{w=1}^\infty \int_{v=0}^{2\pi} h(v) \exp(iav) g(w) (w^2 - 1)^{a/2} D^a P_{-1/2+it}(w) \, dv \, dw$$

$$= (-1)^a \int_{w=1}^\infty \int_{v=0}^{2\pi} h(v) \exp(iav) P_{-1/2+it}(w) D^a[g(w)(w^2 - 1)^{a/2}] \, dv \, dw.$$

(b) Use Exercise 12 to show that

3.2. Harmonic Analysis on H

$$(2\pi)^{-1} \sum_{a \in \mathbb{Z}} (-1)^a \int_{t \in \mathbb{R}^+} \tilde{f}(a,t) \varepsilon_{-a,t}(z) t \tanh \pi t \, dt = \frac{1}{2\pi} \sum_{a \in \mathbb{Z}} (-1)^a e^{-iau}$$

$$\times \int_{t \in \mathbb{R}^+} (-1)^a \int_{w=1}^{\infty} \int_{v=0}^{2\pi} h(v) e^{iav} D^a [g(w)(w^2-1)^{a/2}]$$

$$\times P_{-1/2+it}(w) \, dw (x^2-1)^{-a/2} I^a P_{-1/2+it}(x) t \tanh \pi t \, dt.$$

(c) Finish the proof of Theorem 2 by using the Mehler-Fock inversion formula and the fact that $h(u)$ is represented by its Fourier series.

(d) Show that $\overline{\varepsilon_{a,t}(z)} = \varepsilon_{-a,-t}(z) = \varepsilon_{-a,t}(z)$.

Exercise 15. Can you find a proof of the Mehler-Fock inversion formula which is analogous to the proof of the Kontorovich-Lebedev inversion formula in Exercise 7 of §2.2?

Exercise 16 (Another Derivation of the Kontorovich-Lebedev Inversion Formula).

(a) Use the integral formula in Exercise 1 to see that $K_{it}(y)$ is the Fourier transform on $\mathbb{R}$ of $h_y(x) = \frac{1}{2}\exp(-y\cosh x)$; i.e., $K_{it}(y) = \hat{h}_y(t/2\pi)$.

(b) Use the multiplication formula for the Fourier transform on $\mathbb{R}$ (see Theorem 1 of §1.2) to show that if $f(x)$ is extended to $\mathbb{R}$ by setting $f(x) = f(-x)$, then

$$2\pi^{-2} x \sinh \pi x \int_{y=0}^{\infty} K_{ix}(y) y^{-1} \int_{u=0}^{\infty} f(u) K_{iu}(y) \, du \, dy$$

$$= 2^{-1} \pi^{-2} x \sinh \pi x \int_{y=0}^{\infty} K_{ix}(y) y^{-1} \int_{u=-\infty}^{+\infty} \hat{f}(u/2\pi) \exp(-y \cosh u) \, du \, dy.$$

(c) Show that the Kontorovich-Lebedev inversion formula (3.15) follows from Fourier inversion on $\mathbb{R}$ (§1.2) and

$$\int_{y=0}^{\infty} \exp(-y \cosh u) K_{ix}(y) y^{-1} \, dy = \pi \cos(xu)/(x \sinh \pi x).$$

This last integral formula follows from part (a) of Exercise 13, since

$$P_v^{1/2}(z) = (2\pi)^{1/2}(z^2-1)^{-1/4}((z+\sqrt{z^2-1})^{v+1/2} + (z+\sqrt{z^2-1})^{-v-1/2})$$

(see Erdélyi et al. [1, Vol. I, p. 150]). You also need formula (3.17).

Exercise 17 (Another Derivation of the Mehler-Fock Inversion Formula). Use Exercise 13(a) to see that

$$\int_{u \geq 1} \int_{t \geq 0} P_{-1/2+iw}(u) P_{-1/2+it}(u) g(t) \, dt \, du$$

$$= 2^{1/2} \pi^{-3/2} \int_{v \geq 0} v^{-1/2} \int_{u \geq 1} e^{-uv} P_{-1/2+iw}(u) \int_{t \geq 0} K_{it}(v) g(t) \cosh \pi t \, dt \, du \, dv.$$

Then use Exercise 13(b) and the Kontorovich-Lebedev inversion formula to prove the Mehler-Fock inversion formula.

Exercise 18 (Some Mehler-Fock Transforms).

(a) Show that if $\operatorname{Re} a > \frac{1}{2}$, then

$$\int_1^\infty w^{-a} P_{-1/2+it}(w)\, dw = 2^{a-2}\pi^{-1/2}\Gamma\left(\frac{a+it-\frac{1}{2}}{2}\right)\Gamma\left(\frac{a-it-\frac{1}{2}}{2}\right)\bigg/\Gamma(a).$$

(b) Prove that if $\operatorname{Re} a > 0$, $\operatorname{Re} b > 0$, then

$$\int_1^\infty (a^2+b^2+2abw)^{-1/2}\exp[-(a^2+b^2+2abw)^{1/2}]P_{-1/2+it}(w)\,dw$$

$$= 2\pi^{-1}(ab)^{-1/2}K_{it}(a)K_{it}(b).$$

There are many other examples of Mehler-Fock transforms in Erdélyi et al. [2, Vol. II, pp. 320–326].

Next we give our third and last version of non-Euclidean harmonic analysis on H, that of Helgason [1, 5]. Suppose that $f \in C_c^\infty(H)$. Define the *Helgason transform* of f for $s \in \mathbb{C}$, $k \in SO(2)$, by

$$\mathcal{H}f(s,k) = \int_H f(z)\overline{\operatorname{Im}(k(z))^s}\, y^{-2}\, dx\, dy. \tag{3.24}$$

Set $B = K/M$, where $M = \{-I, I\}$. Then B is called the *boundary* of H. This is reasonable since B can be identified with the circle, which is just the one point compactification of the real line. Note that $\mathcal{H}f(s,k)$ depends only on the coset kM of k in B. We will see in Lemma 3 of §3.7 that the transform (3.24) is the same transform which number theorists call the Selberg transform, when f is K-invariant (see Kubota [1, p. 56]).

The boundary of H has the non-Euclidean analogues of properties of the boundary of the unit disc. For example, it is possible to generalize classical potential theory in the framework of symmetric spaces (see Helgason [1], Koranyi [1], and Chapter 5). And eigenfunctions of invariant differential operators on H can be shown to be given by a Poisson integral over the boundary (even in the higher-rank case) as was shown by Kashiwara et al. [1]. The precise statement of this result requires Sato's theory of analytic functionals or hyperfunctions. These are elements of the dual space $\mathcal{A}'(B)$ to the space $\mathcal{A}(B)$ of analytic functions on the boundary. The space $\mathcal{A}(B)$ has a natural topology which is described for B = the circle as follows. Let U be an open annulus containing B and $\mathcal{A}(U)$ = the holomorphic functions on U topologized by uniform convergence on compact subsets. Identify $\mathcal{A}(B)$ with the union of all the $\mathcal{A}(U)$ (with the inductive limit topology). The eigenfunctions are then just these analytic functionals acting on the Poisson kernel, which is just the power function.

3.2. Harmonic Analysis on H

Theorem 3 (Helgason's Version of Harmonic Analysis on H).

(1) Fourier Inversion on H.
   If $f \in C_c^\infty(H)$, then
   $$f(z) = \frac{1}{4\pi} \int_{t \in \mathbb{R}} \frac{1}{2\pi} \int_{u=0}^{2\pi} \mathscr{H}f(\tfrac{1}{2} + it, k_u) \operatorname{Im}(k_u(z))^{1/2+it} t \tanh \pi t \, du \, dt,$$
   where $k_u \in SO(2)$ is defined in formula. (3.19).

(2) The Paley-Wiener Theorem.
   The map $f \mapsto \mathscr{H}f$ takes $C_c^\infty(H)$ one-to-one, onto the space of C^∞ functions $G(s, k)$ on $\mathbb{C} \times SO(2)$ which are holomorphic in s and have properties a, b below:

   (a) G is of uniform exponential type R:
   $$\sup_{s \in \mathbb{C}, k \in SO(2)} (\exp(-R|\operatorname{Re} s|)(1 + |s|)^N |G(s, k)|) < \infty \qquad \text{for each } N \in \mathbb{Z}^+.$$

   (b) G satisfies the functional equation
   $$\int_{u=0}^{2\pi} \operatorname{Im}(k_u(z))^{1/2+it} G(\tfrac{1}{2} + it, k_u) \, du$$
   $$= \int_{u=0}^{2\pi} \operatorname{Im}(k_u(z))^{1/2-it} G(\tfrac{1}{2} - it, k_u) \, du.$$

(3) The Plancherel Formula.
   The map $f \mapsto \mathscr{H}f$ extends to an isometry of $L^2(H, d\mu)$ onto $L^2(\mathbb{R} \times K, \tfrac{1}{2}(2\pi)^{-2} t \tanh \pi t \, dt \, du)$, where $K = SO(2)$ is identified with the interval $(0, 2\pi]$, using the map sending k_u to u.

PROOF. We shall only discuss part (1). See Helgason [1, 5] for an alternate derivation of (1), as well as proofs of (2) and (3). We will follow Helgason's method for $SL(n, \mathbb{R})/SO(n)$ in Chapter 4.

First let us relate the transform $\tilde{f}(a, t)$ of Theorem 2 to $\mathscr{H}f(\tfrac{1}{2} + it, k)$. Set $c(a, t) = \Gamma(\tfrac{1}{2} + it + a)[2\pi \Gamma(\tfrac{1}{2} + it)]^{-1}$. Then, by the integral formula in Exercise 9, we have

$$\tilde{f}(a, t) = \int_{u=0}^{2\pi} \int_{r=0}^{\infty} f(k_{-u} e^{-r} i) \exp(iau) P_{-1/2+it}^a(\cosh r) \sinh r \, dr \, du$$

$$= c(a, t) \int_{u=0}^{2\pi} \int_{r=0}^{\infty} f(k_{-u} e^{-r} i) e^{iau} \int_{\theta=0}^{2\pi} e^{ia\theta} [\operatorname{Im} k_\theta(e^{-r} i)]^{1/2-it} \sinh r \, d\theta \, dr \, du$$

$$= c(a, t) \int_{\varphi=0}^{2\pi} e^{ia\varphi} \int_{u=0}^{2\pi} \int_{r=0}^{\infty} f(k_{-u} e^{-r} i) \operatorname{Im}(k_\varphi k_{-u} e^{-r} i)^{1/2-it} \sinh r \, dr \, du \, d\varphi.$$

To obtain the last equality, we substituted $\varphi = u + \theta$. Thus

$$\tilde{f}(a, t) = c(a, t) \int_{\varphi=0}^{2\pi} \mathscr{H}f(\tfrac{1}{2} + it, k_\varphi) e^{ia\varphi} \, d\varphi.$$

By Theorem 2, Exercises 9 and 10, we have

$$f(k_{-u}e^{-r}i)$$

$$= \frac{1}{4\pi} \sum_{a \in \mathbb{Z}} \int_{t=-\infty}^{+\infty} (-1)^a c(a,t) \int_{\varphi=0}^{2\pi} e^{ia\varphi} \mathcal{H}f(\tfrac{1}{2} + it, k_\varphi)$$

$$\times e^{-iau} P_{-1/2+it}^{-a}(\cosh r) t \tanh \pi t \, d\varphi \, dt$$

$$= \frac{1}{4\pi} \sum_{a \in \mathbb{Z}} (-1)^a \int_{t=-\infty}^{+\infty} c(a,t) c(-a,-t) \int_{\varphi=0}^{2\pi} e^{ia\varphi} \mathcal{H}f(\tfrac{1}{2} + it, k_\varphi) e^{-iau} \int_{\theta=0}^{2\pi} e^{-ia\theta}$$

$$\times \operatorname{Im}(k_\theta e^{-r}i)^{1/2+it} t \tanh \pi t \, d\theta \, d\varphi \, dt.$$

It follows from $\Gamma(z)\Gamma(1-z) = \pi \csc \pi z$ that $c(a,t)c(-a,-t) = (-1)^a(2\pi)^{-2}$ (see Lebedev [1, p. 3]). Therefore

$$f(k_{-u}e^{-r}i) = (2\pi)^{-3} \frac{1}{2} \sum_{a \in \mathbb{Z}} \int_{t=-\infty}^{+\infty} t \tanh \pi t \int_{\varphi=0}^{2\pi} \int_{\theta=0}^{2\pi} \exp[ia(\varphi - u - \theta)]$$

$$\times \operatorname{Im}(k_\theta e^{-r}i)^{1/2+it} \mathcal{H}f(\tfrac{1}{2} + it, k_\varphi) \, d\theta \, d\varphi \, dt.$$

Now $\operatorname{Im}(k_\theta e^{-r}i)^{1/2+it}$, as a function of θ, is represented by its Fourier series, which implies

$$f(k_{-u}e^{-r}i)$$

$$= (2\pi)^{-2} \frac{1}{2} \int_{\varphi=0}^{2\pi} \int_{t=-\infty}^{+\infty} t \tanh \pi t \operatorname{Im}(k_\varphi k_{-u}e^{-r}i)^{1/2+it} \mathcal{H}f(\tfrac{1}{2} + it, k_\varphi) \, dt \, d\varphi.$$

This completes the proof of part (1) of Theorem 3. $\square$

Helgason [1, pp. 9–10] gives a much more elegant reduction of the proof of Theorem 3, part (1), to the case of $K = SO(2)$-invariant functions $f \in C_c^\infty(H)$. The idea is to imitate the discussion of Fourier inversion on $\mathbb{R}^m$ which was given in §1.2. Define the *inverse transform* $\mathscr{S}$ for functions $h : \mathbb{R} \times B \to \mathbb{C}$ by

$$\mathscr{S}h(z) = (8\pi^2)^{-1} \int_{t \in \mathbb{R}} \int_{u=0}^{2\pi} h(t, k_u) \operatorname{Im}(k_u z)^{1/2+it} t \tanh \pi t \, dt \, du. \quad (3.25)$$

Then Helgason proves that

$$\int_H f \overline{\mathscr{S}\mathcal{H}g} \, d\mu = \int_H (\mathscr{S}\mathcal{H}f) \bar{g} \, d\mu, \quad (3.26)$$

and that $\mathscr{S}\mathcal{H}$ commutes with the action of $G = SL(2, \mathbb{R})$ on H. We shall discuss this proof in detail for $G = SL(n, \mathbb{R})$ in Chapter 4.

Next we want to consider the *Helgason transform of $K = SO(2)$-invariant functions* $f(z) = f(k(z))$, for all k in $SO(2)$ and z in H:

$$\begin{aligned}\hat{f}(s) = \mathscr{H}f(s,k) = \mathscr{H}f(s,I) &= \int_H f(z)y^{\bar{s}-2}\,dx\,dy \\ &= 2\pi\int_{r=0}^{\infty} f(e^{-r}i)P_{-1/2+it}(\cosh r)\sinh r\,dr,\end{aligned} \tag{3.27}$$

if $s = \frac{1}{2} + it$, and $I =$ the 2×2 identity matrix. The *inverse transform* $\mathscr{S}$ for $\hat{f}(s)$ is a transform which takes functions $F: \mathbb{R} \to \mathbb{C}$ onto functions $\mathscr{S}F$ with domain H, defined by

$$\mathscr{S}F(ke^{-r}i) = (4\pi)^{-1}\int_{t\in\mathbb{R}} F(t)P_{-1/2+it}(\cosh r)t\tanh \pi t\,dt \tag{3.28}$$

for k in $SO(2)$ and $r > 0$. Then if $f \in C_c^{\infty}(H)$ and $f(k(z)) = f(z)$ for all $k \in SO(2)$ and $z \in H$, the *Fourier inversion formula* of Theorem 3 is

$$f = S(\hat{f}). \tag{3.29}$$

This is nothing else but the Mehler-Fock inversion formula (3.22).

Exercise 19 (Convolution).

(a) Let $f, g: SL(2,\mathbb{R}) = G \to \mathbb{C}$. Define the *convolution* $f * g$ by

$$(f * g)(a) = \int_G f(b)g(b^{-1}a)\,db,$$

where $da =$ a right- and left-invariant Haar measure on G. Suppose that either f or g is really a function on $K\backslash G/K$; i.e., it is a K-invariant function on H, where $K = SO(2)$, as usual. Show that $\mathscr{H}(f * g) = \mathscr{H}f \cdot \mathscr{H}g$.

(b) Show that you should not expect the convolution property to hold unless either f or g is K-invariant.

Hint. In Chapter 4, we generalize this exercise to $SL(n,\mathbb{R})$. The proof may be slightly clearer in the general context. It uses a property of the power function. See Exercise 2 of §4.3.

Exercise 20 (Writing the Geodesic Radial Coordinate r in Terms of the Rectangular Coordinates x, y). Let $x + iy = k_u e^{-r}i$ as in formula (3.19) above. Note that this means

$$M = \begin{pmatrix} 1 & x \\ 0 & 1 \end{pmatrix}\begin{pmatrix} \sqrt{y} & 0 \\ 0 & 1/\sqrt{y} \end{pmatrix} = k_u\begin{pmatrix} \exp(-r/2) & 0 \\ 0 & \exp(r/2) \end{pmatrix}k_v \qquad \text{for some } v.$$

Deduce that

$$M\,{}^tM = k_u\begin{pmatrix} e^{-r} & 0 \\ 0 & e^{r} \end{pmatrix}k_{-u}$$

and that

$$y + 1/y + x^2/y = 2\cosh r.$$

The following Exercise concerns a transform that will appear again as part of the Selberg trace formula in §3.7.

Exercise 21 (The Helgason Transform of K-Invariant Functions is a Composition of Harish and Mellin Transforms).

(a) Define the *Harish transform* Tf of $f : H \to \mathbb{C}$ by

$$Tf(y) = y^{-1/2} \int_{x \in \mathbb{R}} f(x + iy)\,dx \qquad \text{(see Lang [4, pp. 69ff])}.$$

Let Mg denote the Mellin transform of $g : \mathbb{R}^+ \to \mathbb{C}$ as in §1.4. Show that the Helgason transform of a K-invariant function f, $K = SO(2)$ (i.e., $f(kz) = f(z)$ for all $k \in K$ and $z \in H$) is given by

$$\hat{f}(s) = MTf(\bar{s} - \tfrac{1}{2}).$$

(b) Show that if $g : \mathbb{R}^+ \to \mathbb{C}$ and

$$G(v) = \int_{x \in \mathbb{R}} g(v + x^2/2)\,dx.$$

then

$$g(v) = -(2\pi)^{-1} \int_{w \in \mathbb{R}} G'(v + w^2/2)\,dw.$$

(c) Set $f(x + iy) = g(\cosh r)$, when $x + iy$ and r are related as in Exercise 20. Show that $Tf(y) = G((y + y^{-1})/2)$, where G is the transform of g defined in part (b).

(d) Use (a) and (b) to deduce the inversion formula for the Helgason transform of a K-invariant function at $z = i$ (see Helgason [5, pp. 79–82]).

Hint. Mellin inversion implies that

$$G\left(\frac{y + 1/y}{2}\right) = (2\pi)^{-1} \int_{t \in \mathbb{R}} \hat{f}(it)y^{it}\,dt.$$

Set $t = e^v$ and show that

$$G'(\cosh v) = -(2\pi)^{-1} \int_{t \in \mathbb{R}} \hat{f}(it)t \sin(tv)/\sinh(v)\,dt.$$

Then use part (b) to see that

$$f(i) = g(1) = -(2\pi)^{-1} \int_{w \in \mathbb{R}} G'(1 + w^2/2)\,dw$$

$$= (2\pi)^{-2} \int_{v \in \mathbb{R}} \int_{t \in \mathbb{R}} \hat{f}(it)t \sin(tv)[\cosh(v/2)/\sinh v]\,dt\,dv$$

Here we changed variables according to $\cosh v = 1 + w^2/2$, $dw = \cosh(v/2)\,dv$.

Table 3.1. Short Table of Helgason Transforms

$f(z) = f(k(z)) = f(e^{-r}i)$	$\hat{f}(s), \quad s = \frac{1}{2} + it$
$e^{-T/4}\sqrt{2}(4\pi T)^{-3/2}\displaystyle\int_r^\infty \frac{be^{-b^2/4T}\,db}{\sqrt{\cosh b - \cosh r}}$	$e^{s(s-1)T}$
$(\cosh r)^{-a}, \quad a > 0$	$\sqrt{\pi}\,2^{a-1}\left\lvert\Gamma\left(\frac{a-\frac{1}{2}+it}{2}\right)\right\rvert^2 \Big/ \Gamma(a)$
$\exp(-a\cosh r), \quad a > 0$	$2(2\pi/a)^{1/2}K_{s-1/2}(a)$

Exercise 22. Check Table 3.1 which lists three Helgason transforms.

Hint. For the first line of Table 3.1 use Exercise 21 to see that if $f(s) = e^{s(s-1)T}$, $s = \frac{1}{2} + it$, then

$$F(\cosh r) = (2\pi)^{-1}\int_{t\in\mathbb{R}} \exp[-(\tfrac{1}{4} + t^2)T]\exp(irt)\,dt = \frac{1}{2\sqrt{\pi T}}e^{-T/4-r^2/4T}.$$

Also

$$\begin{aligned} f(e^{-r}i) &= -(2\pi)^{-1}\int_{w\in\mathbb{R}} F'(\cosh r + w^2/2)\,dw \\ &= -(\pi)^{-1}\int_r^\infty \frac{F'(\cosh v)\sinh v\,dv}{\sqrt{2(\cosh v - \cosh r)}}. \end{aligned}$$

No one seems to have evaluated this integral beyond what is in Table 3.1 (a result that would be of interest for work on the central limit theorem on H and it applications).

For the second line, see Exercise 18.

For the third line, see Exercise 13.

Exercise 23 (A Conductor Which is the Surface of Two Intersecting Spheres).

(a) Toroidal coordinates (a, b, ϕ) are defined as follows in terms of cylindrical coordinates (r, ϕ, z) (see Lebedev [1, p. 222]):

$$r = c\frac{\sinh a}{\cosh a - \cos b}, \qquad z = c\frac{\sin b}{\cosh a - \cos b},$$

where c is a constant. Show that the surface $b =$ constant is a sphere

$$(z - c\cot b)^2 + r^2 = \left(\frac{c}{\sin b}\right)^2.$$

(b) Show that

$$\begin{aligned} \Delta f &= f_{xx} + f_{yy} + f_{zz} \\ &= \frac{a}{\partial a}\left(\frac{\sinh a}{\cosh a - \cos b}\frac{\partial f}{\partial a}\right) + \frac{\partial}{\partial b}\left(\frac{\sinh a}{\cosh a - \cos b}\frac{\partial f}{\partial b}\right) \\ &\quad + \frac{1}{(\cosh a - \cos b)\sinh a}\frac{\partial^2 f}{\partial \phi^2}. \end{aligned}$$

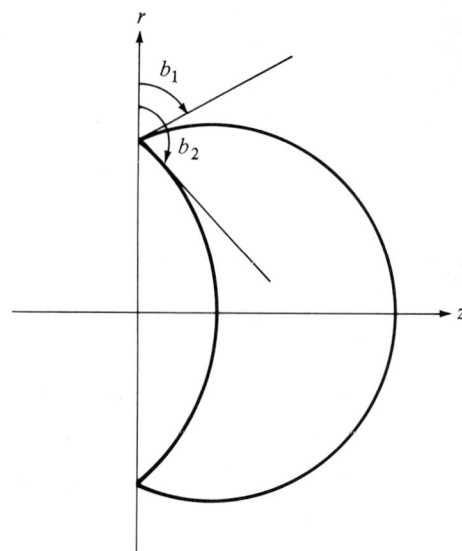

Figure 3.10. A slice of two intersecting spheres.

(c) Consider the Dirichlet problem on the domain in Fig. 3.10. Here the z-axis passes through the center of the two spheres. Choose c equal to the radius of the circle of intersection. The two spheres are given by $b = b_1$ and $b = b_2$ in toroidal coordinates. Then the Dirichlet problem is to find u with

$$\begin{cases} \Delta u = 0, & b_1 < b < b_2, \\ u|_{b=b_j} = f_j, & j = 1, 2. \end{cases}$$

Show that separation of variables leads to a solution which has the form

$$u = \sqrt{2(\cosh a - \cos b)}$$

$$\times \int_0^\infty \frac{F_2 \sinh(b - b_1)t + F_1 \sinh(b_2 - b)t}{\sinh(b_2 - b_1)t} P_{-1/2+it}(\cosh a)\, dt,$$

where

$$F_j(t) = t \tanh \pi t \int_0^\infty \frac{f_j(a)}{\sqrt{2(\cosh a - \cos b_j)}} P_{-1/2+it}(\cosh a) \sinh a\, da,$$

$$j = 1, 2.$$

Hint. See Lebedev [1, pp. 227–230].

Application. The Heat Equation on H in Rectangular Coordinates

The problem is to find $u = u(z, t) =$ the temperature at z in H and time t, if $u(z, 0) = f(z)$ is the initial heat distribution. That is, we want to solve

3.2. Harmonic Analysis on H

$$u_t = \Delta_z u = y^2\left(\frac{\partial^2}{\partial x^2} + \frac{\partial^2}{\partial y^2}\right)u(x+iy, t)$$

$$u(z, 0) = f(z) \tag{3.30}$$

Separation of variables leads us to consider elementary solutions of the form $u(z, t) = Z(z)T(t)$, where

$$\frac{\Delta Z}{Z} = k = \frac{T'}{T}(t).$$

Set $k = -(v^2 + \frac{1}{4}) = s(s-1)$, if $s = \frac{1}{2} + iv$. Then $T(t) = \exp[-(v^2 + \frac{1}{4})t]$ and $Z(z) = e_{a,s}(z)$, as in Theorem 1. Then we can use the spectral decomposition of Δ in Theorem 1 to obtain

$$u(z, t) = \pi^{-2} \int_{b\in\mathbb{R}} \int_{v\in\mathbb{R}} v \sinh \pi v A(b, v) e_{b, 1/2+iv}(z) e^{-(v^2+1/4)t}\, dv\, db,$$

where $\tag{3.31}$

$$A(b, v) = \int_H f(z)\overline{e_{b, 1/2+iv}(z)}\, y^{-2}\, dx\, dy.$$

Then Exercises 24 and 25 below show that this solution can be rewritten in the form

$$u(z, t) = f * G_t,$$

$$G_t(ke^{-r}i) = (4\pi)^{-1}\int_{b\in\mathbb{R}} b \tanh(\pi b) P_{-1/2+ib}(\cosh r) e^{-(b^2+1/4)t}\, db \tag{3.32}$$

where convolution of functions on H is induced from convolution of functions on $SL(2, \mathbb{R})$ as in Exercise 19. A function $f: H \to \mathbb{C}$ induces a function $f: SL(2, \mathbb{R}) \to \mathbb{C}$ by writing $f(g) = f(g(i))$ for g in $SL(2, \mathbb{R})$, $i = \sqrt{-1}$. The function G_t in (3.32) is the fundamental solution of the heat equation and thus gives rise to the analogue of the normal distribution discussed in Example 3 of §1.1 and Example 1 of §1.2, as well as in the section on the central limit theorem at the end of §1.2. We will obtain a more direct treatment of formula (3.32) using Helgason's transform, after the next two exercises. Then we will develop the non-Euclidean central limit theorem for $SO(2)$-invariant random variables.

Exercise 24.

(a) Prove that if $z = x + iy$ and $w = \mu + iv$ in H, then we can write (3.31) as

$$u(z, t) = \int_{w\in H} f(w) G_t(z, w) v^{-2}\, d\mu\, dv,$$

where

$$G_t(z, w) = (vy)^{1/2} \pi^{-2} \int_{r \in \mathbb{R}} \exp[-(r^2 + \tfrac{1}{4})t] r \sinh \pi r$$

$$\times \int_{b \in \mathbb{R}} \exp[2\pi i b(x - \mu)] K_{ir}(2\pi |b| v) K_{ir}(2\pi |b| y) \, db \, dr.$$

(b) Show that

$$\int_{b=0}^{\infty} \cos(xc) K_s(ax) K_s(bx) \, dx$$

$$= \frac{\pi^2}{4} (ab)^{-1/2} \sec(\pi s) P_{s-1/2}[(a^2 + b^2 + c^2)(2ab)^{-1}].$$

(c) Use (a) and (b) to prove

$$G_t(z, w) = (4\pi)^{-1} \int_{b \in \mathbb{R}} e^{-(b^2 + 1/4)t} b \tanh \pi b \, P_{ib-1/2}(\cosh r(z, w)) \, db,$$

where $r(z, w)$ is the geodesic radial coordinate (see formula (3.19)) of the point $M_w^{-1} M_z i$, if $M_z i = z$ and $M_w i = w$ for M_z, M_w in $SL(2, \mathbb{R})$. In particular, we have

$$M_z = \begin{pmatrix} 1 & x \\ 0 & 1 \end{pmatrix} \begin{pmatrix} \sqrt{y} & 0 \\ 0 & 1/\sqrt{y} \end{pmatrix} = k \begin{pmatrix} e^{-r/2} & 0 \\ 0 & e^{r/2} \end{pmatrix} k', \qquad k, k' \in K,$$

$y + y^{-1} + y^{-1} x^2 = 2 \cosh r$, as in Exercise 20. Then use Exercise 25 to see that $2 \cosh r(z, w) = (y^2 + v^2 + (x - \mu)^2)/(vy)$, if z and w are as in part (a).

(d) Prove formula (3.32).

Exercise 25.

(a) Suppose that $z = x + iy$ and use the same matrix M_z as in part (c) of Exercise 24. Similarly, define M_w for $w = \mu + iv$. Show that

$$M_z M_w^{-1} i = \left(\frac{xv - \mu y}{v} \right) + \frac{y}{v} i \quad \text{and} \quad M_w^{-1} M_z i = \frac{x - \mu}{v} + \frac{y}{v} i.$$

(b) Then write $M_w^{-1} M_z i = k_u e^{-r} i$, where $r = r(z, w)$, $k \in K = SO(2)$ are the geodesic polar coordinates for $M_w^{-1} M_z i$. And show that

$$2 \cosh r(z, w) = [y^2 + v^2 + (x - \mu)^2]/(vy).$$

The Heat Equation on H Using Helgason's Transform (from Gangolli [1, 2])

We want to consider the non-Euclidean analogue of the method that was used in Example 1 of §1.2 to find the fundamental solution of the Euclidean heat equation. We shall assume that $f(kz) = f(z)$ for all $k \in SO(2)$ and z in H. Then

take the Helgason transform with respect to z of the PDE in (3.30). This gives

$$\mathscr{H}u_t = \mathscr{H}(\Delta u).$$

Integration by parts or Green's theorem says that

$$\frac{\partial}{\partial t}\mathscr{H}u = s(s-1)\mathscr{H}u.$$

Thus $\mathscr{H}u = \mathscr{H}f \cdot e^{s(s-1)t}$. Using Exercise 19, it follows that

$$u(z,t) = f * \mathscr{S}(e^{s(s-1)t})(z),$$

where $\mathscr{S}$ denotes the inverse transform (3.28):

$$\begin{aligned} G_t(e^{-r}i) &= [\mathscr{S}(e^{s(s-1)t}](e^{-r}i) \\ &= (4\pi)^{-1}\int_{v\in\mathbb{R}} e^{-(v^2+1/4)t}P_{-1/2+iv}(\cosh r)v\tanh \pi v\, dv. \end{aligned}$$

This is formula (3.32). Moreover, Exercise 22 shows that

$$G_t(e^{-r}i) = (4\pi t)^{-3/2}\sqrt{2}e^{-t/4}\int_r^\infty \frac{be^{-b^2/4t}\,db}{\sqrt{\cosh b - \cosh r}}. \tag{3.33}$$

Therefore

$$G_t(z) > 0 \qquad \text{for all } t > 0 \text{ and } z \in H. \tag{3.34}$$

Moreover,

$$G_t(z) \to \delta \qquad \text{as } t \to 0+, \tag{3.35}$$

where $\delta =$ the Dirac delta distribution on H; i.e., $f * \delta =$ f. To see (3.35) note that by the Plancherel theorem for H we have

$$\|f * G_t - f\|_2^2 = (4\pi)^{-1}\int_{\substack{s=1/2+iv\\ v\in\mathbb{R}}} |(\widehat{f * G_t} - \hat{f})(s)|^2 v\tanh \pi v\, dv.$$

Now

$$\widehat{f * G_t} = \hat{f}\cdot\hat{G}_t = \hat{f}\cdot e^{s(s-1)t} \to \hat{f} \qquad \text{as } t \to 0+.$$

Thus,

$$\|f * G_t - f\|_2^2 \to 0 \qquad \text{as } t \to 0+,$$

by the Lebesgue dominated convergence theorem.

In fact, G_t has the properties of the Gauss kernel, except that it does not appear to be possible to find a simpler formula for G_t than (3.32) or (3.33). Formula (3.33) was obtained by Karpelevich, Tutubalin, and Shur [1], as well as the corresponding result for the three-dimensional analogue of H to be discussed in Chapter 5 (the fundamental solution of the heat equation on hyperbolic 3-space).

Application. The Central Limit Theorem for K-Invariant Random Variables on H. Transmission Lines with Random Inhomogeneities

There are many versions of the central limit theorem on symmetric spaces such as the upper half-plane. The first papers on the subject were written by mathematicians in the U.S.S.R. during the 1950s and 1960s (see Karpelevich, Tutubalin, and Shur [1], Tutubalin [1], and Virtser [1]). Mathematicians in the U.S. discussed these matters using very general limit theorems on stochastic differential equations in the 1970s (see Burridge and Papanicolaou [1], Keller and Papanicolaou [1], and Papanicolaou [1]). Mathematicians in France have obtained central limit theorems for various sorts of Lie groups (see Bougerol [1], Clerc and Roynette [1], Faraut [1]). Limit theorems are also obtained by Heyer [1] and Trimèche [1]. A similar discussion of Euclidean rotation-invariant random variables appears in Kingman [1]. There are many potential applications for such limit theorems; e.g., in demography (see Cohen [1, especially p. 290]), learning theory (see LePage [1] and the references there), atomic physics (see Hantsch and von Waldenfels [1], as well as Hurt and Hermann [1]). Other related papers are Dudley [1, 2], Furstenberg [1], Furstenberg and Kesten [1], Gangolli [1], Getoor [1], and Letac [1]. The last reference gives analogues of classical probability problems for various symmetric spaces, including the p-adic symmetric spaces.

First, let us consider an engineering problem of Gertsenshtein and Vasil'ev [1]—a problem which requires the non-Euclidean central limit theorem for its solution. We wish to analyze a very long lossless transmission line with random inhomogeneities caused perhaps by tiny defects. Such inhomogeneities produce reflected waves, whose properties are described by the reflection coefficient which we can view as a random variable Z in H, as we saw at the end of §3.1. The *composite* of two random inhomogeneities with reflection coefficients Z_1 and Z_2 in H produces a reflection coefficient

$$Z_1 \circ Z_2 = M_{Z_1} M_{Z_2} i, \quad \text{where } Z_j = M_{Z_j} i \text{ for } M_{Z_j} \in SL(2, \mathbb{R}), j = 1, 2. \quad (3.36)$$

This composition is only well defined when the Z_j are $SO(2)$-invariant, which is the only case to be considered here. The question then arises as to the distribution of $Z_1 \circ \cdots \circ Z_n$, when it is correctly normalized, as $n \to \infty$. Having found the limit distribution, one should be able to compute the mean power output, for example.

To carry out this project, we need a central limit theorem for H. Luckily we already have a candidate density for the non-Euclidean normal distribution, namely the fundamental solution of the heat equation on H given by equations (3.32) and (3.33). We shall attempt to keep our discussion as close as possible to the discussion of the Euclidean central limit theorem (see Theorem 7 of §1.2).

First, we should set down the requisite definitions. A *random variable Z in H* has *distribution function*

$$P(Z \in A) = \int_A f_Z(z) y^{-2}\, dx\, dy = \int_A f_Z\, d\mu,$$

where f_Z is the *density function* and

$$f_Z \geq 0, \int_H f(z)\, d\mu = 1.$$

We shall consider only *$SO(2)$-invariant random variables* on H; this means that the density function must satisfy the invariance condition

$$f_Z(kz) = f_Z(z) \qquad \text{for all } z \in H,\ k \in K.$$

The *sum or composition of two $SO(2)$-invariant random variables* is defined by formula (3.37).

And we shall say that the random variables Z_1 and Z_2 are *independent* if

$$P(Z_1 \in A \text{ and } Z_2 \in A) = P(Z_1 \in A) P(Z_2 \in A).$$

Exercise 26. Show that if Z_1 and Z_2 are $SO(2)$-invariant independent random variables in H with density functions f_1 and f_2, respectively, then the density function for $Z_1 \circ Z_2$ is $f_1 * f_2$, with convolution defined as in Exercise 19.

Hint. You can imitate the proof that works in the Euclidean case, since

$$P(Z_1 \circ Z_2 \in A) = \iint\limits_{M_{Z_2} M_{Z_1} i \in A} f_1(z_1) f_2(z_2)\, d\mu(z_1)\, d\mu(z_2) \qquad \text{with } f_j = f_{Z_j}.$$

Set $w = M_{Z_2} M_{Z_1} i$, and note that

$$\int_{w \in A} f_{Z_1 \circ Z_2}(w)\, d\mu(w) = \int_{z_1 \in H} f_1(z_1) \int_{w \in A} f_2(M_W M_{Z_1}^{-1} i)\, d\mu(w)\, d\mu(z_1).$$

The *characteristic function* φ_Z of an $SO(2)$-invariant random variable Z in H is the *Helgason or non-Euclidean Fourier transform* from (3.24):

$$\varphi_Z(p) = \int_H f_Z(z) \overline{\operatorname{Im} z^s}\, d\mu(z), \qquad s = \tfrac{1}{2} + ip,\ p \in \mathbb{R}. \tag{3.37}$$

These non-Euclidean characteristic functions possess many (but unfortunately not all) of the properties of Euclidean characteristic functions described in Theorem 6 of §1.2.

Exercise 27 (Properties of Characteristic Functions). Which properties of Euclidean characteristic functions in Theorem 6 of §1.2 fail for non-Euclidean characteristic functions?

Hint. The main problem lies with part (2) of Theorem 6 §1.2.

In the preceding Exercise it was noted that φ_Z lacks one important property of the Euclidean characteristic function. For φ_Z has domain $\mathbb{R}$ or $\mathbb{C}$ and not H itself. This means, for example, that we cannot say that if $a \in \mathbb{R}$, then $\varphi_{aZ}(p) = \varphi_Z(ap)$ for $p \in \mathbb{R}^+$. This property was quite important in our proof of the Central Limit Theorem (Theorem 7 of §1.2). And the lack of this property seems to be the cause of some non-Euclidean trouble. Clerc and Roynette [1] meet this issue head on by considering random variables in the domain of the characteristic function rather than H. Karpelevich, Tutubalin, and Shur [1] do not mention this problem and only sketch the beginnings of a proof of their central limit theorem. We shall combine ideas from both these papers.

The *density function for the non-Euclidean normal distribution* is defined to be $G_c(z)$, the fundamental solution of the non-Euclidean heat equation given by formulas (3.32) and (3.33). Then if N_c is a normally distributed random variable in H, the characteristic function is

$$\varphi_{N_c}(p) = \exp[-c(p^2 + \tfrac{1}{4})].$$

In the non-Euclidean case, there are many possible analogues of the mean and the variance. We choose the most direct analogues and define for an $SO(2)$-invariant random variable Z in H, the *mean* m_Z and the *variance* or, better perhaps, the *second moment* d_Z by

$$m_Z = 2\pi \int_{r>0} f_Z(e^{-r}i) r \sinh r \, dr,$$

$$d_Z = 2\pi \int_{r>0} f_Z(e^{-r}i) r^2 \sinh r \, dr. \qquad (3.38)$$

Before proceeding with our discussion of the non-Euclidean central limit theorem, we need another asymptotic property of Legendre functions.

Exercise 28 (An Asymptotic Relation Between Legendre Functions and J-Bessel Functions).

(a) Show that

$$P_{-1/2+ip}(\cosh r) = \frac{\sqrt{2}}{\pi} \int_0^r \frac{\cos(pu)\,du}{\sqrt{\cosh r - \cosh u}}.$$

(b) Show that $P_{-1/2+ip}(\cosh r) \sim J_0(pr)$, as $r \to 0$.

Note. the formula in (b) is a special case of a very general phenomenon relating the spherical function on G/K to the spherical function on the tangent space to G/K at a point. We will consider the analogue for $SL(n, \mathbb{R})$ in Chapter 4.

The complete asymptotic expansion of $P_{-1/2+ip}(\cosh r)$, as $r \to 0$, can be found in Szegö [1] or Fock [1]. This has been generalized to rank 1 symmetric spaces by Stanton and Tomas [1]. Clerc and Roynette [1] use a similar result,

3.2. Harmonic Analysis on H

which is analogous to

$$P_{-1/2+ipm}\left(\cosh\frac{r}{m}\right) \sim J_0(pr) \quad \text{as } m \to \infty.$$

In the classical case, the mean and standard deviation of a random variable are (essentially) the first and second derivatives of the characteristic function evaluated at 0. However, in the non-Euclidean case, this produces different integrals.

Exercise 29 (Other Analogues of the Mean and the Variance).

(a) Show that if Z is an $SO(2)$-invariant random variable in H, then

$$\varphi'_Z(0) = 0.$$

This can be viewed as a non-Euclidean analogue of the mean, which differs from that in equation (3.38).

(b) Define the *dispersion* D_Z of an $SO(2)$-invariant random variable in H by

$$D_Z = -\frac{\varphi''_Z(0)}{\varphi_Z(0)} = \frac{\displaystyle\int_{r \geq u \geq 0} \frac{u^2 f_Z(e^{-r}i) \sinh r}{\sqrt{\cosh r - \cosh u}} \, du \, dr}{\displaystyle\int_{r \geq u \geq 0} \frac{f_Z(e^{-r}i) \sinh r}{\sqrt{\cosh r - \cosh u}} \, du \, dr}.$$

Hint. Use Exercise 28(a).

Note that D_Z can be viewed as a non-Euclidean analogue of the variance, which differs from that in equation (3.38).

(c) Show that if Z_1 and Z_2 are independent $SO(2)$-invariant random variables in H, then

$$D_{Z_1 \circ Z_2} = D_{Z_1} + D_{Z_2}.$$

(d) Show that $D_{N_c} = 2c$, if N_c has the Gaussian or normal distribution for H.

Exercise 30. Graph $G_c(e^{-r}i)$ for various of c. Then graph the mean and second moment defined by formula (3.38). Show that the mean approaches infinity as c approaches infinity. This can be derived from the following formula of Gertsenshtein and Vasil'ev [1]:

$$\int_1^\infty G_c(e^{-r}i) r \, dr = e^{2c}.$$

To discuss the non-Euclidean central limit theorem, we suppose that we are given a sequence $\{Z_n\}_{n \geq 1}$ of independent, $SO(2)$-invariant, random variables in H, each having the same density function f. We want to find some way to normalize the random variable

$$S_n = Z_1 \circ \cdots \circ Z_n \tag{3.39}$$

in order to be able to say that the normalized variable, which we shall call $S_n^\#$, approaches the random variable with density G_c = the fundamental solution of the non-Euclidean heat equation given by formulas (3.32) or (3.33) above. Now Karpelevich, Tutubalin, and Shur [1] normalize S_n by noting for $A \subset H$,

$$P(S_n \in A) = 2\pi \int_{e^{-r}i \in A, r > 0} f_{S_n}(e^{-r}i) \sinh r \, dr. \tag{3.40}$$

Thus we really have densities on $\mathbb{R}^+$, using the non-Euclidean radial variable (r = non-Euclidean distance of $z \in H$ to $i = \sqrt{-1}$). We want to normalize S_n by dividing the radial coordinate random variable by $\sqrt{n}$. Thus the *characteristic function of the normalized random variable* $S_n^\#$ is

$$\varphi_{S_n^\#}(p) = 2\pi \int_{r>0} f_{S_n}(e^{-r}i) \sinh r \, P_{-1/2 + ip}(\cosh(r/\sqrt{n})) \, dr, \tag{3.41}$$

In the classical case, one could easily move the $n^{-1/2}$ over to the p-variable, since if X is a random variable in $\mathbb{R}$,

$$\varphi_{X/\sqrt{n}}(p) = \int_{x \in \mathbb{R}} f(x) \exp(ipx/\sqrt{n}) \, dx = \varphi_X(p/\sqrt{n}).$$

But the relation between $P_{-1/2 + ip/\sqrt{n}}(\cosh r)$ and $P_{-1/2 + ip}(\cosh(r/\sqrt{n}))$ appears to be rather bizarre at first sight.

However, we can use Exercise 28(b) to rectify this situation and find that

$$\varphi_{S_n^\#}(p) \sim \left(2\pi \int_{r>0} f(e^{-r}i) \sinh r \, J_0\left(\frac{rp}{\sqrt{n}}\right) dr\right)^n \quad \text{as } n \to \infty.$$

And we have the power series

$$J_0(x) = \sum_{k \geq 0} \frac{(-1)^k}{(k!)^2} \left(\frac{x}{2}\right)^{2k}.$$

This implies that

$$\varphi_{S_n^\#}(p) \sim \left(2\pi \int_{r>0} f(e^{-r}i) \sinh r \, dr - \frac{2\pi}{4n} p^2 \int_{r>0} f(e^{-r}i) r^2 \sinh r \, dr\right)^n$$

$$\sim \left(1 - \frac{dp^2}{4n}\right)^n \sim e^{-dp^2/4} \quad \text{as } n \to \infty,$$

$$d = d_{Z_n} = 2\pi \int_{r>0} f(e^{-r}i) r^2 \sinh r \, dr,$$

as in formula (3.38).

In order to complete the discussion of the central limit theorem for rotation-invariant random variables on H, one must imitate the argument given in the proof of the Euclidean central limit theorem (Theorem 7 of §1.2). We have proved that

3.2. Harmonic Analysis on H

$$\lim_{n\to\infty} \varphi_{S_n^\#}(p) = \exp(-dp^2/4). \tag{3.42}$$

Note that the density function for the random variable $S_n^\#$ is

$$f_n^\#(ke^{-r}i) = \sqrt{n}(f* \cdots *f)(e^{-\sqrt{n}r}i)\sinh(r\sqrt{n})/\sinh r, \tag{3.43}$$

where $k \in SO(2)$ and $r > 0$. Next let $\alpha \in C_c^\infty(H)$. And set

$$d\sigma(p) = \frac{1}{4\pi} p \tanh \pi p \, dp.$$

By formula (3.42), the Plancherel theorem (Theorem 3), and the Lebesgue dominated convergence theorem, we have

$$\lim_{n\to\infty} \int_H f_n^\#(z)\alpha(z)\,d\mu = \lim_{n\to\infty} \int_{s=1/2+ip,\,p\in\mathbb{R}} \mathcal{H}f_n^\#(s)\mathcal{H}\alpha(s)\,d\sigma(p)$$

$$= \int_{s=1/2+ip,\,p\in\mathbb{R}} \exp(-dp^2/4)\mathcal{H}\alpha(s)\,d\sigma(p) \tag{3.44}$$

$$= e^{d/4} \int_H G_{d/4}(z)\alpha(z)\,d\mu.$$

We can approximate the indicator function of a measurable set in H by α in $C_c^\infty(H)$ to complete the proof of the following theorem.

Theorem 4 (A Non-Euclidean Central Limit Theorem for Rotation-Invariant Random Variables). *Suppose that $\{Z_n\}_{n\geq 1}$ is a sequence of independent, $SO(2)$-invariant random variables in H, each having the same density function $f(z)$. Let $S_n = Z_1 \circ \cdots \circ Z_n$ be normalized as in formulas (3.41) and (3.43) above. Suppose that $f_n^\#$ is the density function for the normalized random variable $S_n^\#$. Then for measurable sets $A \subset H$ we have*

$$\int_A f_n^\#(z)\,d\mu \sim e^{d/4} \int_A G_{d/4}(z)\,d\mu \qquad \text{as } n \to \infty.$$

Here G_c denotes the fundamental solution of the non-Euclidean heat equation given by formulas (3.32) and (3.33). And d is defined by formula (3.38).

Exercise 31.

(a) Prove that $|P_s(\cosh r)| \leq 1$, if $-1 \leq \operatorname{Re} s \leq 0$.
(b) Use (a) to show that if φ_Z is the characteristic function of an $SO(2)$-invariant random variable Z with density function $f(z)$, then $|\varphi_Z(p)| \leq 1$ for all $p \in \mathbb{R}$.
(c) Justify formula (3.44).

Finally, let us return to the discussion of the lossless transmission line with random inhomogeneities. Exercise 30 showed that the mean distance from i to a normally distributed random variable with density $G_c(z)$ increases exponentially as c approaches infinity. And we can conclude from parts (c) and (d) of Exercise 29 that c approaches infinity as the length of the transmission line increases. Recalling the discussion in §3.1 which showed that the transmitted power decreases as the distance of the reflection coefficient from the origin increases (measuring distance using the non-Euclidean metric), we conclude that a long transmission line reflects almost all of the incoming power. More precise calculations might allow an engineer to do something about this (see Feller [1, Vol. II, pp. 208–209]).

Some Remarks on the Uses of Harmonic Analysis on Groups Such as $SL(2, \mathbb{R})$ in Theoretical Physics

We shall keep our remarks very brief. Detailed discussions of the representations of $SL(2, \mathbb{R})$ can be found in Lang [4], Sugiura [1], and Vilenkin [1]. Applications of representations of noncompact groups like $SL(2, \mathbb{R})$ to physics are considered, for example, in Barut and Rączka [1], Mackey [1, 2, 3], and Wybourne [1].

Physicists are interested in representations of groups such as $SO(3, 1)$ and $SO(4, 2)$ which leave invariant various PDEs describing, for example, electromagnetic waves or elementary particles. Actually, the full description of the representations of the groups $SO(p, q)$ is not complete, in general. These are the real $n = p + q$ by n matrices which leave invariant the quadratic form $x_1^2 + \cdots + x_p^2 - x_{p+1}^2 \cdots - x_n^2$. We saw in Exercise 13 of §3.1 that $SL(2, \mathbb{R})$ and $SO(2, 1)$ are closely related. In fact, they have the same Lie algebra.

The Fourier transform of a function on $G = SL(2, \mathbb{R})$ involves various series of representations of G—the principal continuous series and the discrete series (see Vergne [1] or Lang [4]). And Fourier analysis on $SL(2, \mathbb{R})$ involves a mixture of Fourier series and integrals. The inversion formula goes back to Bargmann [1] and Harish-Chandra [1]. We shall not state it here. Instead, let us just say a few words about the physical relevance of such expansions. A good introduction to this subject is Chapter 18 of Wybourne [1], where it is shown that many physics problems lead to the Lie algebra of $SL(2, \mathbb{R})$. In particular, one can find the Lie algebra of $SL(2, \mathbb{R})$ or equivalently, $SU(1, 1)$, in the ODE:

$$f''(y) + q(y)f(y) = 0, \qquad q(y) = ay^{-2} + by^2 + c.$$

In fact, this sort of argument allows one to being $SO(2, 1)$ into the discussion of the hydrogen atom. Then one can bring in $SO(4)$ and $SO(4, 2) \cong SU(2, 2)/\pm I$. In this way more accurate information on the spectrum of hydrogen has been obtained.

Clearly, solutions of non-Euclidean wave equations and their properties such as the truth or falsity of the Huygens' principle are of interest in general relativity (see Friedlander [2] and Helgason [1]).

3.3. Fundamental Domains for Discrete Subgroups Γ of $G = SL(2, \mathbb{R})$.

A 2-dimensional smooth orientable, but not compact space of constant negative curvature with the topology of a torus [or a sphere] is investigated. It contains an open end; i.e. an exceptional point at infinite distance, through which a particle or wave can enter or leave, as in the exponential horn of certain antennas or loudspeakers.

From M. Gutzwiller [1, p. 341].
See Exercise 13 for more details.

We saw in §3.2 that it was quite useful to study $\mathbb{R}/\mathbb{Z}$ which we think of as the circle or [0, 1] with 0 and 1 identified. And in §3.4, we saw that it is valuable to study crystals $E(3)/\Gamma$, where $E(3)$ is the Euclidean motion group and Γ is a discrete subgroup of $E(3)$. Similarly, there are higher-dimensional crystals $E(n)/\Gamma$, with Γ any discrete subgroup of the n-dimensional Euclidean motion group $E(n)$, which is a semidirect product of $O(n)$ and the translation group $\mathbb{R}^n$. Bieberbach proved in 1910 that $E(n)$ has only a finite number of different subgroups with compact fundamental domains. This answered part of the 18th of Hilbert's famous problems (see Milnor [1, pp. 491–497]).

In this section, we shall consider discrete subgroups Γ of $SL(2, \mathbb{R})$. It was proved by Poincaré that, unlike $\mathbb{R}^n$, the upper half-plane has an infinite number of essentially different kinds of compact fundamental domains $\Gamma\backslash H$. Thus Hilbert called these domains Poincaré polygons. We shall be most interested, however, in the noncompact fundamental domain for the (*elliptic*) *modular group* $\Gamma = SL(2, \mathbb{Z})$ of 2×2 integer matrices of determinant 1. A fundamental domain for this group is pictured in Fig. 3.11. The main goal of this chapter is to study harmonic analysis on $SL(2, \mathbb{Z})\backslash H$ (see §7). In later chapters we aim to discuss the analogue for $SL(n, \mathbb{Z})$ and other more complicated discrete groups.

General references for this section are Apostol [3], Ford [1], Gunning [1], Hecke [1, 2], Klein and Fricke [1], Knopp [1], Lang [6], Lehner [1], Maass [1], Rankin [1], Schoeneberg [1], Serre [1], Shimura [1], and Siegel [1].

One can cover all of H with Γ-translates of a fundamental domain $\Gamma\backslash H$. This produces what is called a tessellation of H and an interesting picture. In 1958 the artist M.C. Escher saw such a picture in a book of H.S.M. Coxeter (Fig. 3.12) and was inspired to create the various circle limit drawings (see Ernst [1, p. 108]). In Escher's circle limit pictures (see Ernst [1]), the upper half-plane is replaced by the unit disc, using the Cayley transform, as in Exercise 12 of §3.1. Fig. 3.14 is the tessellation of H for $\Gamma = SL(2, \mathbb{Z})$. It was drawn by computer (as directed by Mark Eggert). In Chapter 5, we will find some beautiful three-dimensional analogues of this tessellation, also drawn by computer with Mr. Eggert's direction. You can find other two-dimensional

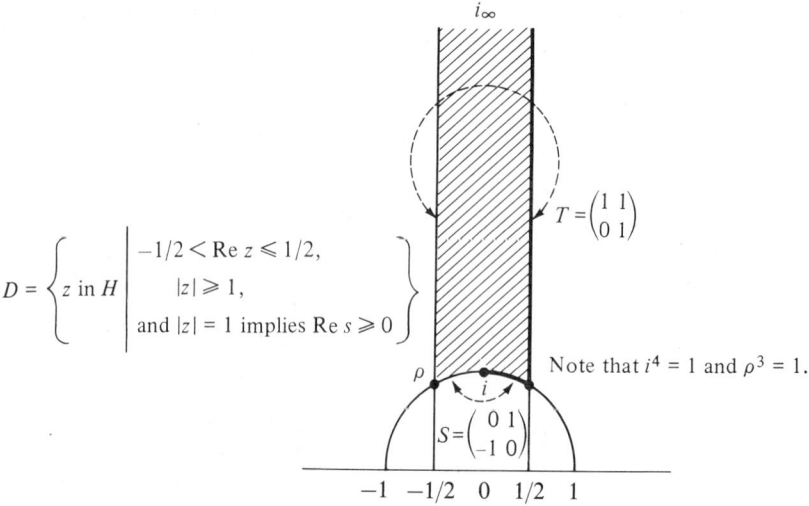

Figure 3.11. A noneuclidean triangle D through the points $\rho, \rho + 1, i\infty$, which is a fundamental domain for H mod $SL(2, \mathbb{Z})$. The domain D is shaded. Arrows show boundary identifications by the fractional linear transformations from S and T which generate $SL(2, \mathbb{Z})/\pm 1$.

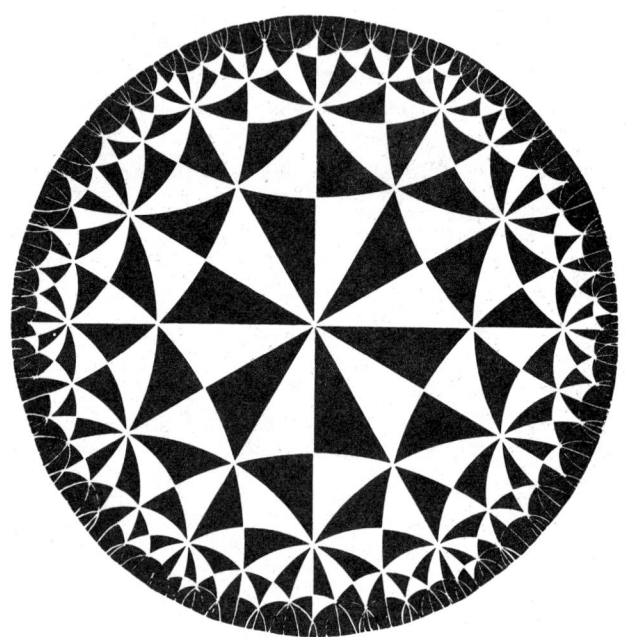

Figure 3.12. The Coxeter illustration that inspired Escher. (From Coxeter [2, p. 285]. Reprinted by permission of John Wiley & Sons.)

3.3. Fundamental Domains for Discrete Subgroups Γ of $G = SL(2, \mathbb{R})$

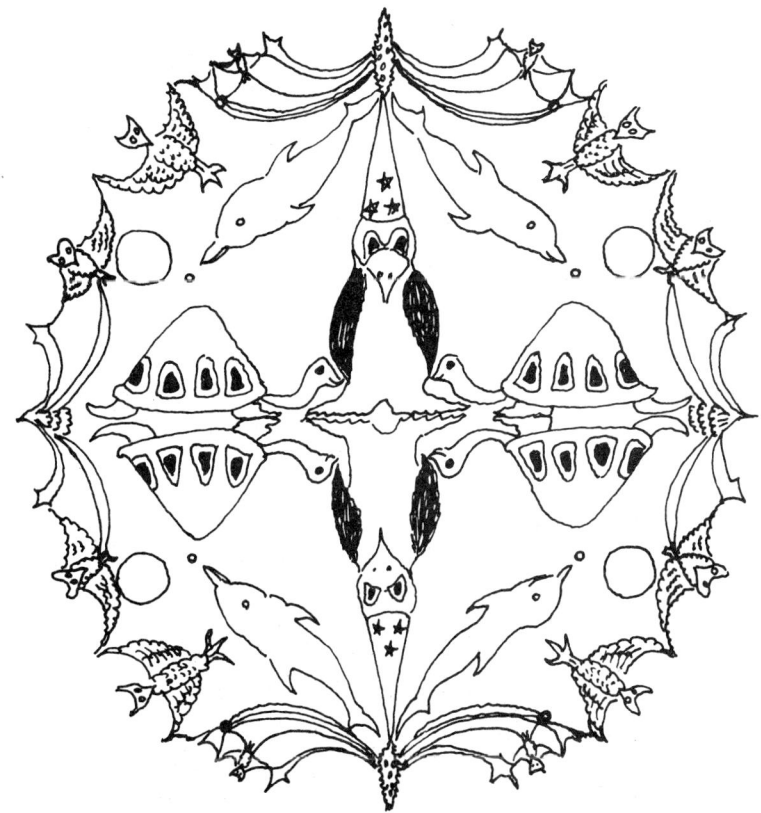

Figure 3.13

tessellations in Ford [1, pp. 305–309], Hurwitz and Courant [1, pp. 430–445], Klein and Fricke [1], Lehner [1, pp. 11–17], and Nehari [1, pp. 308–316].

Why study harmonic analysis on $\Gamma \backslash H \cong \Gamma \backslash G/K$, $G = SL(2, \mathbb{R})$, $K = SO(2)$, and the analogue for other Lie groups? One justification comes from the following dialogue which appeared in the 1953 Scientific American (see P. Le Corbeiller [1]):

> "In the last 60 years, however, a new revolution has taken place, and everywhere we look we find that what seems to be continuous is really composed of atoms...."
> "But are not modern mathematicians interested in such things?" asked Empeiros [a physicist].
> "They are," I answered, "but they give them other names. They call them Number Theory and the Theory of Discontinuous Groups. Actually they have found much more than we can use as yet in physics, but we have in crystals illustrations of some of their simpler theorems...."

Functions on $\Gamma \backslash H$ (or equivalently, on the fundamental domain with boundaries identified) are called *automorphic functions*. Surprisingly, *every*

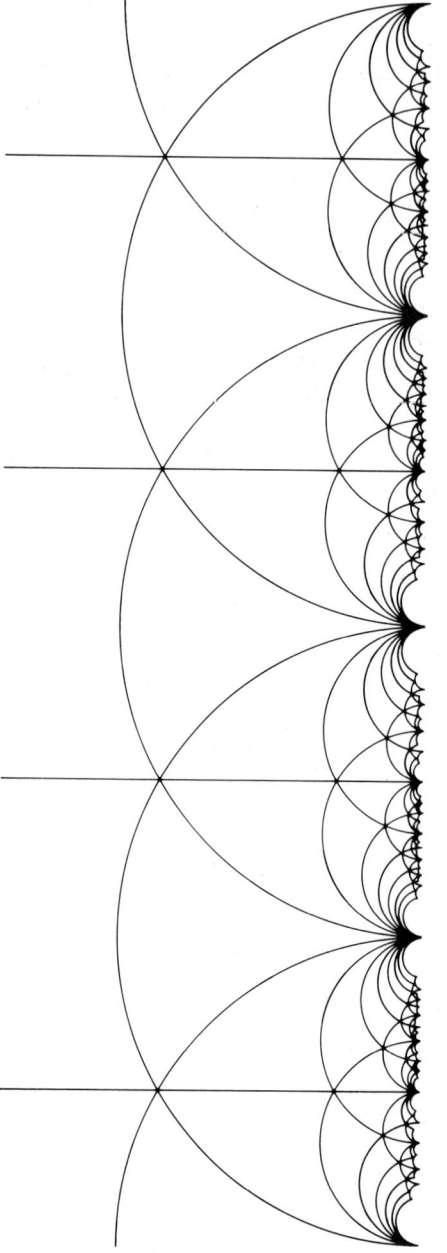

Figure 3.14. Tessellation of H for $SL(2, \mathbb{Z})$. (Computer drawing by the UCSD VAX and Mark Eggert.)

3.3. Fundamental Domains for Discrete Subgroups Γ of $G = SL(2, \mathbb{R})$

analytic function $f(z)$ can be considered as an automorphic function. For f lives on a Riemann surface S which has a (simply connected) universal covering surface $\hat{S}$. The Riemann mapping theorem identifies $\hat{S}$ with either the Riemann sphere, the complex plane or H. The group of isometries of S contains a subgroup Γ isomorphic to the group of covering transformations of $\hat{S}$ over S. Thus f is automorphic with respect to Γ. For the details of this argument, see Siegel [1, Vol. I, Ch. 2] or Hurwitz and Courant [1, pp. 453–550].

Historically, the study of automorphic functions arose with the attempt to solve various problems arising in applied mathematics, as well as number theory. For example, the problem of heat diffusion leads to *theta functions* (see Exercise 7 of §1.3). The search for a proof of the prime number theorem, giving the asymptotic behavior of the number of primes less than or equal to x as $x \to \infty$, leads to the study of Riemann's zeta function. The latter function is connected with the theta function by Mellin transform (see Exercise 7 of §1.4). The simplest theta function is

$$\theta(z) = \sum_{n \in \mathbb{Z}} \exp(\pi i n^2 z) \quad \text{for } z \in H. \tag{3.45}$$

Theta is not quite an automorphic function. It is what is known as a *modular* or *automorphic form*, since theta has the following transformation properties:

$$\theta(z + 2) = \theta(z),$$
$$\theta(-1/z) = (z/i)^{1/2} \theta(z).$$

The last equation results from Poisson summation (see Exercises 6 of §1.3 and 1.4). The two maps $z \mapsto z + 2$ and $z \mapsto -1/z$ generate a discrete subgroup of $SL(2, \mathbb{R})$ called the *theta group* Γ_θ. Also, $\theta(2z)$ is a modular form of weight $\frac{1}{2}$ for the congruence group $\Gamma_0(4)$ (see Pfetzer [1] and Serre and Stark [1]). The *congruence subgroup* $\Gamma_0(4)$ of $SL(2, \mathbb{Z})$ is defined by

$$\Gamma_0(4) = \left\{ \begin{pmatrix} a & b \\ c & d \end{pmatrix} \middle| a, b, c, d \text{ in } \mathbb{Z}, ad - bc = 1, c \equiv 0 \pmod{4} \right\}.$$

Many problems of mathematics and physics lead to the *elliptic integrals*, for example the lemniscate integral:

$$w = f(z) = \int (1 - z^4)^{-1/2} dz.$$

The *Dirichlet problem* for various domains in the plane can often by solved via such integrals, thanks to the Schwarz-Christoffel transformation (see Carrier, Krook, and Pearson [1, Ch. 4]). It turns out that the inverse function $z = g(w)$ is doubly periodic (see Ahlfors [1, p. 232]). The theory of such functions leads one quickly to automorphic functions and forms. In fact, one is often able to compute elliptic integrals via theta functions (see §3.4). Other references for these things are Lehner [1], Nehari [1, pp. 280–296, 308–316], and Siegel [1, Vols. I, II].

The J-function is another example of an automorphic function that can be viewed as giving a certain conformal mapping. It can also be used to prove *Picard's theorem* that an entire function takes on every finite value with one possible exception (see §3.4). The harder Picard theorem says that a function holomorphic in a punctured disc with an essential singularity at the center has the property that there is at most one finite number a such that the function takes on the value a only finitely many times. There is also a way of dealing with the harder result via automorphic forms (see Lehner [2]).

There are many applications of automorphic forms to algebra and number theory. The computation of a fundamental domain for $SL(2,\mathbb{Z})\backslash H$ yields an *algorithm for the computation of class numbers of imaginary quadratic fields* (see Exercise 5, which follows). This is just the classical correspondence between ideal classes of imaginary quadratic fields and reduced integral positive binary quadratic forms (see Borevitch and Shafarevitch [1, pp. 149ff] or Davenport [1, p. 195ff]). The same fundamental domain can also be considered a fundamental domain for the *reduction theory of positive definite quadratic forms of determinant 1* by Exercise 9 of §3.1 and Siegel [2, pp. 68–74]. Hermite, Kronecker, and Brioschi used automorphic functions to *solve the general algebraic equation of degree 5* (see Lehner [1, p. 10]). Automorphic functions are also necessary for *explicit versions of class field theory*—the study of normal extensions of algebraic number fields with abelian Galois group. The J function is useful here, since it has the property that $J(a)$ is an algebraic integer when a lies in the upper half-plane and an imaginary quadratic field. More details on this subject can be found in Borel, Chowla, et al. [1]. *Hilbert's problem 12* asks for functions whose special values generate abelian extensions of number fields, just as values of the exponential function generate abelian extensions of the rational numbers. The answer for imaginary quadratic base fields requires automorphic functions and was found in the late 1800s and early 1900s by H. Weber, Kronecker, Fueter, and others. Much of the recent work on the subject is due to G. Shimura (see Shimura [1]). See also Stark [3, 4].

The study of the *representations of $SL(2,\mathbb{R})$* also leads to classical automorphic forms in the case of discrete series representations (see Gelfand, Graev, and Piatetskii-Shapiro [1, pp. 43–48]). So far the only functions on H that we have discussed are complex-analytic or meromorphic functions of z in H (see §3.4). However, there are other interesting automorphic functions. Their study is rather recent. They correspond to the continuous series representations of $SL(2,\mathbb{R})$ and they are eigenfunctions of the non-Euclidean Laplacian invariant under $SL(2,\mathbb{Z})$. Such functions were first studied systematically in 1949 by Maass [3]. We will consider these Maass wave forms in §3.5. They are essential for harmonic analysis on $\Gamma\backslash H$ to be discussed in §3.7.

Automorphic forms and harmonic analysis on fundamental domains of discrete subgroups of Lie groups have appeared in many papers in physics journals recently (see Gutzwiller [1–3] and Hurt [1, 2]). For example, Monastyrsky and Perelomov [1] consider automorphic forms on general symmetric

3.3. Fundamental Domains for Discrete Subgroups Γ of G = SL(2, ℝ)

spaces and give several references to applications, one of which concerns the problem of boson pair creation in alternating external fields. These applications are basically a reflection of the usefulness of group representations in physics as we mentioned at the end of §3.2, plus the fact that automorphic forms are involved in the description of various series of representations of the Lie groups like $SL(2, \mathbb{R})$. The terminology used in the quantum mechanics work is that automorphic forms appear in *systems of coherent states*.

After this, perhaps overlong, introduction to our subject, let us get down to the study of $\Gamma \backslash H$, for discrete subgroups Γ of $SL(2, \mathbb{R})$. For concreteness, we shall mostly restrict our study to $\Gamma = SL(2, \mathbb{Z})$. A *fundamental domain D* for Γ is a subset D of H which behaves like the quotient space $\Gamma \backslash H$ (at least up to boundary identifications). Thus for every z in H, there is a γ in Γ with γz in D. Moreover, if z and w lie in the interior of D and z = γw, then γ is either the identity matrix I or −I. We shall say that z and γz are *equivalent under* Γ if γ ∈ Γ. As in §3.1, the action of γ ∈ Γ on z in H is by fractional linear transformation.

Exercise 1 (A Fundamental Domain for $SL(2, \mathbb{Z})$). Show that a fundamental domain for $SL(2, \mathbb{Z})$ can be taken as the region pictured in Fig. 3.11; that is,

$$D = \{z \in H \mid -\tfrac{1}{2} < \text{Re } z \leq \tfrac{1}{2}, |z| \geq 1, \text{ and if } |z| = 1, \text{ then Re } z \geq 0\}.$$

Hints.

(1) *The Highest Point Method.*

   Suppose $z \in H$. You must find γ in $\Gamma = SL(2, \mathbb{Z})$ so that $\gamma z \in D$. Clearly you can translate z to make $\text{Re}(\gamma z) \in (-\tfrac{1}{2}, +\tfrac{1}{2}]$. Next note that $\text{Im}(\gamma z) = y|cz + d|^{-2}$, if $\gamma z = (az + b)/(cz + d)$, with $a, b, c, d \in \mathbb{Z}$ and $ad - bc = 1$. Now you can choose γ so that $|cz + d|$ is minimal; i.e., $\text{Im}(\gamma z)$ is maximal, so that γz is a "highest point" equivalent to z. Why? To prove that $|z| \geq 1$, use the existence of $Sz = -1/z$ to obtain a contradiction to the maximality of $\text{Im}(\gamma z)$ otherwise.

   Next you must prove that $z, \gamma z \in D$ for $\gamma \in \Gamma$, implies that $\gamma = \pm I$. This can be done by brute force. Suppose $\gamma z = (az + b)/(cz + d)$. Then

   $$|cz + d|^2 = c^2 z \bar{z} + cd(z + \bar{z}) + d^2 \geq (|c| - |d|)^2 + |cd|.$$

   Therefore $c \neq 0$ implies that $|cz + d| \geq 1$. It follows that $\text{Im } z = \text{Im } \gamma z$ if $c \neq 0$, since you can also consider γ^{-1}. Then $(|c| - |d|)^2 + |cd| = 1$ and only a few possible matrices γ need to be considered; $S, STS, ST^{-1}S$, with S, T as in Fig. 3.11. It is easy to see what these matrices do to the fundamental domain (see Fig. 3.15). The case $c = 0$ is even easier.

(2) *The Generators and Relations Method of Poincaré.*

   If you compute the tessellation of H by $SL(2, \mathbb{Z})$ as in Fig. 3.15, you may notice, as Poincaré did, that the fundamental domain gives generators and defining relations of $PSL(2, \mathbb{Z}) = SL(2, \mathbb{Z})/\{+I, -I\}$. The generators are the maps identifying the sides of the fundamental region. To have pairs of

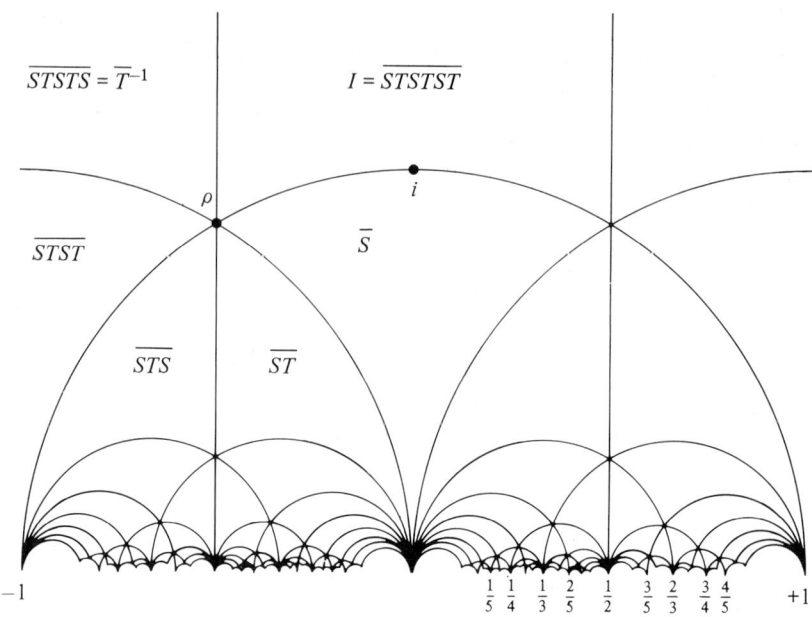

Figure 3.15. Poincaré's method of generators and relations illustrated for $SL(2, \mathbb{Z})/\{+I, -I\}$. $\bar{S}z = -1/z$; $\bar{T}z = z + 1$; $\rho = (-1 + \sqrt{-3})/2$; $i = \sqrt{-1}$. Generators of $SL(2, \mathbb{Z})/\{+I, -I\}$ are $\bar{S}$ and $\bar{T}$. Defining relations are $\bar{S}^2 = \bar{I}$ and $(\bar{S}\bar{T})^3 = \bar{I}$.

sides you need four sides, and not three. So cut the bottom circle in half at the point $i = \sqrt{-1}$. Then the maps identifying the sides of D in pairs are $\bar{T}z = z + 1$ and $\bar{S}z = -1/z$. The defining relations for $PSL(2, \mathbb{Z})$ are obtained by picture. To do this, follow the mappings that circle the two vertices $\rho = (-1 + \sqrt{-3})/2$ and $i = \sqrt{-1}$. Since these mappings must lead to the identity when composed, you find that $(\bar{S}\bar{T})^3 = \bar{I}$ when you follow the copies of D around the vertex ρ and you find that $\bar{S}^2 = \bar{I}$ when you follow the copies of D around the point i.

Poincaré showed in fact that one can start with a geodesic polygon P with certain properties and obtain a discrete group Γ for which the polygon is a fundamental domain. The sides of P must be geodesics arranged in pairs equivalent under certain fractional linear maps and equal in non-Euclidean length. The sums of angles at equivalent vertices in the fundamental domain which are fixed by order k elliptic fractional linear transformations must be $2\pi/k$. See the end of this section for the definition of "elliptic." Thus, in the case under consideration, the sum of the angles of the vertices of the fundamental domain D at ρ and $\rho + 1$ is $2\pi/3$ and 3 is just the order of the fractional linear map fixing ρ. And the sum of the angles at i is just $\pi = 2\pi/2$, and 2 is again the order of the fractional linear map fixing i.

3.3. Fundamental Domains for Discrete Subgroups Γ of $G = SL(2, \mathbb{R})$

Stark [5] has shown how to generalize these conderations to higher-dimensional cases in order to obtain analogues of Poincaré's results and to compute the fundamental domains. Other references for Poincaré's theory are Lehner [1] and Poincaré [1].

In order to complete the exercise using Poincaré's ideas you must show that the generators and relations theorem actually gives an algorithm for finding fundamental domains.

Exercise 2.

(a) Use the Euclidean algorithm to show that $PSL(2, \mathbb{Z}) = SL(2, \mathbb{Z})/\{+I, -I\}$ is generated by $Sz = -1/z$ and $Tz = z + 1$.

Hint. Note that
$$\begin{pmatrix} 1 & t \\ 0 & 1 \end{pmatrix} \begin{pmatrix} a & b \\ c & d \end{pmatrix} = \begin{pmatrix} a + tc & * \\ * & * \end{pmatrix}.$$

You can choose t to make $0 \leq a < c$ by the Euclidean algorithm. Then you can use S to flip a and c. Finally, you can obtain $c = 0$ in a finite number of steps, since a and c are relatively prime.

(b) Show that the algorithm in (a) also gives the continued fraction expansion of a/c (see Hardy and Wright [1]).

Exercise 3. Which of Figs. 3.16, 3.17, or 3.18 is the image of the tessellation of H which was given in Fig. 3.14 under the Cayley transform $z \mapsto i(z - i)/(z + i)$? Why? In Figs. 3.16 and 3.17, a shaded and a light region together make up a fundamental domain.

Exercise 4. The fundamental domain for $SL(2, \mathbb{Z})$ in Fig. 3.11 is a non-Euclidean triangle with angles $\pi/3$ at $\rho = (-1 + \sqrt{-3})/2$ and $\rho + 1$, and angle 0 at ∞. Show that the non-Euclidean area of the fundamental domain for $SL(2, \mathbb{Z})$ is $\pi/3$.

The fundamental domain for $SL(2, \mathbb{Z})$ was used by Gauss in his research on quadratic forms. We will consider Minkowski's generalization of this result to $SL(n, \mathbb{Z})$ in Chapter 4. The analogue for more general discrete groups will be considered in Chapter 5.

Application to Number Theory. Computation of Class Numbers of Imaginary Quadratic Fields

Suppose that $K = \mathbb{Q}(\sqrt{d})$, $d < 0$, with $d =$ the discriminant of K. There is a one-to-one correspondence between ideal classes $C \in I_K$ (as defined in §1.4) and points z in the fundamental domain D for $SL(2, \mathbb{Z})$ in Fig. 3.11. For we can

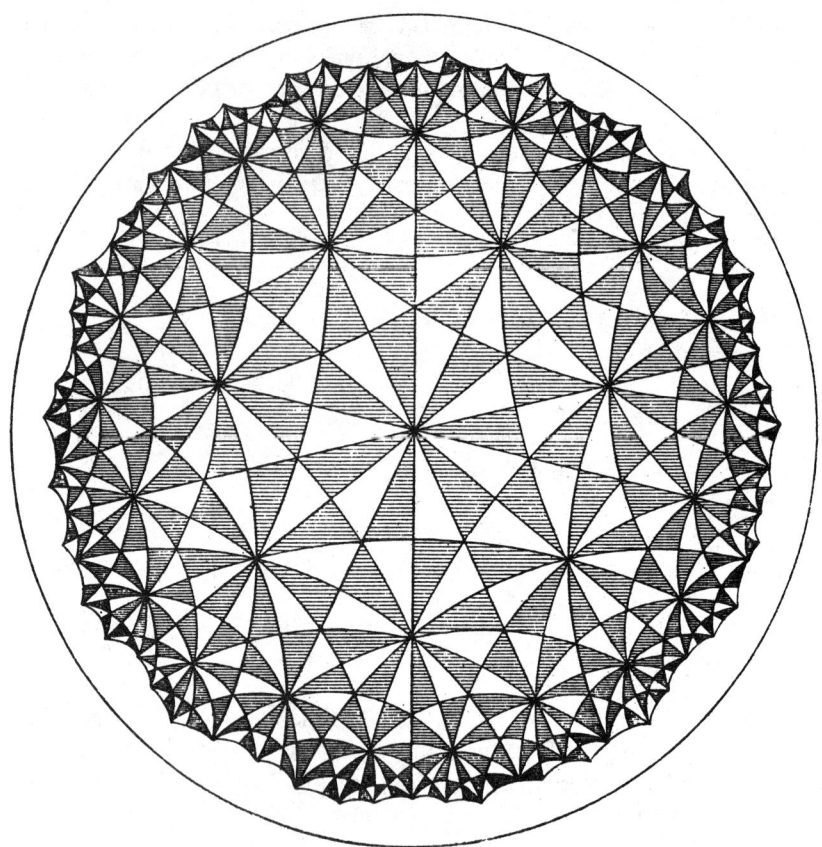

Figure 3.16. Tessellation of the unit disc. (From Klein and Fricke [1]. Reprinted by permission of Teubner.)

choose an ideal $\mathfrak{A}$ in D of the form

$$\mathfrak{A} = \mathbb{Z} \oplus \mathbb{Z}z, \qquad z = (-b + \sqrt{d})/2a,$$

with

$$d = b^2 - 4ac, \qquad a > 0, \qquad (a, b, c) = 1.$$

Here a, b, c are integers in $\mathbb{Z}$. So we need $-a < b \leq a$ and $1 \leq |z|^2 = c/a$. It follows from the definition of the fundamental domain that the class number of K is

$$h_K = \#\left\{(a,b,c) \in \mathbb{Z}^3 \,\middle|\, \begin{array}{l}(a,b,c) = 1, \quad a > 0, \quad -a < b \leq a, \quad c \geq a, \\ c > a \quad \text{if} \quad b < 0, \quad d = b^2 - 4ac \end{array}\right\}.$$

Note that you need only count the pairs (a, b), since they determine c. Also, the inequalities force $a < \sqrt{|d|/3}$. Thus the set of pairs to be considered is at most

3.3. Fundamental Domains for Discrete Subgroups Γ of $G = SL(2, \mathbb{R})$ 173

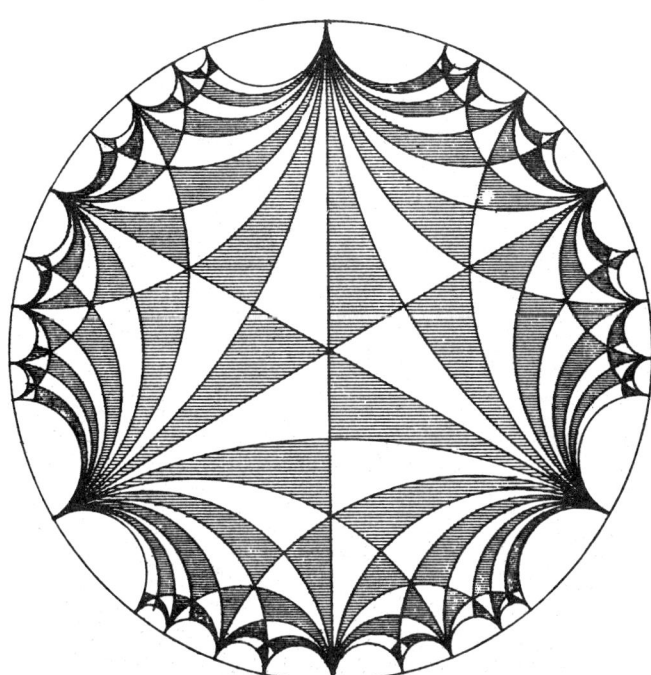

Figure 3.17. Another tessellation of the unit disc. (From Klein and Fricke [1]. Reprinted by permission of Teubner.)

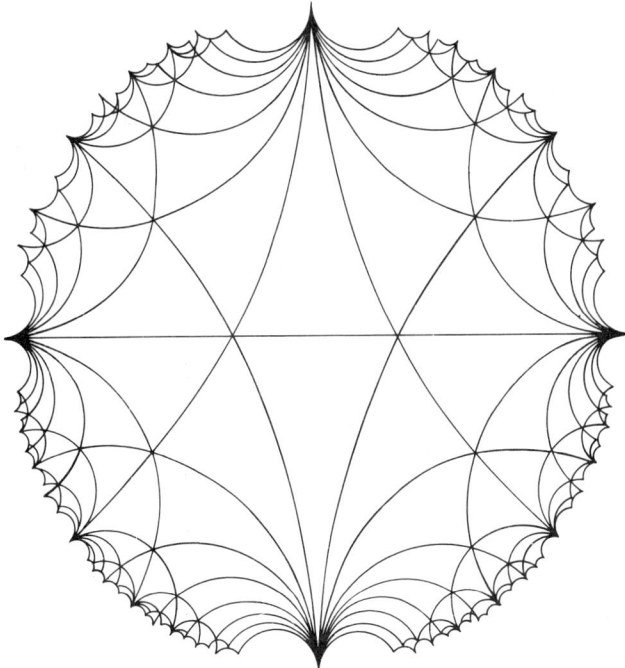

Figure 3.18. Yet another tessellation of the unit disc. (Drawn by the USCD VAX computer and Mark Eggert.)

$|d|$ in number. It is an easy matter to program a calculator or computer to find the class number of an imaginary quadratic field K with d near -10^6. Below is a table of some class numbers in this range. Fancier programming is called for if one wants to deal with monster discriminants. Shanks [1] uses Gaussian composition of the quadratic forms $ax^2 + 2bxy + cy^2$ (which corresponds to multiplication of ideal classes) to speed the algorithm and finds, for example, that if the discriminant is -4722366483281962074113, then $h_K = 50866650112$.

Exercise 5. Make a table of class numbers of imaginary quadratic fields similar to Table 3.2 which follows.

Table 3.2. Class Numbers of Imaginary Quadratic Fields

Discriminant	Class Number	Discriminant	Class Number
−1000003	105	−1000052	306
−1000007	630	−1000055	828
−1000011	368	−1000056	364
−1000015	430	−1000059	240
−1000019	342	−1000063	394
−1000020	320	−1000067	318
−1000023	706	−1000068	372
−1000024	274	−1000072	264
−1000027	168	−1000079	974
−1000031	928	−1000083	184
−1000036	192	−1000084	300
−1000039	877	−1000087	366
−1000040	688	−1000088	372
−1000043	192	−1000091	342
−1000047	508	−1000095	720
−1000051	276	−1000099	187

Exercise 6. Construct fundamental domains for some congruence subgroups of $SL(2, \mathbb{Z})$. These subgroups are defined by formula (3.46), which follows.

Exercise 7 (Another Construction of a Fundamental Domain for $\Gamma \backslash H$). Suppose that Γ is a discrete subgroup of $SL(2, \mathbb{R})$.

(a) Show that Γ is countable.
(b) Show that there is a point $w \in H$ which is not fixed by any element of Γ.
(c) Let $d(z_1, z_2)$ be the non-Euclidean distance between z_1 and z_2. Set

$$D = \{z \in H \mid d(z, w) \leq d(z, \gamma w) \text{ for all } \gamma \in \Gamma\},$$

where w is the point from part (b). Show that D is a fundamental domain for Γ.

3.3. Fundamental Domains for Discrete Subgroups Γ of $G = SL(2, \mathbb{R})$

Hints.

(a) Any neighborhood of the identity in $SL(2, \mathbb{R})$ contains only finitely many points of Γ.

(c) Using the invariance of the non-Euclidean distance under elements of Γ, it is easy to show that H is a union of images of D under elements of Γ. Secondly, one must show that z and γz in D imply that z and γz lie on the boundary of D. Now the boundary of D consists of points z such that $d(z, w) = d(z, \lambda w)$ for some $\lambda \in \Gamma$, while the inequality holds for other $\gamma \in \Gamma$. In fact, if we set $D_\gamma = \{z \mid d(z, w) \leq d(z, \gamma w)\}$, then D_γ is a hyperbolic half-plane whose boundary is the non-Euclidean perpendicular bisector of the geodesic through w and γw. Since D is the intersection of all the D_γ, $\gamma \in \Gamma$, the fundamental domain is a convex set in the sense of the non-Euclidean metric. If z and γz both lie in D, then $d(z, w) = d(z, \gamma^{-1} w)$ and z is a boundary point unless $\gamma = \pm I$. For more hints, see Maass [1, pp. 12–15] or Siegel [1, Vol. II, pp. 35–38]. It turns out that D may not have a finite number of sides. If it does, then it has finite area and conversely (see Siegel [1, Vol. II., pp. 39–46]).

The fundamental region in Exercise 7 is called the *normal fundamental region*. In the special case of $\Gamma = SL(2, \mathbb{Z})$, the normal fundamental region with center iy_0, $y_0 > 1$, is just the fundamental domain in Fig. 3.11. The non-Euclidean perpendicular bisectors of the lines joining iy_0 and $\pm 1 + iy_0$ are the vertical lines bounding this fundamental domain. The non-Euclidean perpendicular bisector of the geodesic joining iy_0 and i/y_0 is the circle of center 0 and radius 1, the other boundary of the fundamental domain. We mentioned an application of the computation of this fundamental domain in Exercise 5. There is also an application to the theory of Riemann surfaces of genus 1. For this fundamental domain is in correspondence with the classes of conformally equivalent Riemann surfaces of genus 1 (see Maass [1, pp. 51ff]).

There is an overabundance of terminology for groups of fractional linear transformations mapping the inside of a fixed circle onto itself. We owe this perhaps to the rivalry of Poincaré and Klein (see their letters in Acta Math. 39 (1923) or Rankin [2]). Poincaré called a group of fractional linear transformations mapping the inside of a fixed circle onto itself a *Fuchsian group*, and Klein called such a group a *Hauptkreisgruppe* (principal circle group). There is a finer distinction, according to whether all points of the principal circle are limit points of the group. By a limit point of Γ, we mean a point of accumulation of the set of centers of the isometric circles

$$\{z \in \mathbb{C} \mid |cz + d| = 1\}, c \neq 0, \text{ of } \begin{pmatrix} a & b \\ c & d \end{pmatrix} \text{ in } \Gamma.$$

If all points of the principal circle are limit points, the group is called a *Fuchsian group of the first kind* in English and a *Grenzkreisgruppe* in German. One can prove that Γ is a finitely generated Fuchsian group of the first kind if and only if Γ is a discontinuous subgroup of the automorphism group of the

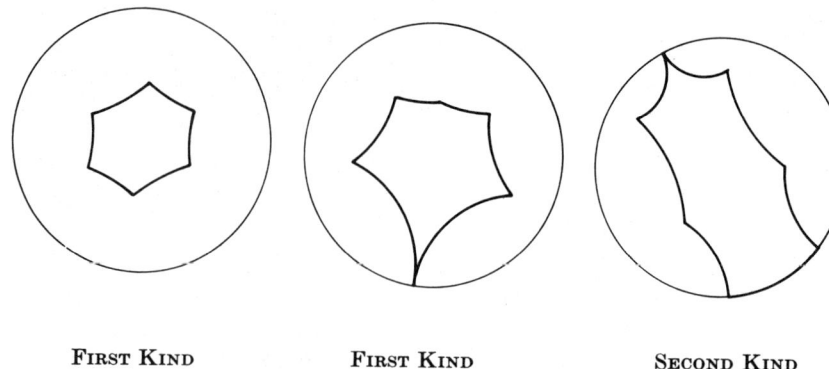

FIRST KIND   FIRST KIND   SECOND KIND

Figure 3.19. Fundamental domains for Fuchsian groups. (Reprinted from *Mathematical Surveys and Monographs*, "Discontinuous Groups and Automorphic Functions", Joseph Lehner, Volume 8, pg. 22, by permission of the American Mathematical Society.)

interior of its principal circle having a fundamental domain of finite invariant volume. Poincaré called a subgroup of $SL(2, \mathbb{C})$ *Kleinian* if it had no limit circle. Klein apparently did not like this use of his name. Figure 3.19 shows some possible fundamental domains for Fuchsian groups.

We have already noted that the sides of our fundamental domain for $SL(2, \mathbb{Z})$ are identified by the generators of $SL(2, \mathbb{Z})$. This is also true in the general case of a fundamental domain with a finite number of sides (see Maass [1, p. 26]).

Another important example of a discrete subgroup of $SL(2, \mathbb{R})$ is the *principal congruence subgroup* $\Gamma(N)$ of level N defined by

$$\Gamma(N) = \left\{ \begin{pmatrix} a & b \\ c & d \end{pmatrix} \middle| a, b, c, d \in \mathbb{Z}, a \equiv d \equiv 1 \pmod{N} \text{ and } b \equiv c \equiv 0 \pmod{N} \right\}.$$
(3.46)

A reference on generators of this group for $N \equiv$ prime is Frasch [1]. A subgroup Γ of $SL(2, \mathbb{R})$ is called a *congruence subgroup* if and only if Γ contains $\Gamma(N)$ for some positive integer N. It was discovered in the 1880s that there are an infinite number of examples of noncongruence subgroups (see Maass [1, pp. 76–78]). It turns out that $SL(n, \mathbb{Z})$ behaves quite differently for $n \geq 3$. For it has been proved by Mennicke as well as Bass, Lazard, and Serre that every subgroup of $SL(n, \mathbb{Z})$ of finite index is a congruence subgroup, if $n \geq 3$ (see Bass [1]).

Exercise 8. Show that a fundamental domain for a subgroup Γ_1 of $\Gamma = SL(2, \mathbb{Z})$ can be obtained by taking translates of a fundamental domain for Γ by representatives of the quotient Γ/Γ_1; i.e., if D is a fundamental domain for Γ, then a fundamental domain for Γ_1 is

3.3. Fundamental Domains for Discrete Subgroups Γ of $G = SL(2, \mathbb{R})$

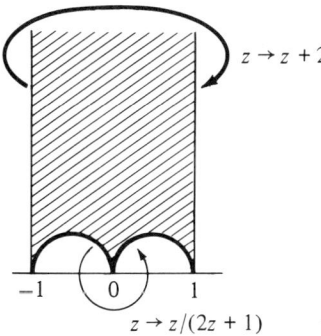

Figure 3.20. A fundamental domain for $\Gamma(2)$.

$$\bigcup_{\bar{\gamma} \in \Gamma/\Gamma_1} \gamma D \qquad \text{where } \bar{\gamma} = \gamma \Gamma_1.$$

Exercise 9. Show that if $\Gamma_1 = \Gamma(N)$, then it is a normal subgroup of $\Gamma = SL(2, \mathbb{Z})$ and the quotient Γ/Γ_1 is isomorphic to $SL(2, \mathbb{Z}/N\mathbb{Z})$. The only nontrivial part of this problem is the proof that for each solution $a, b, c, d \in \mathbb{Z}$ of $ad - bc \equiv 1 \pmod{N}$, there are integers $a_1 \equiv a, b_1 \equiv b, c_1 \equiv c, d_1 \equiv d \pmod{N}$ such that $ad - bc = 1$.

It is also possible to show that the index of $\Gamma(N)$ in $SL(2, \mathbb{Z})$ is

$$[SL(2, \mathbb{Z}) : \Gamma(N)] = N^3 \prod_{p | N}(1 - p^{-2}).$$

This result is proved in Shimura [1, pp. 21–22] and Schoeneberg [1, pp. 74–75]. A fundamental domain for $\Gamma(2)$ is pictured in Fig. 3.20. Since $[SL(2, \mathbb{Z}) : \Gamma(2)] = 6$, we can take our fundamental domain for $\Gamma(2)$ to be the union of six appropriate translates of a fundamental domain for $SL(2, \mathbb{Z})$. More information on fundamental domains for $\Gamma(2)$ can be found in Maass [1, pp. 71–73] and Schoeneberg [1, pp. 83–84].

Recall from the formulas following (3.45) that the theta group is the group generated by $z \mapsto z + 2$ and $z \mapsto -1/z$. A fundamental domain for the theta group is pictured in Fig. 3.21 (see Schoeneberg [1, pp. 84–86] and Maass [1, pp. 74–75]).

The final step in the study of fundamental domains for discrete subgroups of $SL(2, \mathbb{R})$ is to note that they can be made into *Riemann surfaces* with the upper half-plane as a (usually branched) covering surface.

First, we need to classify the fractional linear maps on the upper half-plane by classifying the corresponding elements of $SL(2, \mathbb{R})$ according to their Jordan form type. The Jordan form of such a matrix is one of the following:

$$\begin{pmatrix} a & 1 \\ 0 & a \end{pmatrix} \quad \text{or} \quad \begin{pmatrix} a & 0 \\ 0 & b \end{pmatrix}.$$

The corresponding fractional linear map will be one of four types:

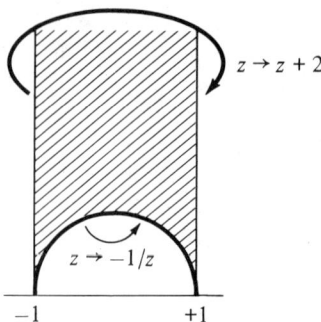

Figure 3.21. Fundamental domain for the theta group.

(1) $z \to z$;
(2) $z \to cz$, $\quad |c| = 1, c \neq 1$;
(3) $z \to cz$, $\quad c > 0, c \neq 1$;
(4) $z \to z + a$, $\quad a \neq 0$.

In case 1, the element of $SL(2, \mathbb{R})$ is $\pm I$, $I =$ the *identity*. In case 2, the element is called *elliptic*. In case 3, the element is *hyperbolic*. In case 4, the element is *parabolic*.

Exercise 10. Show that if $\sigma \in SL(2, \mathbb{R})$, $\sigma \neq \pm I$,

(a) σ is parabolic $\Leftrightarrow |\operatorname{Tr} \sigma| = 2$;
(b) σ is elliptic $\Leftrightarrow |\operatorname{Tr} \sigma| < 2$;
(c) is hyperbolic $\Leftrightarrow |\operatorname{Tr} \sigma| > 2$.

More details on the classification of elements of $SL(2, \mathbb{R})$ and $SL(2, \mathbb{C})$ can be found in Schoeneberg [1, pp. 2–4] and Shimura [1, pp. 5–10], for example.

A point $z \in H$ is called *elliptic* (with respect to Γ) if there is an elliptic element of Γ fixing z. A point s of $\mathbb{R} \cup \{\infty\}$ is called a *cusp* (with respect to Γ) if there is a parabolic element of Γ fixing s. These will be the interesting points on the Riemann surface of a fundamental domain for Γ. The elliptic points in the fundamental domain for $SL(2, \mathbb{Z})$ given in Fig. 3.11 are the points $i = \sqrt{-1}, \rho = (-1 + i\sqrt{3})/2$, and $-\bar{\rho}$. The only cusp of this fundamental domain for $SL(2, \mathbb{Z})$ is the cusp at infinity. For p in $H \cup \{\infty\}$, let S_p (the *stabilizer* of p) denote the group of maps $z \mapsto \gamma z$ which fix the point p, with $\gamma \in \Gamma$. For $\Gamma = SL(2, \mathbb{Z})$, $S_p =$ the identity, unless $p = \infty, i, \rho,$ or $\rho + 1$. In these four cases one has the following: S_∞ is infinite cyclic generated by $z \mapsto z + 1$; S_i is cyclic of order 2 generated by $z \mapsto -1/z$; S_ρ and $S_{\rho+1}$ are cyclic of order 3 generated by $-1/(z + 1)$ and $1 - 1/z$, respectively.

Exercise 11. Check the preceding statements about S_p for $\Gamma = SL(2, \mathbb{Z})$.

3.3. Fundamental Domains for Discrete Subgroups Γ of $G = SL(2, \mathbb{R})$

The cusps and elliptic points often make trouble in calculations. Thus some computations have been done for congruence subgroups $\Gamma(N)$ rather than for $SL(2, \mathbb{Z})$ itself, since these subgroups do not have elliptic points when $N > 1$ (see Exercise 12 and Fig. 3.20).

One can show that compactness of the fundamental domain $\Gamma \backslash H$ for Γ a discrete subgroup of $SL(2, \mathbb{R})$ implies that Γ has no parabolic elements. Examples of such groups having arithmetic interest come from quaternion algebras (see Gelfand, Graev, and Piatetskii-Shapiro [1, pp. 116–119], Marie-France Vignéras [4], Magnus [1], or Fricke and Klein [1, pp. 502–634]).

If Γ does have parabolic elements, one can compactify the fundamental domain by adding the cusps. The result can be made into a compact branched Riemann surface. Let us consider only the case that $\Gamma = SL(2, \mathbb{Z})$.

Exercise 12. Show that the congruence subgroups $\Gamma(N)$ of $SL(2, \mathbb{Z})$, with $N > 1$, have no elliptic elements.

Hint. See Shimura [1, p. 22].

The only problem in describing the *local coordinates for the Riemann surface* $D^* = D \cup \{\infty\}$, *where D is the fundamental domain of Fig. 3.11 for $SL(2, \mathbb{Z})$*, occurs at the elliptic points and cusps. At any point $p \in D^*$ such that p is not an elliptic point or cusp, define the local coordinate t_p by $t_p = (z - p)/(z - \bar{p})$. When $p = i$, define the coordinate to be $t_i = [(z - i)/(z + i)]^2$. There is a double branching here. When p is one of the other two elliptic points, define $t_p = [(z - p)/(z - \bar{p})]^3$. There is a triple branching here. Then finally when $p = \infty$, define $t = \exp(2\pi i z)$. There is an infinite branching here. If you think about the fundamental domain and its identifications and try to draw neighborhoods of each point you will believe these "weird" coordinates are necessary and perhaps not even "weird." More details can by found in Maass [1, pp. 30–37], Schoeneberg [1, p. 27], and Shimura [1, p. 18].

One can now compute the genus of the branched Riemann surface D^* formed from the fundamental domain in Fig. 3.11 for $SL(2, \mathbb{Z})$. The genus is the number of handles when you view the Riemann surface as a sphere with handles. Now the fundamental domain is a non-Euclidean triangle. Thus you can use Euler's formula for the genus to see that D^* *has genus zero*. It is also possible to use a formula of Hurwitz to compute the genus of other fundamental domains of interest (see Lehner [1, Ch. 6], Maass [1, pp. 30–32], Schoeneberg [1, pp. 93–103], and Shimura [1, pp. 18–23]).

Exercise 13 (The Leaky Torus) (from Gutzwiller [1]). Consider the fundamental domain D in Fig. 3.22 with its edges identified by

$$A = \begin{pmatrix} 1 & 1 \\ 1 & 2 \end{pmatrix} \quad \text{and} \quad B = \begin{pmatrix} 1 & -1 \\ -1 & 2 \end{pmatrix}.$$

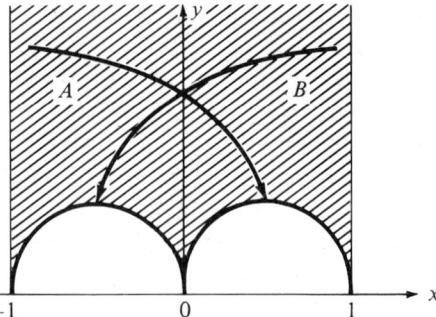

Figure 3.22. Fundamental domain for the group in Exercise 13—the leaky torus. (From Gutzwiller [1, pp. 1–3]. Reprinted by permission of Elsevier Science Publishers.)

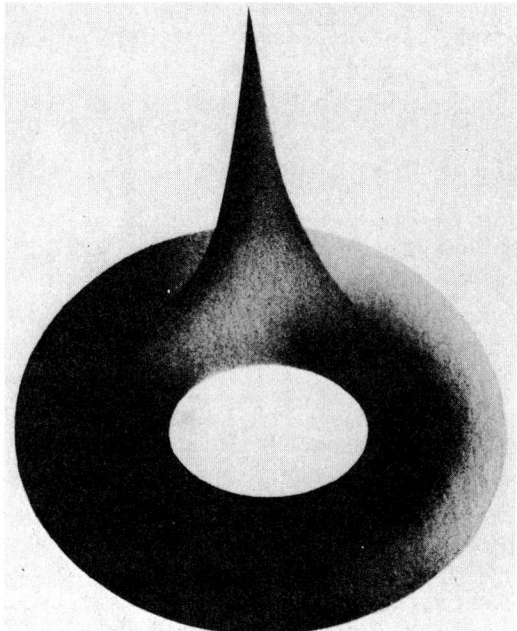

Figure 3.23. Another view of the leaky torus. (From Gutzwiller [1, pp. 1–3]. Reprinted with permission by Elsevier Science Publishers.)

(a) Show that the domain D, with sides identified as above, is equivalent to that given by Fig. 3.24, with boundary edges identified by A and B as above, plus

$$K = \begin{pmatrix} -1 & -6 \\ 0 & -1 \end{pmatrix} = B^{-1}A^{-1}BA \quad \text{and} \quad C = \begin{pmatrix} 3 & -1 \\ 1 & 0 \end{pmatrix}.$$

Does it bother you that the group Γ generated by A and B is not the congruence group $\Gamma(2)$ although the pictures of the fundamental domains

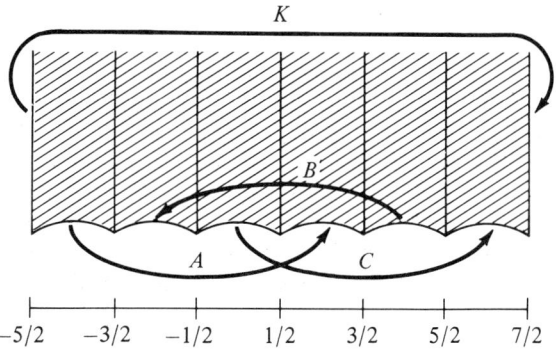

Figure 3.24. Another fundamental domain for the leaky torus.

in Figs. 3.20 and 3.22 look the same? Can you explain how different discrete groups can have the same fundamental domain except for the boundary identifications?

(b) Show that the genus of the domain D in Fig. 3.22 is 1, while that in Fig. 3.20 is 0. Thus the fundamental domain for Γ generated by A and B above is topologically a torus with one cusp. This is pictured in Fig. 3.23. On the other hand, the fundamental domain for $\Gamma(2)$ is topologically a sphere with two cusps.

(c) Does the group Γ generated by A and B above contain some congruence group $\Gamma(N)$? This would make Γ a congruence group of level N.

Hint. See Schoeneberg [1] for hints on part (c).

Gutzwiller [1, pp. 345–346] notes: "The leaky torus is topologically different from the ordinary torus which physicists like to use in Euclidean space. The four corners of the domain D become one point as usual, but this point is now infinitely removed. A path which goes around this exceptional point cannot be contracted to zero, because that would require moving it over an infinite distance."

3.4. Automorphic Forms—Classical

The abundance of topics in this subject is so large that I have investigated only a few problems. I think that these facts could be put in order only by using notions from different arithmetical theories which are perhaps not sufficiently developed.

From E. Hecke [2].

There are many books on classical automorphic forms. To the list at the beginning of §3.3, you can add Ogg [1, 2], for example. Thus we do not feel compelled to write a book on this subject and we shall only present *some* of the classical theory in this section. Moreover, we will usually consider only the discrete group $SL(2,\mathbb{Z})$. For a treatment from the point of view of adelic representation theory, see Gelbart [1] or Jacquet-Langlands [1].

A function $f: H \to \mathbb{C}$ is said to be an (*entire*) *modular or automorphic form of weight k for the discrete group* $\Gamma = SL(2,\mathbb{Z})$ if

(1) f is entire on H; i.e., $f(z)$ is holomorphic for every $z \in H$;
(2) $f(gz) = (cz + d)^k f(z)$, for any $g = \left(\begin{smallmatrix} a & b \\ c & d \end{smallmatrix}\right)$ in Γ;
(3) f is holomorphic at infinity; i.e., f has the Fourier series expansion

$$f(z) = \sum_{n \geq 0} a_n \exp(2\pi i n z).$$

In order to check (2), it suffices to check that $f(z+1) - f(z)$ and $f(-1/z) = z^k f(z)$. Property (3) is equivalent to the requirement that f be bounded in the fundamental domain for $SL(2,\mathbb{Z})$.

The automorphic forms of weight k for $SL(2,\mathbb{Z})$ constitute a vector space which we will denote by $\mathcal{M}(SL(2,\mathbb{Z}),k)$. *In order that this space be nonzero, the weight k must be a nonnegative even integer.* To see that k must be even, replace g by $-g$ in property (2). To see that k must be nonnegative, note that the function $y^{k/2}|f(z)|$ is invariant under $SL(2,\mathbb{Z})$. If $k < 0$, $y^{k/2}$ and thus $y^{k/2}|f(z)|$ is bounded on the fundamental domain and thus on all of the upper half-plane. Thus $|f(z)| \leq My^{-k/2}$ for some constant M. But then the coefficients in the Fourier series for f satisfy

$$a_n \exp(-2\pi n y) = \int_0^1 f(x+iy)\exp(-2\pi i n x)\,dx.$$

Thus $|a_n| \leq My^{-k/2}\exp(2\pi n y)$. If k is negative, it follows that $a_n = 0$ for all n, since you can let y approach zero (from above).

In order to obtain nonconstant modular forms of weight $k = 0$, one must relax the definition and allow the function f to have poles. Automorphic forms of weight zero are called *automorphic functions*. For groups Γ with more than one cusp, one must consider Fourier expansions at each cusp in part (3) of the definition of automorphic form. And in order to call the theta function in formula (3.45) of §3.3 an automorphic form, one must introduce the concept of *multiplier system* in part (2) of the definition of automorphic form. A multiplier system $v: \Gamma \to \mathbb{C}$, corresponding to the weight k, has the properties $|v(\gamma)| = 1$ for all $\gamma \in \Gamma$, and

$$v(\gamma_1\gamma_2)(c_3 z + d_3)^k = v(\gamma_1)v(\gamma_2)(c_1\gamma_2 z + d_1)^k(c_2 z + d_2)^k \tag{3.47}$$

for

$$\gamma_i = \begin{pmatrix} a_i & b_i \\ c_i & d_i \end{pmatrix}, \qquad i = 1, 2, 3 \quad \text{and} \quad \gamma_3 = \gamma_1\gamma_2.$$

If k is an integer, then equation (3.47) reduces to $v(\gamma_1\gamma_2) = v(\gamma_1)v(\gamma_2)$. We say

3.4. Automorphic Forms—Classical

that $f: H \to \mathbb{C}$ is an automorphic form of weight k and multiplier system v for Γ if f satisfies the conditions (1) and (3) of the preceding definition (perhaps modified somewhat as we mentioned above), and condition (2) is replaced by

(2')
$$f(\gamma z) = v(\gamma)(cz + d)^k f(z) \quad \text{for} \quad \gamma = \begin{pmatrix} * & * \\ c & d \end{pmatrix} \in \Gamma.$$

Here, when $k \notin \mathbb{Z}$, one must fix the branch of $(cz + d)^k$ by assuming that

$$z^k = |z|^k e^{ik \arg z}, \quad -\pi \leq \arg z < \pi, \quad \text{for } z \in \mathbb{C}.$$

There is an overabundance of terminology with respect to the word "weight." Some replace k by $-k$ (cf. Lehner [1, p. 268]). Some use the word "degree" (cf. Knopp [1, p. 12]). See also Serre [1] for slightly divergent notation, as well as a nice short survey of the subject of this section—a survey which we follow somewhat here.

As an example of an automorphic form of weight $k = 4, 6, \ldots$, consider the Eisenstein series G_k defined by

$$G_k(z) = \sum_{(m,n) \in \mathbb{Z}^2 - 0} (mz + n)^{-k}, \quad k = 4, 6, 8, \ldots. \tag{3.48}$$

Exercise 1.

(a) Set

$$P = \left\{ \begin{pmatrix} a & b \\ 0 & d \end{pmatrix} \in SL(2, \mathbb{Z}) \right\}.$$

Show that

$$G_k(z) = 2\zeta(k) \sum_{g \in P \backslash SL(2,\mathbb{Z})} \left(\frac{d(gz)}{dz} \right)^{k/2},$$

where $\zeta(k)$ is Riemann's zeta function. The sum is over a complete set of representatives g in $SL(2, \mathbb{Z})$ for the equivalence relation

$$g_1 \sim g_2 \quad \text{if} \quad g_1 = pg_2, \quad \text{for some } p \in P.$$

(b) Show that $G_k(z)$ is a nonzero modular form for $SL(2, \mathbb{Z})$ of weight k, if $k = 4, 6, 8, \ldots$.

Hints. For the convergence of (3.48), compare the series with the Epstein zeta function in Exercise 5 of §1.4. To see that G_k has property (2) of modular forms, use part (a) and the chain rule. To see that $G_k \neq 0$ when $k = 4, 6, 8, \ldots$, show that

$$\lim_{y \to \infty} G_k(iy) = 2\zeta(k), \quad k = 4, 6, 8, \ldots.$$

Exercise 2. Obtain the complete Fourier expansion of G_k, $k = 4, 6, 8, 10, \ldots$.

Answer.

$$G_k(z) = 2\zeta(k) + 2\frac{(2\pi i)^k}{(k-1)!}\sum_{n\geq 1}\sigma_{k-1}(n)\exp(2\pi i n z),$$

where

$$\sigma_k(n) = \sum_{o<d|n} d^k \qquad \text{(the sum of the } k\text{th powers of divisors of } n).$$

Hint. One method is to use the Poisson sum formula (see Maass [1, pp. 209–212] for a generalization). Another method starts with the partial fraction expansion of $\pi\cot(\pi z)$ (see Shimura [1, pp. 32–33]).

Exercise 3. Define $\Delta(z) = (60G_4)^3 - 27(140G_6)^2$. Show that Δ is a modular form of weight 12. Then prove that

$$\Delta(\infty) = 0.$$

A modular form that vanishes at infinity is called a *cusp form* (forme parabolique or Spitzenform). We define the space of cusp forms $\mathcal{S}(SL(2,\mathbb{Z}),k)$ to be the automorphic forms for $SL(2,\mathbb{Z})$ of weight k which vanish at infinity. From Exercise 3, we see that $\Delta\in\mathcal{S}(SL(2,\mathbb{Z}),12)$. However, it is possible that Δ is identically zero. We need some information on the order of zeros of automorphic forms to preclude this possibility (see the proof of Theorem 1 below). In fact, we will learn that $\Delta(z)$ does not vanish at all on H. This means that you can divide by Δ and that is a useful fact for class field theory (see Borel and Chowla [1]).

The facts about zeros of automorphic forms can be derived in several ways. First, if p is a point in the usual fundamental domain D for $SL(2,\mathbb{Z})$ as in Fig. 3.11, then let $n = n_p(f)$ denote the *order of the zero* of f at p; i.e., the integer n such that $f(z)(z-p)^{-n}$ is holomorphic and nonzero at p. If f is in $\mathcal{S}(SL(2,\mathbb{Z}),k)$, then $n_p(f) = n_{\gamma p}(f)$ for all γ in $SL(2,\mathbb{Z})$. The *order of the zero of f at infinity* is denoted $n = n_\infty(f)$ and is obtained by looking at the Fourier expansion; i.e.,

$$f(z) = \sum_{k\geq n} a_k \exp(2\pi i k z), \qquad a_n \neq 0.$$

Let e_p denote the order of the group of maps $z\mapsto\gamma z$ fixing $p\neq\infty$. Then $e_p = 1$ when $p\neq i$ or ρ; $e_i = 2$ and $e_\rho = 3$. *The fundamental formula for the orders of zeros of automorphic forms of weight k for $SL(2,\mathbb{Z})$ is*

$$n_\infty(f) + \sum_{p\in D} n_p(f)/e_p = k/12, \tag{3.49}$$

where D is the usual fundamental domain for $SL(2,\mathbb{Z})$ (see Fig. 3.11).

The easiest proof of this formula comes from the Riemann-Roch theorem (see Gunning [1, Ch. II] and Shimura [1]). This method also generalizes to

3.4. Automorphic Forms—Classical

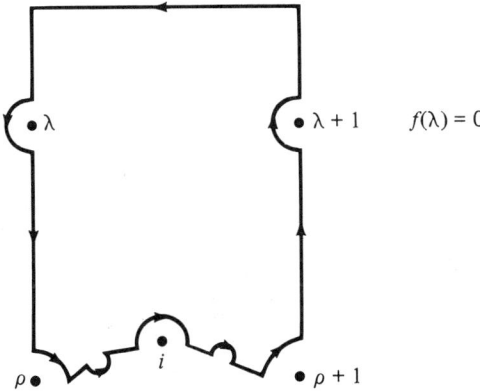

Figure 3.25. The contour C needed for the proof of formula (3.49).

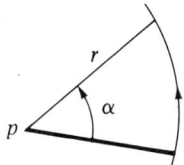

Figure 3.26. The arc A near a zero of $f(z)$.

higher dimensions (see Yamazaki [1]). However, in order to keep this section as elementary as possible, we will present a proof by contour integration.

The proof will be carried out by integrating $d \log f(z)$ over the contour C in Fig. 3.25, taking the contour to enclose all zeros of f in the fundamental domain minus the points $i, \rho, \rho + 1, \infty$. Note that f has only a finite number of zeros in the fundamental domain because f is holomorphic at infinity, so zeros cannot accumulate there.

To do the contour integral over C, first note that if $f(z) = (z - p)^m g(z)$ with $g(p) \neq 0$, then, when you integrate $d \log f(z)$ over the arc A in Fig. 3.26 and let the radius r go to zero, you get $m\alpha i$. There is cancellation of the curved parts of the contour integral of $d \log f(z)$ around the zeros of f on the vertical boundaries. And the curved part of the contour integral of $d \log f(z)$ around i contributes $-n_i \pi i$. Finally, the curved part of the contour integral around ρ contributes $-n_\rho \pi i/3$, with the same contribution for the curved part of the contour integral around $\rho + 1$.

The limit of the part of the contour integral of $d \log f(z)$ along the horizontal contour at the top of C, as the horizontal line moves up to infinity, is $-2\pi i n_\infty$, using the expansion of $f(z)$ at infinity.

The contour integral over the bottom of the contour C is obtained from the fact that f is a modular form of weight k, as follows:

$$\int_{\text{bottom of contour}} d\log f(z) = \int_i^\rho d\log f(z) - d\log f(-1/z)$$

$$= \int_i^\rho (-k)\frac{dz}{z} = k\left(\frac{\pi}{2} - \frac{\pi}{3}\right) = \frac{k}{6}\pi i.$$

Now, since the integrals over the vertical parts of the contour C cancel, one obtains the following formula for the number n of zeros of $f(z)$ in the fundamental domain minus the points $i, \rho, \rho + 1, \infty$:

$$n = (2\pi i)^{-1}\int_C d\log f(z) = -\frac{n_i}{2} - \frac{n_\rho}{6} - \frac{n_\rho}{6} - n_\infty + \frac{k}{12}.$$

Here the zeros are counted with multiplicity. This completes the proof of formula (3.49).

Formula (3.49) leads to the following theorem.

Theorem 1 (Formula for the Dimension of the Spaces of Automorphic Forms for $SL(2,\mathbb{Z})$). *If $[x]$ denotes the greatest integer less than or equal to x, we have the following formula for the dimension of the space of modular forms of weight k for $SL(2,\mathbb{Z})$:*

$$d_k = \dim_{\mathbb{C}} \mathcal{M}(SL(2,\mathbb{Z}),k) = \begin{cases} 0, & \text{if } k < 0 \text{ or } k \notin 2\mathbb{Z}, \\ 1, & \text{if } k = 0, \\ [k/12] + 1, & \text{for } k \not\equiv 2 \pmod{12}, k \in 2\mathbb{Z}, \\ [k/12], & \text{for } k \equiv 2 \pmod{12}, k \in 2\mathbb{Z}. \end{cases}$$

PROOF. We have already proved the first line of the dimension formula (see the discussion immediately following the definition of $\mathcal{M}(SL(2,\mathbb{Z}),k)$).

Using formula (3.49), we find that $d_2 = 0$, since

$$\frac{1}{6} = n + n_\infty + \frac{n_i}{2} + \frac{n_\rho}{3}$$

is not solvable in integers n, n_∞, n_i, n_ρ. It also follows that $d_k \leq 1 + [k/12]$. For suppose that $m > 1 + [k/12]$ and $f_1, \ldots, f_m \in \mathcal{M}(SL(2,\mathbb{Z}),k)$. Then there are constants $c_1, \ldots, c_m$, not all of which are zero, such that

$$\sum_{i=1}^m c_i f_i^{(j)}(\infty) = 0 \qquad \text{for } j = 0, 1, \ldots, m - 2.$$

Thus $f = \sum_{i=1}^m c_i f_i$ is an element of $\mathcal{M}(SL(2,\mathbb{Z}),k)$ with a zero of order $\geq m - 1$ at ∞. But then the formula for the orders of zeros of elements of $\mathcal{M}(SL(2,\mathbb{Z}),k)$ forces a contradiction.

We know from Exercises 1 and 2 that the Eisenstein series $G_k, k = 4, 6, 8, \ldots$, are nonzero elements of $\mathcal{M}(SL(2,\mathbb{Z}),k)$. This implies that $d_k = 1$ for $k = 4, 6, 8, 10$. Then Exercise 3 implies that $d_{12} = 2$, using formula (3.49) again. For we find from formula (3.49) that $n_\infty(\Delta) = 1$ and $n = n_i = n_\rho(\Delta) = 0$. Thus Δ does

3.4. Automorphic Forms—Classical

not vanish at any finite point of H. Note here that Δ is not identically 0, since $G_4(\rho) = 0 = G_6(i)$.

To finish the proof of the dimension formula, note that every form in $\mathcal{M}(SL(2, \mathbb{Z}), k)$ is a unique linear combination of forms $G_{k-12n}\Delta^n$, for $n \geq 0$, $k - 12n \geq 4$. To prove this, use the fact that there is a constant a_0 with $(f - a_0 G_k)(\infty) = 0$. Since Δ is only zero at ∞, you obtain $(f - a_0 G_k)/\Delta \in \mathcal{M}(SL(2, \mathbb{Z}), k - 12)$. The proof is completed by induction. □

It follows from the preceding argument that every modular form of weight k is a unique linear combination of terms of the form $G_4^a G_6^b$, with nonnegative integers a, b such that $4a + 6b = k$. The proof is in Serre [1, p. 89].

More general results on dimensions of spaces of automorphic forms with multiplier systems for $SL(2, \mathbb{Z})$ can be found in Maass [1, p. 144]. Both Eisenstein series and Poincaré series (to be considered later in this section) are used to generate the spaces of such forms. The same result can also be found in Rankin [1, p. 208]. Both of these references show that there are exactly six different multiplier systems for $SL(2, \mathbb{Z})$ for each weight k (see Maass [1, p. 132] and Rankin [1, p. 206]). Rankin obtains the basis elements of the form $G_{r+12s}\Delta^{-s+(k-r)/12}$, $0 \leq s \leq (k-r)/12$, with $r = 0, 4, 6, 8, 10, 14$ if $k \geq r$. Again, it is important that Δ does not vanish on the finite upper half-plane. If $k < r$, above, then there are no forms but 0. If $k \not\equiv r \pmod{12}$, the forms above are all cusp forms. A dimension formula for automorphic forms for more general discrete groups is to be found in Shimura [1, pp. 46–47]. The computation is a nice application of the Riemann-Roch theorem for curves.

Next we want to consider some examples of modular forms—the discriminant Δ, the modular invariant J, theta functions, etc.

The Discriminant

The *discriminant* $\Delta(z)$ is defined by the formula in Exercise 3. It will be convenient to use the notation

$$g_2 = 60 G_4 \quad \text{and} \quad g_3 = 140 G_6. \tag{3.50}$$

The function is called the discriminant because it is the discriminant of the cubic on the right-hand side of the equation

$$u'^2 = 4u^3 - g_2 u - g_3, \tag{3.51}$$

satisfied by the *Weierstrass elliptic function* $u = \wp(t)$ (see Ahlfors [1, pp. 264–268]). Note that the differential equation for $\wp(z)$ shows that it is the inverse function for the *elliptic integral* in Weierstrass normal form, i.e.,

$$z - z_0 = \int_{\wp(z_0)}^{\wp(z)} (4w^3 - g_2 w - g_3)^{-1/2} \, dw.$$

(see Siegel [1, Vol. I] and Lang [5]). The fact that the discriminant does not vanish on the upper half-plane shows that the cubic (5) has three distinct roots e_1, e_2, e_3 such that

$$\wp'(z)^2 = 4(\wp(z) - e_1)(\wp(z) - e_2)(\wp(z) - e_3).$$

Then one can construct the function λ called the *modulus* by

$$\lambda(z) = (e_3 - e_2)/(e_1 - e_2), \tag{3.52}$$

for a specific normalization of the roots e_j. It turns out that this function is an automorphic function for the congruence subgroup $\Gamma(2)$ (see Ahlfors [1, pp. 269–274]), and gives a useful conformal mapping. We will consider λ again later in this section.

The *Ramanujan τ-function* is defined on the positive integers as follows:

$$(2\pi)^{-12}\Delta(z) = \sum_{n\geq 1} \tau(n)\exp(2\pi i n z). \tag{3.53}$$

It is possible to use the Fourier expansion of the Eisenstein series given in Exercise 2 to see that the Ramanujan numbers $\tau(n)$ are all integers. Table 3.3 gives the first 11 of them.

Table 3.3. Ramanujan's Tau Function

n	1	2	3	4	5	6	7	8	9	10	11
$\tau(n)$	1	-24	252	-1472	4830	-6048	-16744	84480	-113643	-115920	534612

The Ramanujan conjecture states that

$$|\tau(p)| < 2p^{11/2} \qquad \text{for all primes } p. \tag{3.54}$$

This implies, using the multiplicative properties of the τ-function, that

$$|\tau(n)| \leq n^{11/2}\sigma_0(n) \qquad \text{where} \quad \sigma_0(n) = \sum_{o<d|n} 1, \qquad n = 1, 2, 3, \ldots.$$

The multiplicative properties of the τ-function will be discussed in §6, using the theory of Hecke operators. Recently Deligne proved the Ramanujan conjecture and its generalization to forms of weight k as a consequence of the Weil conjectures in algebraic geometry (see Deligne [1, 2] and the expository article of Katz [1]). The generalization is called the Ramanujan-Petersson conjecture (see Theorem 4 and the discussion following it). There are also analogues for nonholomorphic Maass waveforms which have not been proved yet (see the discussion of the tables in §3.5 as well as that following Theorem 3 of §3.5).

Another question concerning $\tau(n)$ is still open: Can $\tau(n)$ vanish? Long computer tables exist showing that $\tau(p) \neq 0$ for $p \leq 10^{15}$, but no one knows how to answer the question about the possible vanishing of $\tau(n)$ (see Lehmer [1–3]).

Write $\tau(p) = 2p^{11/2}\cos\varphi_p$. Inspired by the Sato-Tate conjecture for elliptic

3.4. Automorphic Forms—Classical

curves, Serre conjectured that the angles φ_p are distributed in $[0, \pi]$ with respect to the measure

$$\frac{2}{\pi} \sin^2 \varphi \, d\varphi.$$

The conjecture is open. Lehmer [4] did a numerical study. Serre [5] and Ogg [3] showed that the Sato-Tate conjecture follows from nonvanishing properties of the analytic continuation of L-functions associated to the mth symmetric power representation:

$$L_m(s) = \prod_{\substack{p \\ \text{prime}}} \prod_{i=0}^{m} (1 - \alpha_p^i \bar{\alpha}_p^{m-i} p^{-s})^{-1} \quad \text{if} \begin{cases} \tau(p) = \alpha_p + \bar{\alpha}_p, \; \alpha_p \bar{\alpha}_p = p^{11}, \\ \alpha_p = p^{11/2} \exp(i\varphi_p). \end{cases}$$

So far the necessary results about $L_m(s)$ are only known for $m = 1$ (Hecke), $m = 2$ (Rankin and Shimura), and $m = 3, 4$ (Jacquet and Shahidi). Additional references for this subject are Serre [4], M. Ram Murty [1, 2], Katz [2, p. 14], Leveque reviews [1, Vol. II, p. 578; pp. 626–627], and Moreno and Shahidi [1]. Note that Lehmer's problem mentioned in the preceding paragraph is the question, Can $\varphi_p = \pi/2$?

Another useful fact about $\Delta(z)$ is *Jacobi's identity*:

$$\Delta(z) = (2\pi)^{12} q \prod_{n \geq 1} (1 - q^n)^{24} \quad \text{for } q = \exp(2\pi i z). \tag{3.55}$$

Proofs of (3.55) can be found in Maass [1, p. 108], Ogg [1, pp. I-43–44], Rankin [1, pp. 196–197], Serre [1, pp. 95–96], and Weil [3, Ch. 4], for example.

The *eta function of Dedekind* is defined by

$$\eta(z) = q^{1/24} \prod_{n \geq 1} (1 - q^n), \quad q = \exp(2\pi i z). \tag{3.56}$$

Like the theta function in formula (3.45), the eta function is a modular form of weight $\frac{1}{2}$; but eta is a form for the whole modular group $SL(2, \mathbb{Z})$. In particular, eta satisfies

$$\eta(z + 1) = \eta(z) \quad \text{and} \quad \eta(-1/z) = (z/i)^{1/2} \eta(z). \tag{3.57}$$

Note that, like theta, the eta function is an automorphic form with a multiplier system. If you can prove the last transformation formula in (3.57) for eta, then you can prove the product formula for the discriminant (3.55) since there is only one cusp form of weight 12. The transformation formula of eta is proved in many of the basic references on modular forms as well as in de la Torre and Goldstein [1] and Goldstein [2]. The complete multiplier system for eta involves Dedekind sums.

One of the main applications of the eta function to number theory comes from the fact that

$$q^{1/24} \eta(z)^{-1} = \sum_{n \geq 0} p(n) q^n,$$

where $p(n)$ is the number of *partitions* of n; i.e., the number of ways of writing

$$n = n_1 + \cdots + n_r, \qquad n_j \in \mathbb{Z}, \; n_j > 0, \; j = 1, 2, \ldots, r.$$

Rademacher used this fact to obtain an exact formula for $p(n)$ as an infinite series with terms involving certain finite trigonometric sums $A_k(n)$ and I-Bessel functions. The method is a further development of a technique of Hardy and Ramanujan from 1917, which begins with the Cauchy formula

$$p(n) = \frac{1}{2\pi i} \int_C x^{-n-1} f(x) \, dx,$$

with C a closed curve inside the unit circle and containing the origin. Here $f(q) = q^{1/24} \eta(z)^{-1}$. The path C is then taken as a circle around 0 and close to the unit circle. It is then cut into parts corresponding to neighborhoods of just one root of unity. For a discussion of the rather intricate arguments that ensue, see Rademacher [1, Ch. 14] or Lehner [1, p. 351]. The method is called the "circle method." For some history of other work on formulas for $p(n)$, see D.H. Lehmer's review of a paper of Whiteman in Leveque [1, Vol. IV, pp. 510–511]. Selberg (age 19 or so) obtained a formula for $A_k(n)$ as a finite Fourier series at around the same time that Rademacher had obtained his result. Lehmer had factored $A_k(n)$ to make computations.

The eta function and its generalizations have also been used to obtain *class number formulas* (see Goldstein and Razar [1]).

The Modular Invariant

Our next example of a modular form is *the modular invariant $J(z)$* defined by

$$J = (60G_4)^3 / \Delta. \tag{3.58}$$

This function was first constructed by Dedekind in 1877 and Klein in 1878. Here, once again, the definition seems mysterious, but we shall see that J is of fundamental importance, as is $j = 1728J$. The number 1728 gives j integer coefficients in its Fourier expansion. On the other hand, J has nice values at the elliptic and parabolic points in a fundamental domain for $SL(2, \mathbb{Z})$.

Theorem 2 (The Main Properties of the Modular Invariant).

(1) $J(z)$ is a modular function; i.e., an automorphic form of weight 0, which may have poles. J is holomorphic in H with a simple pole at ∞, $J(i) = 1$ and $J(\rho) = 0$, where $i = \sqrt{-1}$ and $\rho = (-1 + \sqrt{-3})/2$.

(2) J defines a conformal mapping which is one-to-one from D onto $\mathbb{C}$, where D is the fundamental domain for $SL(2, \mathbb{Z})$ in Fig. 3.11. And thus J provides an identification of $H/SL(2, \mathbb{Z}) \cup \{\infty\}$ with the Riemann sphere $\mathbb{C} \cup \{\infty\}$.

(3) The following are equivalent for a function f which is meromorphic on H:
   (a) f is a modular function;
   (b) f is a quotient of two modular forms of the same weight;
   (c) f is a rational function of J (i.e., a quotient of polynomials in J) and thus is caled the **Hauptmodul** or *fundamental function*.

PROOF.

(1) This property is clear from the properties of the discriminant and the Eisenstein series. The last two statements are proved by noting that $G_4(\rho) = 0$ and $G_6(i) = 0$, from formula (3.49). Thus $\Delta(i) = [60G_4(i)]^3$.

(2) To see this property, use formula (3.49) to show that $f_s = G_4^3 - s\Delta$ can only be zero at one point.

(3) To prove this property, first note that clearly (c) ⇒ (b) ⇒ (a). In order to prove (a) ⇒ (c), suppose that $f(z)$ is a modular function. Then multiply f by a polynomial in $J(z)$ to make f holomorphic on H. Then $g = \Delta^n f$ is holomorphic at infinity for some $n \geq 0$, to put g in $\mathcal{M}(SL(n, \mathbb{Z}), 12n)$. Thus g is a linear combination of $G_4^a G_6^b$, with $4a + 6b = 12n$. It suffices to show (c) for $f = G_4^a G_6^b/\Delta^n$. Now $p = a/3$ and $q = b/2$ are integers and $f = G_4^{3p} G_6^{2q}/\Delta^{p+q} = (G_4^3/\Delta)^p (G_6^2/\Delta)^q$. Finally, it is easy to see that G_4^3/Δ and G_6^2/Δ are rational functions of J, completing the proof. □

Exercise 4. Fill in the details in the proof of Theorem 2.

One can use the automorphic function J to prove the *small Picard theorem* very quickly. The inverse function $w = J^{-1}(z)$ is infinite valued with branch points at 0, 1, ∞ (the values of J at the vertices of the fundamental domain). Suppose that f is an entire function omitting the values 0, 1, ∞. Then $J^{-1} \circ f = g$ is holomorphic in the z-plane and single valued. Since g lands in the upper half-plane, $\exp(ig)$ is bounded and thus constant. It follows that f must be constant. Still another, short proof of the little Picard theorem can be found in Kobayashi [1]. Moreover, Kobayashi's methods generalize to prove the *big* Picard theorem in higher-dimensional cases.

Another application of the J-function is in number theory. For the maximal unramified abelian extension of an imaginary quadratic number field is produced by a value of j, also called *the modular invariant*. An explicit reciprocity law shows how the Galois group of the extension is isomorphic to the ideal class group of the quadratic field. This is *explicit class field theory* (see our discussion of number fields in §1.4). In the ramified case, other functions than j are needed. A reference for this is Borel, Chowla, et al. [1].

It turns out that the Fourier or q-expansion of j involves only integers:

$$j(z) = q^{-1} + 744 + 196884q + 21493760q^2 + \cdots, \tag{3.59}$$

where $q = \exp(2\pi iz)$. This is useful in the applications to number theory. There is also a surprising connection of the coefficients in (3.59) with the representations of the largest of the 26 sporadic finite simple groups which is known as the *Fischer-Griess monster group M*. All of the early Fourier coefficients in (3.59) are simple linear combinations of degrees of characters of M. This was first noticed by John McKay and John Thompson. No one has explained the phenomenon yet, but some interesting analogues for Lie groups are explained by V. Kac [1]. See also Conway [1] and Conway and Norton [1]. Conway reports on the situation as follows: "Because these new links are still almost completely unexplained, we refer to them collectively as the *'moonshine'* properties of the MONSTER, intending the word to convey our feelings that they are seen in a dim light, and that the whole subject is rather vaguely illicit!"

The modular invariant j is important to algebraic geometers because it characterizes elliptic curves (see Shimura [1, p. 97]). An elliptic curve is a projective nonsingular variety with a structure of algebraic group of dimension 1. Such a group must be commutative. This is, after all, the reason that the modular group is so called; the moduli variety classifying elliptic curves is $SL(2, \mathbb{Z}) \backslash H$.

Theta Functions

The last example of a modular form that we want to consider in some detail is the *theta function*. It is an important example for us (like G_k and unlike J and Δ) because generalizations to higher dimensions are known. The *theta function associated to a positive definite symmetric $n \times n$ real matrix P and a point z in the upper half-plane* is

$$\theta(P, z) = \sum_{a \in \mathbb{Z}^n} \exp(\pi i P[a] z). \tag{3.60}$$

Here $P[a]$ denotes the quadratic form associated to P; i.e., $P[a] = {}^t a P a$, thinking of a as a column vector in $\mathbb{Z}^n$, as usual. Note that if $z = it$, for $t > 0$, then $\theta(P, z) = \theta(P, t)$ as defined in Exercise 6 of §1.4. Note also that if $n = 1$, $P = I$, $\theta(I, z) = \theta(z)$, the function defined by formula (3.45).

Exercise 5. Prove the transformation formula

$$\theta(P, z) = |P|^{-1/2}(z/i)^{-n/2} \theta(P^{-1}, -1/z).$$

Hint. Recall Exercise 6 of §1.4.

In order for the theta function defined by (3.60) to be a modular form as a function of z (with trivial multiplier system for the group $SL(2, \mathbb{Z})$), we need to place two restrictions on P.

Theorem 3. *If P is an even integral positive symmetric $n \times n$ matrix of determinant 1, then 8 divides n and $\theta(P, z) \in \mathcal{M}(SL(2, \mathbb{Z}), n/2)$. Here we say P is even integral if $P[a] \in 2\mathbb{Z}$ for all a in $\mathbb{Z}^n$. And $\theta(P, z)$ is defined by formula (3.60).*

PROOF. First we need an exercise.

Exercise 6. Show that under the hypotheses on P in Theorem 3, the size n of the matrix P must be divisible by 8.

Hint. (See Serre [1, pp. 53 and 109].)
    You can assume that $n \equiv 4 \pmod 8$ by replacing P by

3.4. Automorphic Forms—Classical

$$\begin{pmatrix} P & 0 \\ 0 & P \end{pmatrix} \quad \text{or} \quad \begin{pmatrix} P & & 0 \\ & P & \\ & & P \\ 0 & & P \end{pmatrix}.$$

Then let $V = ST$, with $Sz = -1/z$ and $Tz = z + 1$. Compute $\theta(P, V(iy))$ for $y > 0$, as well as $\theta(P, V^3(iy))$. Obtain a contradiction from $V^3 = I$.

From Exercises 5 and 6, it follows that $\theta(P, z + 1) = \theta(P, z)$ and $\theta(P, z) = z^{-n/2}\theta(P, -1/z)$, if P satisfies the hypotheses of Theorem 3. For if $|P| = 1$ and P is integral, then, upon setting

$$S_p = \{P[a] | a \in \mathbb{Z}^n\},$$

we see that $S_p = S_{p^{-1}}$. This is derived from the fact that we can write

$$P^{-1}[b] = P[a] \quad \text{for } b = Pa \in \mathbb{Z}^n. \qquad \square$$

Exercise 7. Show that the following 8×8 matrix, which comes out of the theory of the exceptional Lie group E_8, gives an example of a matrix satisfying the hypotheses of Theorem 3. It can, in fact, be proved that this is the only possible example, up to transformation by $GL(8, \mathbb{Z}) = $ the group of 8×8 integral matrices of determinant $+1$ or -1 (see the definition of equivalence below).

$$P_8 = \begin{pmatrix} 2 & 0 & -1 & 0 & 0 & 0 & 0 & 0 \\ 0 & 2 & 0 & -1 & 0 & 0 & 0 & 0 \\ -1 & 0 & 2 & -1 & 0 & 0 & 0 & 0 \\ 0 & -1 & -1 & 2 & -1 & 0 & 0 & 0 \\ 0 & 0 & 0 & -1 & 2 & -1 & 0 & 0 \\ 0 & 0 & 0 & 0 & -1 & 2 & -1 & 0 \\ 0 & 0 & 0 & 0 & 0 & -1 & 2 & -1 \\ 0 & 0 & 0 & 0 & 0 & 0 & -1 & 2 \end{pmatrix}.$$

For $n = 16$, there are two nonequivalent examples modulo $GL(16, \mathbb{Z})$. For $n = 24$, there are 24 nonequivalent examples modulo $GL(24, \mathbb{Z})$ (see Serre [1, Chs. 5 and 7]). We call two positive definite real $n \times n$ matrices P and Q *equivalent* if there is a matrix A in $GL(n, \mathbb{Z}) = $ the integral $n \times n$ matrices of determinant $+1$ or -1, such that $P = Q[A] = {}^tAQA$. In fact, this equivalence relation will be quite important in Chapter 4, as an analogue of the equivalence relation defining the quotient space $SL(2, \mathbb{Z})\backslash H$.

If one considers relaxing the hypotheses on P in Theorem 3; e.g., allowing P to have arbitrary determinant, then one obtains a modular form attached to *a congruence subgroup*

$$\Gamma_0(N) = \left\{ \begin{pmatrix} a & b \\ c & d \end{pmatrix} \in SL(2, \mathbb{Z}) \,\middle|\, N \text{ divides } c \right\}.$$

And the multiplier system will not be trivial for such theta functions. The theory is harder for odd n, even if $n = 1$, which is the most widely used special case—the theta function corresponding to the Riemann zeta function by Mellin transform. The *level* N of the congruence subgroup is called the level of the form P and defined to be the least positive integer such that NP^{-1} is even integral. For n even, it was proved by Hermite that

$$\theta(P, \gamma z) = \varepsilon(d)(cz + d)^k \theta(P, z) \quad \text{for } \gamma = \begin{pmatrix} a & b \\ c & d \end{pmatrix} \in \Gamma_0(N), \, k = n/2,$$

where the multiplier system is given by

$$\varepsilon(d) = (\operatorname{sign} d)^k \left(\frac{(-1)^k |P|}{|d|} \right). \tag{3.61}$$

The symbol $(-)$ in the multiplier system is the *Kronecker symbol*, which is a generalization of the *Legendre symbol* $\left(\frac{a}{p}\right)$ defined, for a, p integers with p an odd prime not dividing a, by

$$\left(\frac{a}{p}\right) = \begin{cases} +1 & \text{if } a \equiv x^2 \pmod{p} \text{ has a solution } x \text{ in } \mathbb{Z}, \\ -1 & \text{otherwise.} \end{cases}$$

The *quadratic reciprocity law*, first proved by Gauss (eight times), says that if p and q are distinct odd primes then

$$\left(\frac{p}{q}\right)\left(\frac{q}{p}\right) = (-1)^{(p-1)/2 \,(q-1)/2}. \tag{3.62}$$

It is possible to prove the quadratic reciprocity law (one of the most basic truths in number theory) using theta functions (see Dym and McKean [1, p. 225]). Hecke generalized this argument to obtain the quadratic reciprocity law for algebraic number fields, using theta functions (see Hecke [3] and Siegel [5, Vol. III, Paper 74]). The definition of the Kronecker symbol can be found, for example, in H. Cohn [1], Grosswald [1], or Siegel [3]. Proofs of formula (3.61) can be found in Eicher [1, pp. 48–52], Ogg [1, Ch. 9], and Schoeneberg [1, Ch. 9]. The proof given by Eichler is quite interesting because it uses theta functions on the Siegel upper half-space (see Chapter 5 for an introduction to such Siegel modular forms). Moreover, Eichler's proof works for an odd number of variables. The latter case is due to Pfetzer [1]. There is also a very interesting article by Eichler [2], which includes another proof of the result for an even number of variables plus some history and connections with other areas; e.g., the representations of $SL(2, \mathbb{Z}/n\mathbb{Z})$. Actually, one must generalize the theta functions defined in this section and introduce spherical harmonics in the sums (the same spherical harmonics that we considered in §2.1) in order to obtain cusp forms and bases for spaces of modular forms for $\Gamma_0(N)$ (see also

3.4. Automorphic Forms—Classical

Hijikata, Pizer, and Schemanske [1], Serre and Stark [1]). For other discussions of theta functions and multiplier systems, see Knopp [1], Maass [1], Mumford [4], and Rankin [1].

Exercise 8. Prove the following formula of Landsberg and Schaar:

$$p^{-1/2} \sum_{n=0}^{p-1} e^{2\pi i n^2 q/p} = (2q)^{-1/2} e^{i\pi/4} \sum_{n=0}^{2q-1} \exp\left(\frac{-\pi i n^2 p}{2q}\right)$$

for any integers p and q.

Hint. Use the transformation formula for theta (for $z \in H$):

$$\sum_{n \in \mathbb{Z}} \exp(i\pi n^2 z) = (z/i)^{-1/2} \sum_{n \in \mathbb{Z}} \exp(i\pi n^2 (-1/z)).$$

Let $z = 2q/p + it$ and let t approach zero. It is possible to use this exercise to prove the quadratic reciprocity law (see Dym and McKean [1, pp. 222–226] and Bellman [1, §29]).

For the applications to elliptic integrals, it is useful to define slightly more general theta functions, similar to that in Exercise 6 of §1.3. These functions will be studied in the following pair of exercises. There are multitudes of theta functions, and the notation for these functions is definitely not standardized. So be careful if you consult references. Here are some books to consult, in addition to our standard list from the beginning of §3.3 and the books just mentioned: Bellman [1], Erdélyi et al. [1], Hurwitz and Courant [1], Igusa [1], Lion and Vergne [1], Rauch and Lebowitz [1], and Whittaker and Watson [1].

Exercise 9 (Another Theta Function).
Define

$$\vartheta(z, s) = \sum_{n \in \mathbb{Z}} \exp(\pi i z n^2 + 2\pi i s n) \qquad \text{for } z \in H \text{ and } s \in \mathbb{C}.$$

Prove the following statements about this theta function.

(a) $\vartheta(z, s)$ is an entire function of $z \in H$ and of $s \in \mathbb{C}$.
(b) ϑ is quasi-periodic as a function of s, in the following sense:

$$\vartheta(z, s + n) = \vartheta(z, s) \qquad \text{for all } n \in \mathbb{Z};$$

$$\vartheta(z, s + zn) = \exp[-\pi i(zn^2 + 2ns)]\vartheta(z, s) \qquad \text{for all } n \in \mathbb{Z}.$$

(c) $\vartheta(z, s)$ satisfies the transformation formula

$$\vartheta(z, s) = (z/i)^{1/2} \sum_{n \in \mathbb{Z}} \exp\left(\frac{-\pi i(n - s)^2}{z}\right).$$

(d) $\vartheta(z, (1 + z)/2) = 0$.

(e) The only zero of $\vartheta(s) = \vartheta(z,s)$ as a function of s, holding z fixed, for s in the period parallelogram on 1 and z is $s = (1+z)/2$. Moreover, this zero is simple.

Hints. For (c) recall Poisson summation. For (e) imitate the proof of formula (3.49). This theta function is considered by Siegel [1, Vol. II, pp. 158–172] and Mumford [4, Vol. I] along with generalizations to the Siegel upper half-plane.

Exercise 10 (Jacobi's Four Theta Functions).
Define for $q = \exp(\pi i z)$, $z \in H$,

$$\theta_1(u,q) = 2 \sum_{n=0}^{\infty} (-1)^n q^{(n+1/2)^2} \sin[(2n+1)u];$$

$$\theta_2(u,q) = 2 \sum_{n=0}^{\infty} q^{(n+1/2)^2} \cos[(2n+1)u];$$

$$\theta_3(u,q) = 1 + 2 \sum_{n=1}^{\infty} q^{n^2} \cos(2nu);$$

$$\theta_4(u,q) = 1 + 2 \sum_{n=1}^{\infty} (-1)^n q^{n^2} \cos(2nu).$$

Prove that

$$\theta_3(u,q) = \vartheta\left(z, \frac{u}{\pi}\right); \quad \theta_4(u,q) = \vartheta\left(z, \frac{u}{\pi} + \frac{1}{2}\right);$$

$$\theta_2(u,q) = \theta_1\left(u + \frac{\pi}{2}, q\right); \quad \theta_1(u,q) = -i\exp\left(iu + \frac{\pi i z}{4}\right)\theta_4\left(u + \frac{\pi z}{2}, q\right).$$

We will show later that you can use products and quotients of the theta functions in Exercise 9 to obtain any elliptic function (see the discussion of elliptic integrals which follows). This aids in the computation of elliptic functions, because the series for theta functions converge rapidly, when the variable z is away from the real line. And you can use property (c) of Exercise 9 to deal with other values of z. Siegel extends these results to generalizations of elliptic functions called Jacobi-Abel functions in [1], using Siegel modular forms to be considered in Chapter 5. The Jacobi-Abel functions are related to abelian integrals.

Set $\vartheta_{11}(z,s) = \exp[\frac{1}{4}\pi i z + \pi i(s + \frac{1}{2})]\vartheta(z, s + \frac{1}{2}(1+z))$.

Mumford [4, Vol. I, p. 25] shows that up to a constant the Weierstrass $\wp$-function is obtained from $\vartheta_{11}(z,s)$ by taking the logarithm of theta and then two derivatives in the s-variable. We should caution the reader that our notation is the mirror image of Mumford's. Mumford [loc. cit., p. 64] proves Jacobi's derivative formula which shows that taking derivatives of $\vartheta_{11}(z,s)$ with respect to s and then evaluating at $s = 0$ leads to cusp forms of weight 3 and level 4 (after squaring the result). And Mumford [loc. cit., p. 72] shows that $\Delta(z)$ can be written as a 24th power of a value of $\vartheta(z,s)$.

Theta functions are viewed by algebraic geometers as providing "nice projective embeddings of a polarized abelian variety" (see Igusa [1], Mumford [4]). There are connections between this view and the application to elliptic integrals which follows. A historical sketch explaining some of these things can be found in Shafarevitch [1, pp. 411–430 and Chapter 11]. Another reference is Hoobler and Resnikoff [1]. Much of the modern algebraic–geometric theory of theta functions (of higher genus) has been developed by Mumford (see Tate [1] and Mumford [1–4]). The higher-genus theta functions will be considered in Chapter 5. The book of Mumford [4] is particularly recommended as a nice introduction to theta functions which includes several applications to the solution of partial differential equations.

There are numerous applications of theta functions in number theory; e.g., in order to obtain correspondences between various spaces of automorphic forms (see Kudla [1, 2], Stark [5], and Marie-France Vignéras [1]), to study representations of integers by quadratic forms and related questions (see Exercises 17 and 18, and Siegel [5, Vol. I, pp. 326–405, 410–443, 469–548; Vol. II, pp. 1–7, 20–40]). Siegel's work on quadratic forms finds a new adelic and representation-theoretic formulation in Weil [5, Vol. III, pp. 1–157] and Tamagawa (see Kneser's article in Cassels and Fröhlich [1]). But these are higher-dimensional matters to be considered in Chapter 5.

Application to Elliptic Integrals

The name "elliptic integral" arises because the arc length of an ellipse is such an integral. An *elliptic integral* is an integral of the form $\int R(x,y)\,dx$, where $R(x,y)$ is a rational function over $\mathbb{C}$ (i.e., a quotient of polynomials with complex coefficients) and x and y are related by an algebraic equation which can be solved by substitutions $x = f(u)$ and $y = g(u)$, with f, g being elliptic functions; e.g., $y^2 = G(x)$, where $G(x)$ is a third- or fourth-degree polynomial over $\mathbb{C}$. Here an *elliptic function* means a meromorphic function of $z \in \mathbb{C}$ which is doubly periodic; i.e., has two linearly independent periods over $\mathbb{R}$. Elliptic functions arise in many ways in applied mathematics; e.g., in potential theory on rectangles, in problems involving pendulums, and in problems involving ellipses (see Jeffreys and Jeffreys [1, Ch. 25]). Perhaps one of the strangest things about elliptic functions is that, unlike the well-known functions of freshman mathematics, they do not satisfy linear differential equations (e.g., see formula (3.51) and Exercise 19). Elliptic integrals arise, for example, when one uses the Schwarz–Christoffel formula to obtain the conformal mapping of the rectangle in Fig. 3.27 onto the upper half-plane. Here the points are mapped according to Table 3.4 and $sn'(0) = 1$.

The integral defining the inverse function for the *Jacobian elliptic function* $w = sn(z, k) = sn(z)$ is

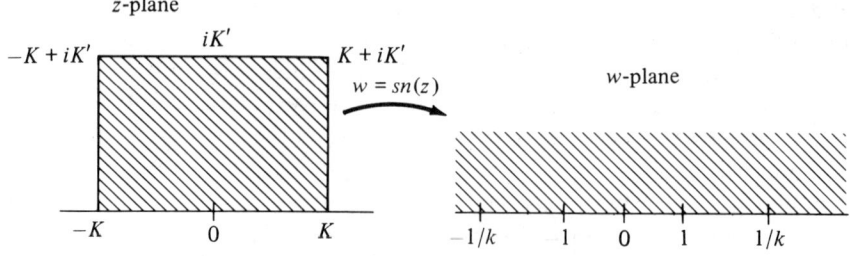

Figure 3.27. Conformal mapping by $sn(z)$.

Table 3.4

z	$-K+iK'$	$-K$	0	K	$K+iK'$	iK'
$sn(z)$	$-k^{-1}$	-1	0	1	k^{-1}	∞

$$z = \int_0^w [(1-w^2)(1-k^2w^2)]^{-1/2}\, dw. \qquad (3.63)$$

Note that the case $k=0$ gives the ordinary trigonometric function $w = \sin z$.

The Jacobian elliptic function is a doubly periodic function with periods $4K$ and $2iK'$. The *modulus* k^2 (also k with $0 < k < 1$) is uniquely determined by $q = \exp(i\pi\tau)$, $\tau = iK/K'$. In the theory of elliptic integrals, k^2 was always called the modulus and it turns out that $k^2(q) = \lambda(\tau)$, thus explaining why we call λ the "modulus." References for these things are Copson [1, Ch. 14], DuVal [1], and Nehari [1, pp. 280–296], for example.

The elliptic function $sn(z)$ is used in applied mathematics to produce the potential distribution inside a rectangle when there is some kind of charge distribution on the boundary (see Morse and Feshbach [1, Vol. II, p. 1252]). Reading this page in Morse and Feshbach, one is struck by the regret with which they turn to higher-dimensional problems from those of the complex plane, for higher-dimensional problems seldom have such elegant solutions. The method of conformal mapping mostly fails. For Liouville proved that the only nontrivial conformal maps in 3-space are "mappings by reciprocal radii" (inversions in spheres)—a fact also bemoaned by Sommerfeld [1, p. 140].

Exercise 11. Use the Schwarz-Christoffel transformation to show that the Jacobian elliptic function has the properties that we claimed above.

Exercise 12. Show that $w = sn\,z$ satisfies
$$w'^2 = (1-w^2)(1-k^2w^2).$$

Exercise 13. Show that if $q = e^{\pi i \tau}$,
$$\frac{\theta_1(z,q)}{\theta_4(z,q)} = A\,sn\frac{2K}{\pi}z.$$

3.4. Automorphic Forms—Classical

Hint. Compare zeros and poles of the two doubly periodic functions. The functions θ_1 and θ_4 are defined in Exercise 10.

Exercise 14. Show that the constant A in Exercise 13 has the value

$$A = \frac{\theta_2(0, q)}{\theta_3(0, q)}.$$

Also show that

$$k^2 = \left(\frac{\theta_2(0, q)}{\theta_3(0, q)}\right)^4.$$

Because the theta series converge rapidly, these formulas can be used to evaluate the functions involved on a computer.

There are many other known and useful formulas for these functions. We mention, for example, the *relation between the modulus and J*:

$$J = \frac{4}{27} \frac{(1 - \lambda + \lambda^2)^3}{\lambda^2 (1 - \lambda)^2}.$$

Jacobi's triple product formula says that if $z = \exp(2iu)$

$$\theta_3(u, q) = \prod_{n=1}^{\infty} (1 - q^{2n})(1 + q^{2n-1} z^2)(1 + q^{2n-1} z^{-2}).$$

For a proof, see Hurwitz and Courant [1], for example. This and related identities for the Dedekind eta function have attracted much attention recently thanks to connections with Lie groups (see V. Kac [2] and MacDonald [2]). For more connections between theta functions and the theory of representations of Lie groups, see Lion and Vergne [1] and Mumford [4].

The Connection between Theta Functions and Coding Theory

References for this subject are Sloane [1, 2], MacWilliams and Sloane [1], and Conway, Odlyzko, and Sloane [1].

Consider a noisy telephone connection transmitting 0's and 1's. Sometimes a 1 is received as a 0. The coding theorists at Bell labs, and elsewhere, correct errors by sending messages encoded into *codewords* which are vectors of 0's and 1's. The *Hamming distance* between two codewords x, y is the number of indices i in the vectors $x = (x_1, \ldots, x_n)$ and $y = (y_1, \ldots, y_n)$ such that $x_i \neq y_i$.

A *binary code* C of type $[n, k, d]$ is a set of 2^k vectors $u = (u_1, \ldots, u_n)$ with $u_i \in \mathbb{Z}/2\mathbb{Z} = \mathbb{F}_2$ for all $i = 1, \ldots, n$ such that

(a) C is closed under addition (componentwise and mod 2);
(b) the Hamming distance between two codewords is $\geq d$.

Coding theorists have the incompatible goals of seeking n to be small for speed of transmission, k to be large for efficiency, and d to be large to correct many errors. It turns out that a code $[n, k, d]$ can correct $[(d - 1)/2]$ errors.

Suppose that C is a *linear* code of type $[n, k, d]$; i.e., C forms a vector space over the finite field $\mathbb{F}_2$. Then it is shown in Sloane [1] that one can obtain a lattice in $\mathbb{R}^n$ from C in several ways (see §1.4 for the definition of lattice and the connection with positive definite symmetric matrices). One way of obtaining a lattice from C (and the only one that we shall discuss) is to set

$$L(C) = \{(u + 2m)/\sqrt{2} | u \in C, m \in \mathbb{Z}^n\}.$$

For example, the code $C = \{000, 011, 101, 110\}$ of type $[3, 2, 2]$ gives rise to the face-centered cubic lattice (see §1.4) in this way and thus to the densest lattice packing of spheres in $\mathbb{R}^3$ (see Ch. 4).

If $u \in C$, the *weight* of $u = wt(u)$ is defined to be the number of components u_i of u such that $u_i \neq 0$. And the *weight enumerator* of C is a homogeneous polynomial of degree n in two indeterminates x and y defined by

$$w_C(x, y) = \sum_{u \in C} x^{n-wt(u)} y^{wt(u)}.$$

It has been proved by Berlekamp, MacWilliams, and Sloane as well as Broue and Enguehard (see Sloane [1]) that the *theta function* associated to $L(C)$ by

$$\theta_{L(C)}(z) = \sum_{a \in L(C)} \exp(\pi i z {}^t a a), \quad z \in H$$

is actually a weight enumerator with Jacobi theta functions from Exercise 10 substituted for x and y. The explicit formula is

$$\theta_{L(C)}(z) = w_C(\theta_3(0, q), \theta_2(0, q)) \quad \text{if } q = e^{2\pi i z}.$$

In particular, it is possible to use a code called the extended Hamming code H_8 to obtain the theta function mentioned earlier, which is associated to the Lie group E_8 (see Exercise 7). To obtain the Hamming code H_8, one writes down a matrix called the parity check matrix:

$$H = \begin{pmatrix} 1 & 1 & 1 & 1 & 1 & 1 & 1 & 1 \\ 0 & 0 & 0 & 1 & 1 & 1 & 1 & 0 \\ 0 & 1 & 1 & 0 & 0 & 1 & 1 & 0 \\ 1 & 0 & 1 & 0 & 1 & 0 & 1 & 0 \end{pmatrix}.$$

Then the code words are obtained by chosing components u_1, u_2, u_3, u_4 arbitrarily in $\mathbb{F}_2$ while chosing the remaining components of u so that the equation $Hu = 0$ holds (modulo 2, of course). This is 4 equations in 4 unknowns, so that the solution is unique. And H_8 is a code of type $(8, 4, 4)$.

Poincaré Series

There is another example of a modular form which has been useful—the *Poincaré series*, which generalizes the Eisenstein series G_k defined in formula (3.48) by replacing the

3.4. Automorphic Forms—Classical

$$(cz+d)^{-k} = \left(\frac{d(\gamma z)}{dz}\right)^{k/2} \quad \text{for } \gamma = \begin{pmatrix} a & b \\ c & d \end{pmatrix} \in SL(2,\mathbb{Z}),$$

by

$$f(\gamma z)(cz+d)^{-k},$$

for some holomorphic bounded function $f: H \to \mathbb{C}$. Usually, a function such as $f(z) = \exp(2\pi i m z)$ for $m \in \mathbb{Z}$ is chosen (see Lehner [1, p. 275] and Rankin [1, p. 136]). Petersson showed that you can obtain a basis for $\mathcal{M}(SL(2,\mathbb{Z}), k)$ of such Poincaré series (see Lehner [1, p. 293] and Rankin [1, pp. 153–159]). These Poincaré series are also useful in the study of automorphic forms for congruence subgroups (see Rankin [1, pp. 172–193]). In his article on the basis problem for modular forms on $\Gamma_0(N)$, Eichler uses the Poincaré series with $f_w(\gamma, z) = \chi(\gamma)(\gamma(z) - w)^{-1}$, where $\chi(\gamma) = \chi(a)$, for χ a character of the group $(\mathbb{Z}/N\mathbb{Z})^* =$ the multiplicative group of integers $a \bmod N$ with $(a, N) = 1$ (see Eichler [2, p. 122]). These Poincaré series are used to construct a "Green's function" to be used in the trace formula. The Fourier coefficients of Poincaré series involve interesting arithmetical sums called Kloosterman sums (see Rankin [1]). We should note that one of the big problems with Poincaré series is that *they may be identically zero* (see Lehner [1, p. 25, pp. 290–291]).

Finally, we should mention that Poincaré series (like Eisenstein series) are capable of extensive generalization. For example, Svetlana Katok [1] considers *relative Poincaré series attached to hyperbolic elements* $\gamma_0 \in \Gamma$, where Γ is a Fuchsian group of the first kind with compact fundamental domain. Equivalently, one can say that these series are attached to closed or periodic geodesics in the fundamental domain of Γ. She proves that these series generate the whole space of cusp forms of weight greater than or equal to 4. These series are sums over $\Gamma_0 \backslash \Gamma$, where Γ_0 is the centralizer in Γ of γ_0. If

$$\gamma_0 = \begin{pmatrix} a_0 & b_0 \\ c_0 & d_0 \end{pmatrix},$$

set

$$Q_{\gamma_0}(z) = c_0 z^2 + (d_0 - a_0)z - b_0.$$

Then the relative Poincaré series for γ_0 is essentially the sum over γ representing $\Gamma_0 \backslash \Gamma$ of

$$Q_{\gamma_0}(\gamma z)^{-k/2}(cz+d)^{-k} \quad \text{with } \gamma = \begin{pmatrix} a & b \\ c & d \end{pmatrix}.$$

The construction works for groups with cusps also (see Exercise 15). Furthermore, Katok shows that the periods over closed geodesics play as important a role as the Fourier coefficients play for the group $SL(2,\mathbb{Z})$ and congruence subgroups.

Exercise 15 (from Katok [1]) (Relative Poincaré Series for $SL(2,\mathbb{Z})$). Let $\gamma_0 \in SL(2,\mathbb{Z})$ be a hyperbolic element. With the notation above, consider the relative Poincaré series defined by:

$$\sum_{\Gamma_0\backslash\Gamma} Q_{\gamma_0}(\gamma z)^{-k/2}(cz+d)^{-k} \quad \text{with } \gamma = \begin{pmatrix} a & b \\ c & d \end{pmatrix} \in \Gamma = SL(2,\mathbb{Z}).$$

Show that the series represents a modular form of weight k for $k \geq 4$.

Let us close this section with some estimates for the Fourier coefficients of modular forms.

Theorem 4 (Estimates for the Fourier Coefficients of Modular Forms). *The Fourier coefficients of the Eisenstein series $G_k(z)$ defined by equation* (3.48)

$$G_k(z) = \sum_{n \geq 0} a_n \exp(2\pi i n z) \quad (k = 4, 6, 8, 10, 12, \ldots)$$

satisfy

$$An^{k-1} \leq |a_n| \leq Bn^{k-1} \quad \text{for some positive constants } A, B.$$

The Fourier coefficients of a cusp form $f(z)$ of weight k for $SL(2,\mathbb{Z})$ with

$$f(z) = \sum_{n \geq 1} a_n \exp(2\pi i n z)$$

satisfy the Ramanujan-Petersson conjecture (proved by Deligne [1,2]*):*

$$|a_n| \leq Cn^{(k-1)/2+\varepsilon} \quad \text{for any } \varepsilon > 0.$$

PROOF (of a weaker result). To prove the estimate for the coefficients of G_k, note that Exercise 2 implies that

$$a_n = C\sigma_{k-1}(n)$$

for some constant $C \neq 0$, where

$$\sigma_s(n) = \sum_{\substack{d \text{ divides } n \\ 0 < d}} d^s.$$

We can estimate $\sigma_s(n)$, using the Riemann zeta function, as follows:

$$\sigma_s(n) = \sum_{0 < d | n} d^s = n^s \sum_{0 < d | n} d^{-s} \leq n^s \zeta(s),$$

if $s > 1$.

The estimate for cusp forms comes from Deligne's proof of the Weil conjectures on zeta functions of n-dimensional projective nonsingular varieties over finite fields, as we mentioned earlier in our discussion of Ramanujan's tau function. Deligne's proof cannot be discussed here. Instead we give an exercise proving an easier result.

Exercise 16 (An Estimate for the Fourier Coefficients of Cusp Forms). Show that if $f(z)$ is a cusp form of weight k for $SL(2,\mathbb{Z})$ as in Theorem 4, we have the inequality $|a_n| \leq Cn^{k/2}$ for some positive constant C.

Hint. Since $f(z)$ vanishes at infinity, we can factor $\exp(2\pi i z)$ out of the Fourier expansion and obtain $|f(z)| \leq C\exp(-2\pi y)$. Since the function $g(z) = |f(z)| y^{k/2}$ is invariant under $SL(2,\mathbb{Z})$ and approaches zero as y approaches infinity, it is bounded by some constant M. It follows that the Fourier coefficient a_n satisfies $|a_n| \leq My^{-k/2}\exp(2\pi ny)$ for all $y > 0$. Take $y = 1/n$ to complete the proof. □

The idea of Exercise 16 is Hecke's. Classical improvements in the estimate were obtained by Rankin [3] and Selberg [2], among others. Note that the proof in Exercise 16 applies also to noncongruence subgroups and nonintegral weights, as well as to the nonholomorphic modular forms to be considered in the next section, while Deligne's method does not apply to these other situations.

There is a interesting question raised by Atkin and Swinnerton-Dyer. The Fourier coefficients for modular forms on congruence subgroups have "bounded denominators" in the sense that the forms are linear combinations of functions whose coefficients are integers in some algebraic number field (of finite degree over $\mathbb{Q}$). This is not the case for noncongruence subgroups.

Question. do *all* of the modular forms for noncongruence subgroups have this "bad" property unless they are really modular forms on congruence subgroups?

Application to the Representation of Integers by Quadratic Forms

Let P be a positive definite even integral matrix P of determinant one as in Theorem 3. Define the representation numbers $r_P(m)$ by

$$r_P(m) = \#\{a \in \mathbb{Z}^n | P[a] = 2m\}.$$

It follows from what we know about modular forms of weight k for $SL(2,\mathbb{Z})$ that there is a positive constant C such that for every $\varepsilon > 0$,

$$\left| r_P(m) - \frac{2k}{B_k}\sigma_{k-1}(m) \right| \leq Cm^{(k-1)/2+\varepsilon} \quad \text{where } k = n/2. \quad (3.64)$$

Here B_k denotes the kth Bernoulli number defined in §1.4.

Exercise 17. Prove the estimate (3.64).

Hint. From Theorem 3, you know that $\theta(P,z) \in \mathcal{M}(SL(2,\mathbb{Z}), n/2)$. So there is a constant c_k and a cusp form f_k such that

$$\theta(P,z) = c_k G_k(z) + f_k(z).$$

Compare constant terms in the Fourier series for both sides of this equation to see that $c_k = [2\zeta(k)]^{-1}$. This allows you to evaluate the mth Fourier coefficient of $c_k G_k(z)$, using Exercise 2. Use the formula for the Riemann zeta function at even integers, which was mentioned in §1.4 and Theorem 4, to finish the exercise.

It would be more interesting perhaps to obtain formulas for the representation numbers of $P = I =$ the $n \times n$ identity matrix. Then one would have a formula for the number of ways of representing m as a sum of m squares. Such results for certain n are considered in Rademacher [1, Ch. 11] and Knopp [1, Ch. 5]. Siegel answers much more general questions than these in his work on quadratic forms mentioned earlier.

Exercise 18. Show that if $n = 8$ and P is a positive definite 8×8 symmetric matrix satisfying the conditions of Theorem 3, then

$$r_P(m) = 240\sigma_3(m).$$

Exercise 19 (The Korteweg-DeVries Equation). Show that the Weierstrass $\wp$-function which satisfies the nonlinear differential equation given in formula (3.51) also represents a time-independent solution of the Korteweg-DeVries equation for $u = u(x, t)$: $u_t = u_{xxx} - 12uu_x$.

The partial differential equation in Exercise 19 was derived by Korteweg and DeVries in 1895 in a study of long water waves in a rectangular canal. Often the constant 12 is replaced by 6. Of course, in physical problems the constants would reflect the properties of the materials involved. The KdV equation applies to many other situations (e.g., plasma physics) in which solitons occur. This means that waves show a particle-like behavior in interations; that is, the interacting waves keep their shape and amplitude while undergoing a shift. Korteweg and DeVries found a solution involving the hyperbolic secant and one involving elliptic functions. See the book of Lonngren and Scott [1] for various articles on solitons and the KdV equation (particularly the articles of Miura and Hermann).

3.5. Automorphic Forms—Not So Classical— Maass Waveforms

il a fallu Maass pour nous sortir du ghetto des fonctions holomorphes*

From A. Weil [5, Vol. III, p. 463].

Our main goal for the remainder of this chapter is to develop harmonic analysis on $SL(2, \mathbb{Z}) \backslash H$. That is, we seek the spectral decomposition of the non-

---

* It was necessary for Maass to help us get out of the ghetto of automorphic functions.

3.5. Automorphic Forms—Not So Classical—Maass Waveforms 205

Euclidean Laplace operator Δ on H into $SL(2,\mathbb{Z})$-invariant eigenfunctions. This is a non-Euclidean analogue of Fourier series. However, because the quotient $SL(2,\mathbb{Z})\backslash H$ is not compact, there is a continuous spectrum as well as a discrete spectrum, so that one has a mixture of Fourier series and integrals. In this section, we shall study the eigenfunctions of Δ on $SL(2,\mathbb{Z})\backslash H$. We will find that these eigenfunctions have much in common with the classical automorphic or modular forms considered in the preceding section. So we shall also call these eigenfunctions of Δ "automorphic forms." Another name in use is "Maass waveforms," because they were first sytematically considered by Maass [3] in 1949. We will find that the only known example of a Maass waveform for $SL(2,\mathbb{Z})$ is the Epstein zeta function from §1.4, which can be viewed as a nonholomorphic analogue of the Eisenstein series in §3.4. However, analogues of cusp forms are also known to exist, but we cannot produce examples such as the discriminant function in §3.4, except for congruence subgroups (since the product of eigenfunctions of Δ will not usually be an eigenfunction of Δ).

Some references for this section are Borel and Casselman [1], Borel and Mostow [1], Cartier and Hejhal [1], Delsarte [1, Tome II, pp. 599–601, 829–845], Elstrodt [1, 2], Faddeev [1], Fay [1], Gangolli [1], Gangolli and Warner [1, 2], Gelbart [1], Gelfand, Graev, and Piatetskii-Shapiro [1], Goldfeld and Husemoller [1], Hejhal [1, 2, 3], Huber [1, 2], Kubota [1], Lachaud [1], Lang [4], Lax and Phillips [1], Maass [1, 3], Moreno [1], Roelcke [1, 2, 3], Selberg [1, 2, 3, 6], Venkov [1, 2, 3], Weil [4], and Zagier [1].

Definition. A function $f : H \to \mathbb{C}$ is a *nonholomorphic modular (or automorphic) form* or *Maass waveform* if it satisfies the following three conditions:

(1) f is an eigenfunction of the non-Euclidean Laplacian; i.e., $\Delta f = \lambda f$, $\Delta = y^2(\partial^2/\partial x^2 + \partial^2/\partial y^2)$;
(2) f is invariant under the modular group; i.e., $f(\gamma z) = f(z)$ for all $\gamma \in \Gamma = SL(2,\mathbb{Z})$ and all $z \in H$;
(3) f has at most polynomial growth at infinity; i.e., there are constants $C > 0$ and k such that $|f(z)| \leq Cy^k$, as $y \to \infty$ uniformly in x.

We shall denote by $\mathcal{N}(SL(2,\mathbb{Z}),\lambda)$ the vector space of such Maass waveforms. Note that, since $\Delta f = \lambda f$ is an elliptic partial differential equation, the Maass waveform f is automatically real analytic (see Garabedian [1, Ch. 5]). Maass [3] considered such automorphic forms in 1949. There is also a discussion in Maass's lectures [1, Ch. 4] of a generalization of the Maass waveform which contains the classical holomorphic modular form from §1.4 as well.* A fundamental application of Maass waveforms for discrete subgroups Γ of $SL(2,\mathbb{R})$ with compact quotients $\Gamma\backslash H$ was already noted by Delsarte in 1942 [1, Tome II, pp. 599–601]. Then in the 1950s Selberg developed his trace formula, along with many number-theoretical applications, some of which were found inde-

---

* Actually, you must multiply Maass's functions g in $[SL(2,\mathbb{Z}),s,s,1]$ by y^s in order to obtain our functions f in $\mathcal{N}(SL(2,\mathbb{Z}),s(s-1))$.

pendently by Eichler (see Selberg [1] and Eichler [3]). We shall discuss the trace formula in §3.7.

Example (Eisenstein Series (alias Epstein's Zeta Function)). As in §3.2, the easiest way to construct eigenfunctions of Δ is to build them up out of the *power function*

$$p_s(z) = (\text{Im } z)^s = y^s \qquad \text{if } z = x + iy \in H, \tag{3.65}$$

for clearly, $\Delta p_s = s(s-1)p_s$. Formulas (3.14) and (3.21), along with the exercises following those formulas, show that the K-Bessel and associated Legendre functions are built up out of power functions by integration over the appropriate subgroup of $SL(2, \mathbb{R})$. In the present situation we must sum rather than integrate, since $SL(2, \mathbb{Z})$ is a discrete subgroup.

Definition. The *Eisenstein series* $E_s(z)$ for $s \in \mathbb{C}$ with $\text{Re } s > 1$, and $z \in H$, is defined by

$$E_s(z) = \sum_{\gamma \in \Gamma_\infty \backslash \Gamma} p_s(\gamma z), \tag{3.66}$$

where

$$\Gamma_\infty = \left\{ \begin{pmatrix} \pm 1 & * \\ 0 & \pm 1 \end{pmatrix} \in SL(2, \mathbb{Z}) = \Gamma \right\}. \tag{3.67}$$

The notation in (3.66) means that the sum runs over a complete set of representatives $\gamma \in \Gamma$ for the quotient $\Gamma_\infty \backslash \Gamma$. It is necessary to "mod out" Γ_∞, because

$$p_s(\gamma z) = p_s(z) \qquad \text{for all } \gamma \in \Gamma_\infty, z \in H.$$

Note that Γ_∞ is the stabilizer in Γ of the cusp at infinity. When there are more cusps in the fundamental domain for the discrete group Γ, there will be more Eisenstein series (one for each Γ-inequivalent cusp). The convergence in (3.66) is absolute and uniform on compacta in the s-plane, by arguments given in §1.4 for Epstein's zeta function, using the following exercise.

Exercise 1 (The Connection between Eisenstein Series and Epstein Zeta Functions). Let Y in $\mathcal{SP}_2 = \{Y \in \mathbb{R}^{2 \times 2} | Y \text{ positive definite}, |Y| = 1\}$. Define Epstein's zeta function for $\text{Re } s > 1$, $Y \in \mathcal{SP}_2$, as in §1.4, by

$$Z(Y, s) = \frac{1}{2} \sum_{a \in \mathbb{Z}^2 - 0} Y[a]^{-s},$$

where $Y[a] = {}^t a Y a$, viewing $a \in \mathbb{Z}^2 - 0$ as a column vector.

Recall that, according to Exercise 9 of §3.1, we can identify H and $\mathcal{SP}_2$ via $z \mapsto W_z$ where $z = x + iy$ and $W_z \in \mathcal{SP}_2$ is defined by

$$W_z = \begin{pmatrix} 1/y & 0 \\ 0 & y \end{pmatrix} \begin{bmatrix} 1 & -x \\ 0 & 1 \end{bmatrix} = \begin{pmatrix} 1/y & -x/y \\ -x/y & (x^2 + y^2)/y \end{pmatrix}.$$

3.5. Automorphic Forms—Not So Classical—Maass Waveforms

(a) Show that if $a = \begin{pmatrix} n \\ -m \end{pmatrix} \in \mathbb{R}^2$, then $W_z[a] = {}^t a W_z a = y^{-1}|mz + n|^2$.

(b) Show that if $\operatorname{Re} s > 1$, then the Eisenstein series (3.66) is given by

$$E_s(z) = \frac{1}{2} y^s \sum_{(m,n)=1} |mz + n|^{-2s}.$$

Here $(m, n) =$ the greatest common divisor of m and n.

(c) Prove that $\zeta(2s)E_s(z) = Z(W_z, s)$, if $\zeta(s) =$ Riemann's zeta function.

(d) Show that $Z(W, s) = Z(W^{-1}, s)$ if $W \in \mathcal{SP}_2$.

Hints on part (b). You must show that the quotient $\Gamma_\infty \backslash \Gamma$ has as representatives

$$\begin{pmatrix} * & * \\ m & n \end{pmatrix}$$

with $(m, n) = 1$. This is a consequence of the fact that

$$\begin{pmatrix} \pm 1 & q \\ 0 & \pm 1 \end{pmatrix} \begin{pmatrix} a & b \\ c & d \end{pmatrix} = \begin{pmatrix} * & * \\ \pm c & \pm d \end{pmatrix}.$$

You also need to know that a matrix lies in $SL(2, \mathbb{Z})$ if and only if the g.c.d. of its second row is one.

Hints on part (c). In the definition of $Z(W, s)$, the sum runs over all $(m, n) \in \mathbb{Z}^2 - 0$. Factor the greatest common divisor out of m and n.

Hints on part (d). If W is a 2×2 positive matrix of determinant 1, then

$$W^{-1} \begin{bmatrix} c \\ d \end{bmatrix} = W \begin{bmatrix} -d \\ c \end{bmatrix}.$$

From Exercise 1, we see that $E_s \in \mathcal{N}(SL(2, \mathbb{Z}), s(s - 1))$ when $\operatorname{Re} s > 1$, provided that $E_s(z)$ has polynomial growth as $y \to \infty$.

Exercise 2.

(a) Show that $E_s(\gamma z) = E_s(z)$ for all $\gamma \in SL(2, \mathbb{Z})$ and $z \in H$, if $\operatorname{Re} s > 1$.
(b) Show that $E_s(z) \sim y^s$, as $y \to \infty$, for s fixed with $\operatorname{Re} s > 1$.
(c) Conclude that $E_s \in \mathcal{N}(SL(2, \mathbb{Z}), s(s - 1))$ when $\operatorname{Re} s > 1$.

Hint on part (b). Use the formula from part (b) of Exercise 1.

Theorem 1 of §1.4 shows that $E_s(z)$ has an analytic continuation to the entire complex s-plane as a meromorphic function of s. And $E_s(z)$ has a pole at $s = 1$ with residue $3/\pi = (\text{vol}(SL(2, \mathbb{Z}) \backslash H))^{-1}$. Moreover, the Eisenstein series satisfies the *functional equation*

$$\Lambda(s)E_s(z) = \Lambda(1-s)E_{1-s}(z) \quad \text{if} \quad \Lambda(s) = \pi^{-s}\Gamma(s)\zeta(2s). \qquad (3.68)^*$$

There must be a functional equation relating E_s and E_{1-s} because both functions lie in $\mathcal{N}(SL(2,\mathbb{Z}), s(s-1))$, which will be proved to be one dimensional for $\text{Re}\, s \neq \frac{1}{2}$.

Since $E_s(z) = Z(W_z, s)/\zeta(2s)$, any zero of $\zeta(2s)$ will produce a pole of $E_s(z)$ unless $Z(W_z, s)$ vanishes at that value of s. The trivial zeros of $Z(W_z, s)$ and $\zeta(2s)$ are both $s = -1, -2, -3, \ldots$; and both have order 1. Thus the trivial zeros cancel out in $E_s(z)$. Exercise 8 shows $\zeta(2s) \neq 0$ for $\text{Re}\, s > \frac{1}{2}$. This is true in a larger region, but the Riemann hypothesis that the nontrivial zeros of $\zeta(2s)$ must lie on the line $\text{Re}\, s = \frac{1}{4}$ remains unproved after more than 100 years. One of the motivations for the study of harmonic analysis on $\Gamma \backslash H$ is the desire to understand the complex zeros of $\zeta(s)$ (see Hejhal [2, p. 479] or [3], Cartier and Hejhal [1], or Zagier [1, p. 276]). Note that the zeros of $Z(W, s)$ depend on W and, in general, do not satisfy the Riemann hypothesis (see Stark [6] and Titchmarsh [3, p. 244]). They tend, however, to lie on the line $\text{Re}\, s = \frac{1}{2}$, but can also be found in $\text{Re}\, s > 1$.

From Exercise 2 and the functional equation (3.68) we see that

$$\Lambda(s)E_s(z) \sim \Lambda(s)y^s + \Lambda(1-s)y^{1-s} \qquad \text{as } y \to \infty.$$

In fact, it is not hard to obtain the complete Fourier expansion of the Eisenstein series (see Exercises 3 and 4). The result of Exercise 4 is a very useful one in number theory for it reveals much about the Eisenstein series. It is thus reminiscent of the incomplete gamma expansion of Epstein's zeta function (Theorem 1 of §1.4). This might lead one to say that these formulas are the meat and potatoes (quiche & salad) of the subject.

Exercise 3 (Fourier Expansions of Maass Waveforms). Show that an automorphic form $f \in \mathcal{N}(SL(2,\mathbb{Z}), s(s-1))$ has a Fourier series expansion as a periodic function of x given by

$$f(z) = ay^s + by^{1-s} + \sum_{n \neq 0} a_n y^{1/2} K_{s-1/2}(2\pi|n|y) \exp(2\pi i n x),$$

where $K_s(y)$ is the K-Bessel function in §3.2.

Hints. Use separation of variables in $\Delta f = s(s-1)f$, as in Exercise 4 of §3.2, and rule out the second solution $I_s(y)$ of the second-order ordinary differential equation for $K_s(y)$, by the polynomial growth of $f(z)$ as $y \to \infty$ (see Lebedev [1, p. 123]).

Exercise 4 (The Fourier Expansion of the Eisenstein Series). Set $\Lambda(s) = \pi^{-s}\Gamma(s)\zeta(2s)$ and $E_s^*(z) = \Lambda(s)E_s(z)$. Show that E_s^* has the Fourier expansion

---

* We shall use the letter Γ for discrete groups such as $SL(2,\mathbb{Z})$ as well as for the gamma function. Hopefully the meaning will be clear from the context.

3.5. Automorphic Forms—Not So Classical—Maass Waveforms

$$E_s^*(z) = y^s\Lambda(s) + y^{1-s}\Lambda(1-s) + 2\sum_{n\neq 0}|n|^{s-1/2}\sigma_{1-2s}(n)y^{1/2}K_{s-1/2}(2\pi|n|y)e^{2\pi inx}.$$

Here $\sigma_s(n)$ is the *divisor function* defined by

$$\sigma_s(n) = \sum_{0<d|n} d^s,$$

where we mean that the sum runs over all positive divisors d of n.

Hints. Method 1 (Chowla and Selberg [1]). Start with the Mellin transform result in Exercise 7 of §1.4:

$$E_s^*(z) = \Lambda(W_z, s) = \frac{1}{2}\int_0^\infty t^{s-1} \sum_{\binom{a}{b}\in\mathbb{Z}^2 - 0} \exp\left(-\pi t W_z\begin{bmatrix}a\\b\end{bmatrix}\right) dt. \quad (3.69)$$

Now, using the formula for W_z in Exercise 1, we have

$$W_z\begin{bmatrix}a\\b\end{bmatrix} = y^{-1}(a-xb)^2 + yb^2.$$

Thus the $b=0$ term of the sum in (3.69) is integrated to obtain $\Lambda(s)y^s$. When $b\neq 0$, the variable a is summed over all of $\mathbb{Z}$, and thus one can apply the Poisson sum formula and obtain, as in Exercise 9 of §3.4

$$\sum_{a\in\mathbb{Z}}\exp\left[-\pi\frac{t}{y}(a-xb)^2\right] = \sqrt{\frac{y}{t}}\sum_{a\in\mathbb{Z}}\exp\left[-\pi\frac{y}{t}a^2 + 2\pi iabx\right]. \quad (3.70)$$

After pulling out the $b=0$ term, substitute (3.70) into (3.69). Then use the integral formula for $K_s(z)$ in Exercise 1 of §3.1 to complete the proof.

Method 2 (Bateman and Grosswald [1] or A. Terras [4]). Apply the Poisson summation formula directly to the series defining $Z(W,s)$, after taking out the $b=0$ term of the sum over $(a,b)\in\mathbb{Z}^2 - 0$, as in Method 1.

Method 3 (Kubota [1, pp. 13–17]). Note that

$$E_s(z) = y^s + y^s \sum_{\substack{(c,d)=1 \\ c\geq 1}} |cz+d|^{-2s} = y^s + y^s \sum_{c\geq 1} c^{-2s} \sum_{(c,d)=1} |z+d/c|^{-2s}.$$

Let $d = r + cn, 0 \leq r < c, n\in\mathbb{Z}$. Use Poisson summation on the sum over n to complete the proof. You will also need the identity relating the divisor function and the singular series (see Hardy [1, p. 141]):

$$\zeta(2s)\sum_{\substack{c>0, d \bmod c \\ (d,c)=1}} c^{-2s}\exp(2\pi imd/c) = \sigma_{1-2s}(m).$$

The name "singular series" was used by Hardy and Littlewood in their work on Waring's problem (see Hardy [2, Vol. I, pp. 377–532]). It was applied to Fourier coefficients of Eisenstein series by Siegel [5, Vol. 1, p. 329] and Maass [2, pp. 300–313]. This third method is related to the Bruhat decomposition of $SL(2,\mathbb{Q})$ (see Chapters 4 and 5).

Note. The singular series form of the Fourier coefficients of $E_s(z)$ in method (3) involves *Ramanujan sums*:

$$c_r(n) = \sum_{1 \le m < r, (m,r)=1} \exp(2\pi i m n / r).$$

Ramanujan studied various formal trigonometric series involving these sums (including that obtained from $E_s(z)$). Kac [1, pp. 86–96] shows that such series can often be viewed as Fourier expansions of almost periodic functions. See also Delsarte [1, Vol. II, pp. 603–624].

Exercise 5. Show that at $s = \frac{1}{2}$ the poles of the constant term in the Fourier expansion of the Eisenstein series $E_s^*(z)$ in Exercise 4 cancel. Evaluate the resulting constant in terms of known constants like π, $\gamma =$ Euler's constant, etc. That is, evaluate

$$\lim_{s \to 1/2} \{\Lambda(s) y^s + y^{1-s} \Lambda(1-s)\}, \qquad \Lambda(s) = \pi^{-s} \Gamma(s) \zeta(2s).$$

Exercise 6 (Kronecker's Limit Formula). Use Exercise 4 to show that if W_z is as in Exercise 1,

$$\lim_{s \to 1} \left\{ Z(W_z, s) - \frac{\pi}{2} \frac{1}{s-1} \right\} = \pi \{\gamma - \log 2 - \log(y^{1/2} |\eta(x+iy)|^2)\},$$

where $\gamma =$ Euler's constant defined by

$$\gamma = \lim_{n \to \infty} \left(\sum_{m=1}^{n} 1/m - \log n \right) = 0.577215\ldots,$$

and $\eta(z) =$ Dedekind's eta function from §3.4.

Hint. Note that $K_{1/2}(y) = (\pi/2y)^{1/2} e^{-y}$ and use the definition of eta as an infinite product.

The result of Exercise 6 goes back to Kronecker [1, Vol. 4, pp. 222, 347–495, Vol. 5, pp. 1–132], along with a second limit formula for the function

$$\sum_{a \in \mathbb{Z}^2 - 0} W[a]^{-s} \exp(2\pi i {}^t q a) \qquad \text{for } W \in \mathcal{SP}_2, q \in \mathbb{R}^2.$$

Interesting discussions of this result can be found in Siegel [4], Weber [1, Vol. 3], and Weil [3]. The Kronecker limit formulas are of central importance for the construction of class fields of algebraic number fields, as Kronecker had already demonstrated in 1863, by proving that the limit formula gives a solution $(x, y) \in \mathbb{Z}^2$ of Pell's equation

$$x^2 - dy^2 = \pm 1 \qquad \text{(if } d \text{ is a given positive integer)}$$

in terms of elliptic modular functions eta and theta. Some related references are Goldstein [2], Hecke [1, pp. 198–207, 290–312], Katayama [1], Lang [5], Meyer [1], Ramachandra [1], Shintani [2], and Zagier [2]. Stark [3, part 4] uses Kronecker's limit formula to prove Stark's conjectures on values of L-functions when the base field is $\mathbb{Q}$ or an imaginary quadratic field. This gives explicit reciprocity laws in some abelian extensions of the base field. Stark [9] and Gupta [1] use the Kronecker limit formula to obtain an analytic proof of the Coates-Wiles theorem on L functions for elliptic curves with complex multiplication.

There are other sorts of applications of the Fourier expansion in Exercise 4. Stark [2] uses generalizations of that result to show that there are exactly nine imaginary quadratic fields with class number 1. The Fourier expansion can also be used to obtain comparisons of values of the Riemann zeta function at $2n$ and $2n+1$, $n = 1, 2, 3, \ldots$, in terms of rapidly converging series of exponentials (see A. Terras [5] and Exercise 7). One has Euler's formula for $\zeta(2n)$, $n = 1, 2, 3, \ldots$, but there is no simple interpretation for $\zeta(2n+1)$, $n = 1, 2, 3, \ldots$. There are, however, conjectures of Lichtenbaum relating these odd values to orders of K-groups (see Lichtenbaum [1]). Later we will find that $\zeta(3)$ does appear in the volume of the fundamental domain for $SL(3, \mathbb{Z})$. This fact was already noted by Siegel, but the Gauss-Bonnet theorem does not apply in this case (see Weil [5, Vol. I, p. 561]) to relate $\zeta(3)$ to a rational number times some power of π. Recently Apéry showed that $\zeta(3)$ is indeed irrational (see Van der Poorten [1]).

Exercise 7.

(a) Prove Euler's formula:
$$\zeta(2n) = \frac{(-1)^{n-1}}{2(2n)!}(2\pi)^{2n}B_{2n}, \quad n = 1, 2, 3, \ldots.$$

Here $B_n =$ the nth Bernoulli number defined by $x(e^x - 1)^{-1} = \sum_{n \geq 0} B_n x^n/n!$.

(b) Use Exercise 4 to show that
$$\zeta(3) = \frac{2}{45}\pi^3 - 4\sum_{n \geq 1} e^{-2\pi n}\sigma_{-3}(n)\left(2\pi^2 n^2 + \pi n + \frac{1}{2}\right).$$

(c) Show that $\lim_{s \to 1}(\zeta(s) - 1/(s-1)) = \gamma =$ Euler's constant.

Hints.

(a) Look at the contour C_R in Fig. 3.28 and let $R \to \infty$ in $\int_{C_R}(e^z - 1)^{-1}z^{-n}\,dz$, using the Cauchy integral formula and residue calculus.

(b) See Terras [5].

Exercise 8. Show that $\zeta(1 + it) \neq 0$ for all t in $\mathbb{R}$.

Hints. (See Jacquet and Shalika [1] and Zagier [1].) Suppose that $\zeta(1 + it) = 0$. Then $E_{(1+it)/2}(z)$ is of rapid decay as $y \to \infty$, according to Exercise 4. But then if $s = (1 + it)/2$, we have

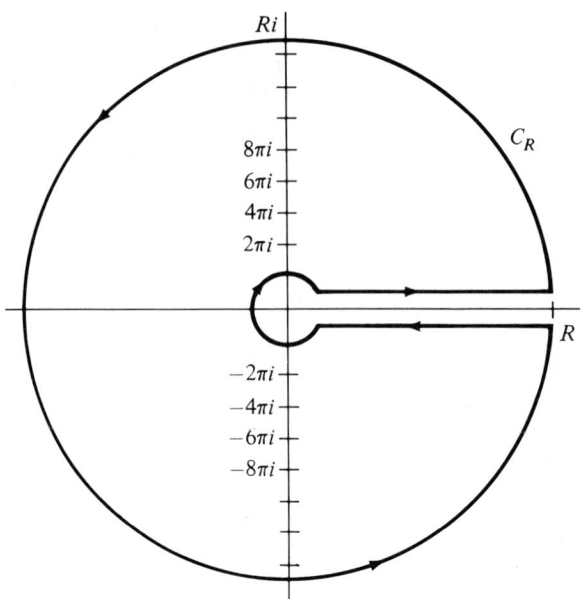

Figure 3.28. Contour for the evaluation of $\zeta(2n)$.

$$C(E_s, y) = \int_{x=0}^{1} E_s(x + iy)\, dx$$

$$= (\text{the constant term in the Fourier expansion of } E_s) = 0.$$

Consider

$$I(E_s, r) = \int_{y=0}^{\infty} C(E_s, y) y^{r-2}\, dy$$

$$= \int_{\Gamma_\infty \backslash H} E_s(z) y^{r-2}\, dx\, dy = \int_{\Gamma \backslash H} E_s(z) E_r(z) y^{-2}\, dx\, dy.$$

Let $r = \bar{s}$. Then $I(E_s, \bar{s})$ is the square of the L^2-norm of E_s and we have a contradiction.

The result in Exercise 8 is quite old and is fundamental for proofs of the prime number theorem using Tauberian theorems (see Wiener [1, pp. 112–121]). As Selberg [3] noticed, the preceding exercise is easy once one has the analytic continuation of $E_s(z)$ to $\operatorname{Re} s \geq \frac{1}{2}$ with the only pole occurring at $s = 1$. For one need only examine a nonconstant Fourier coefficient of $E_s(s)$ (which has $\zeta(2s)$ in the denominator) to see that $\zeta(2s)$ cannot vanish on $\operatorname{Re} s = \frac{1}{2}$. The usual proofs of Exercise 8 are quite different (see Grosswald [1, p. 131]).

Related Fourier expansions of Eisenstein series have been used by Kubota [2] and Heath-Brown and Patterson [1] to study cubic Gauss sums. One can

3.5. Automorphic Forms—Not So Classical—Maass Waveforms

prove Gauss's conjecture that the average order of the number of classes of positive integral binary quadratic forms of discriminant $-D$ is $2\pi D^{1/2}/(7\zeta(3))$ by interpreting class numbers as Fourier coefficients of holomorphic Eisenstein series of weight $\frac{3}{2}$ (see Hecke [1, pp. 499–504]). Goldfeld, Hoffstein, and Patterson [1] use similar ideas to study the arithmetic of elliptic curves.

Exercise 9.

(a) Let $BC^\infty(\Gamma\backslash H)$ denote the set of bounded C^∞ functions on $\Gamma\backslash H$. Show that if f and Δf lie in $BC^\infty(\Gamma\backslash H)$ then (if f is real-valued)

$$(\Delta f, f) = -\int_{\Gamma\backslash H} \left(\left(\frac{\partial f}{\partial x}\right)^2 + \left(\frac{\partial f}{\partial y}\right)^2\right) dx\, dy \leq 0,$$

where the inner product is

$$(f, g) = \int_{\Gamma\backslash H} f(z)\overline{g(z)}\, y^{-2}\, dx\, dy.$$

(b) Deduce that if λ is an eigenvalue of Δ on $L^2(\Gamma\backslash H)$, $\lambda = s(s-1)$, then $\operatorname{Re} s = \frac{1}{2}$ or $s \in [0, 1]$.

(c) Suppose that $f, g \in BC^\infty(\Gamma\backslash H)$ are real functions such that Δf and Δg are also in $BC^\infty(\Gamma\backslash H)$. Then Δ is symmetric; i.e.,

$$(\Delta f, g) = (g, \Delta f).$$

Hint. (See Lang [4, pp. 281–284]). Use Green's theorem on a truncated fundamental domain as in Fig. 3.29.

Using the notation of Exercise 9, let D_Δ be the space of functions f in $BC^\infty(\Gamma\backslash H)$ such that Δf is also in $BC^\infty(\Gamma\backslash H)$. Then D_Δ is a dense subspace of $L^2(\Gamma\backslash H)$ and Δ can be extended to a self-adjoint operator (see Lang [4, pp. 284–287]). This is done using the resolvent $(\Delta - s(s-1))^{-1}$. Exercise 9 shows that the operator is negative.

There are (sometimes) other members of $\mathcal{N}(\Gamma, \lambda)$ besides Eisenstein series.

Definition. We say that $f \in \mathcal{N}(\Gamma, \lambda)$ is a *cusp form* for $\Gamma = SL(2, \mathbb{Z})$ if the constant term in the Fourier expansion in Exercise 3 vanishes; i.e.,

$$\int_{x=0}^{1} f(x+iy)\, dx = 0 = ay^s + by^{1-s} \quad \text{for all } y > 0.$$

Let $\mathcal{SN}(\Gamma, \lambda)$ denote the *vector space of cusp forms* in $\mathcal{N}(\Gamma, \lambda)$.

Exercise 10. Show that if f is a cusp form then $f \in L^2(\Gamma\backslash H)$, using the invariant area element $y^{-2}\, dx\, dy$.

Hint. From the Fourier series for f and the asymptotic behavior of $K_s(y)$ as $y \to \infty$, you can show that a cusp form must be bounded and thus is square-integrable since the fundamental domain has finite area (see §3.3). The Fourier series for f converges uniformly on the fundamental domain by the Weierstrass M-test.

The Eisenstein series E_s is not a cusp form.

Exercise 11. Show that if $\Lambda(W, s) = \pi^{-s}\Gamma(s)Z(W, s)$ for $W \in \mathscr{SP}_2$, $s \neq 0, 1$, then

$\Lambda(W, s) \in L^1(\Gamma\backslash H, y^{-2}\,dx\,dy)$ if $0 < \operatorname{Re} s < 1$ (and not otherwise);

$\Lambda(W, s) \notin L^2(\Gamma\backslash H, y^{-2}\,dx\,dy)$ for all s.

Hint. As in Exercise 10, make use of the Fourier series for $\Lambda(W, s)$ in Exercise 4.

Now we want to study the structure of $\mathscr{N}(SL(2, \mathbb{Z}), s(s-1))$ more closely, recalling what happened for classical modular forms in §3.4. Given f in $\mathscr{M}(SL(2, \mathbb{Z}), k) = $ the classical holomorphic modular forms of weight k, there is always a constant such that $f - cG_k$ is a cusp form, where $G_k = $ the holomorphic Eisenstein series. This is not so clear in the nonholomorphic case, since the Fourier expansion of f in $\mathscr{N}(SL(2, \mathbb{Z}), s(s-1))$ begins with $ay^s + by^{1-s}$, rather than with a constant. Suppose that $\operatorname{Re} s > \frac{1}{2}$, $s \notin [0, 1]$. Then we can find a constant c such that $(f - cE_s)$ has constant term by^{1-s}. This implies that $(f - cE_s)$ is square-integrable over the fundamental domain, using the non-Euclidean invariant area element $y^{-2}\,dx\,dy$. But this contradicts the fact that the Laplace operator is negative on the fundamental domain (see Exercise 9). We have thus proved part (a) of the following theorem.

Theorem 1.

(a) *If $\operatorname{Re} s > \frac{1}{2}$, $s \notin [\frac{1}{2}, 1]$, then $\mathscr{N}(SL(2, \mathbb{Z}), s(s-1)) = \mathbb{C}E_s$.*
(b) *$\mathscr{N}(SL(2, \mathbb{Z}), 0) = \mathbb{C}$.*
(c) *If $\operatorname{Re} s = \frac{1}{2}$ or $s \in [\frac{1}{2}, 1)$, then*

$$\mathscr{N}(SL(2, \mathbb{Z}), s(s-1)) = \mathbb{C}E_s \oplus \mathscr{SN}(SL(2, \mathbb{Z}), s(s-1)).$$

Note. Later we shall prove that if $s \in [\frac{1}{2}, 1)$, $\mathscr{SN}(SL(2, \mathbb{Z}), s(s-1)) = \{0\}$ (see Theorem 3).

PROOF. These results would be easy if the constant term in the Fourier expansion of a Maass waveform did not have two parts. For you could easily subtract a multiple of the Eisenstein series to produce a cusp form. And a harmonic cusp form (i.e., a cusp form f such that $\Delta f = 0$) would have to be identically zero by the maximum principle for harmonic functions (see Garabedian [1]).

In order to circumvent this difficulty, we must show the 1-dimensionality of the vector space V consisting of constant terms in the Fourier expansion of $f \in \mathscr{N}(SL(2, \mathbb{Z}), \lambda)$. The proof of this fact requires Green's theorem on the

3.5. Automorphic Forms—Not So Classical—Maass Waveforms

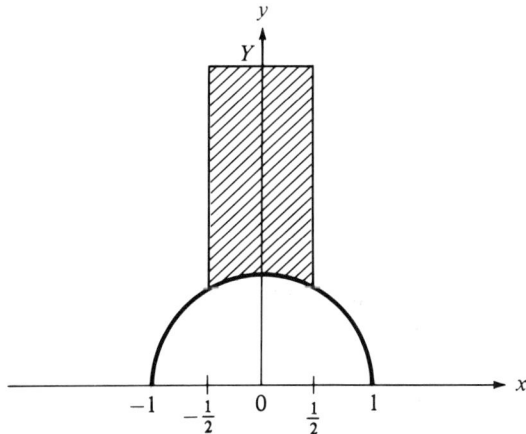

Figure 3.29. Fundamental domain truncated at $y = Y$.

truncated fundamental domain $D_Y = \{z \in H \,|\, |z| \geq 1, \operatorname{Re} z \in [-\frac{1}{2}, \frac{1}{2}], \operatorname{Im} z \leq Y\}$ pictured in Fig. 3.29. Suppose that $f \in \mathcal{N}(SL(2,\mathbb{Z}), \lambda)$ and $g \in \mathcal{N}(SL(2,\mathbb{Z})\mu)$ with Fourier expansions

$$f(z) = \sum_{m \in \mathbb{Z}} a_m(y) \exp(2\pi i m x), \qquad g(z) = \sum_{m \in \mathbb{Z}} b_m(y) \exp(2\pi i m x).$$

Then the Euclidean version of Green's theorem on the region D_Y says that if $\partial/\partial n$ is the normal derivative on the boundary ∂D_Y, and ds the differential of arc length, we have

$$\begin{aligned}
\int_{D_Y} (f \Delta g - g \Delta f) y^{-2} \, dx \, dy &= \int_{\partial D_Y} \left(f \frac{\partial g}{\partial n} - g \frac{\partial f}{\partial n} \right) ds \\
&= \int_{x=-1/2}^{+1/2} \sum_{n, m \in \mathbb{Z}} (a_n(Y) b'_m(Y) \exp[2\pi i(n+m)x] \\
&\quad - a'_n(Y) b_m(Y) \exp[2\pi i(n+m)x]) \, dx \\
&= \sum_{m \in \mathbb{Z}} [a_m(Y) b'_{-m}(Y) - a'_m(Y) b_{-m}(Y)].
\end{aligned}$$

Here we have used the fact that f and g are invariant under $SL(2,\mathbb{Z})$ to see that the integral over the boundary ∂D_Y reduces to the integral over the top horizontal line $\operatorname{Im} z = Y$. For note that the $SL(2,\mathbb{Z})$ identifications make the top the only real boundary of the manifold D_Y. The last formula comes from the orthogonality of the exponentials $\{\exp(2\pi i n x), n \in \mathbb{Z}\}$. The explicit expressions for the Fourier coefficients of f and g from Exercise 3 imply that if $m \neq 0$, then

$$0 = a_m b'_{-m} - a'_m b_{-m} \qquad \text{if } \lambda = \mu = s(s-1).$$

Therefore $\lambda = \mu = s(s-1)$ implies that $0 = a_0 b'_0 - a'_0 b_0$. Moreover, we know that

$$a_0(y) = ay^s + cy^{1-s} \quad \text{and} \quad b_0(y) = by^s + dy^{1-s}.$$

When $s \neq \frac{1}{2}$, it follows that $ad - bc = 0$, which means that the dimension of the vector space V of constant terms is indeed 1. The dimension is obviously 1 in the case that $s = \frac{1}{2}$. □

Note. You can think of the constants in part (b) as residues of the Eisenstein series at $s = 1$.

The following exercise will be useful in the last section of this chapter when we evaluate the parabolic term of the Selberg trace formula. It gives an idea of the independence of the two Eisenstein series as distributions.

Exercise 12.

(a) Show that if $\varphi(s) = \Lambda(1-s)/\Lambda(s)$, $\Lambda(s) = \pi^{-s}\Gamma(s)\zeta(2s)$, then

$$\int_{D_A} E_s E_{s'} y^{-2} \, dx \, dy = \frac{(s'-s)A^{s+s'-1}}{(s-s')(1-s'-s)} - \frac{\varphi(s)\varphi(s')A^{1-s-s'}(s'-s)}{(s-s')(1-s'-s)}$$

$$+ \frac{A^{s-s'}\varphi(s')(1-s-s')}{(s-s')(1-s'-s)} + \frac{A^{s'-s}\varphi(s)(s'+s-1)}{(s-s')(1-s'-s)}$$

$$= \left\{ \frac{A^{s+s'-1}}{s'+s-1} - \frac{A^{1-s'-s}\varphi(s)\varphi(s')}{s'+s-1} \right\} + \frac{A^{s-s'}\varphi(s')}{s-s'}$$

$$- \frac{A^{s'-s}\varphi(s)}{s-s'} + o(A), \text{ as } A \to \infty.$$

Here D_A is the truncated fundamental domain in Fig. 3.29. And E_s denotes the Eisenstein series, as usual. The notation "$o(A)$" stands for a function which, when divided by A, approaches 0, as $A \to \infty$.

(b) Let $s' = \bar{s}$, $s = \sigma + ir$, in part (a) and take the limits as $\sigma \to \frac{1}{2}$ to show that the term in braces from part (a) approaches

$$\frac{1}{2}\left\{ 4\log A - \frac{\varphi'}{\varphi}\left(\frac{1}{2} + ir\right) - \frac{\varphi'}{\varphi}\left(\frac{1}{2} - ir\right) \right\}.$$

The preceding theorem is a special case of a result of Maass [3, pp. 169–170] (see also Maass [1, pp. 195–215]). Formulas such as those in Exercise 12 have been called the *Maass-Selberg relations* (see Kubota [1, pp. 18–20], Harish-Chandra [1, p. 75], Langlands [1, p. 333]). The Maass-Selberg relations will be used in the last section of this chapter to evaluate the parabolic terms in Selberg's trace formula. These relations are used in the last three references mentioned above to obtain the analytic continuation of the Eisenstein series for very general discrete groups, a result which was quite easy for $SL(2, \mathbb{Z})$ (see Theorem 1 of §1.4). It is also possible to give inner-product formulas for truncated Eisenstein series and then no $o(A)$ term appears (see Selberg [3, pp. 183–184]).

3.5. Automorphic Forms—Not So Classical—Maass Waveforms

Now, what can be said of $\mathscr{SN}(SL(2, \mathbb{Z}), s(s-1))$? We know from Exercise 9 and Theorem 1 that this vector space of cusp forms is $\{0\}$ unless s lies in the interval $(0, 1)$ or the real part of s is $\frac{1}{2}$. Furthermore, Theorem 3 will show that, in fact, the space is also $\{0\}$ when $s \in (0, 1)$. We shall show in §7 that $\mathscr{SN}(SL(2, \mathbb{Z}), s(s-1)) \neq \{0\}$ for an infinite number of values s. However, no one has ever produced an exact value of s for which the space of cusp forms is nonzero—much less an example of a cusp form. Discussions of the existence of the discrete spectrum of the Laplacian on $SL(2, \mathbb{Z}) \backslash H$ can be found in many of the references listed at the beginning of this section. We will consider the matter in §7.

However, it *is* possible to obtain computer approximations of cusp forms. Note that a cusp form vanishes like a constant times $\exp(-2\pi y)$ as y approaches infinity. Thus, it is not too far wrong to consider the problem of finding eigenfunctions of the non-Euclidean Laplacian on the compact region D_Y in Figure 3.29. The boundary conditions would be the periodicity coming from $f(z + 1) = f(z)$ and $f(-1/z) = f(z)$, plus the vanishing of $f(z)$ on the horizontal line $y = Y$. Note, further, that the symmetry $u(x + iy) = x - iy$ leaves the domain D_Y and the Laplacian invariant. Thus the symmetry splits the space of solutions to our boundary value problem into even and odd functions. The space of odd functions will satisfy the Dirichlet problem requiring them to vanish on the boundary. The space of even functions will satisfy the Neumann problem requiring that their normal derivatives vanish on the boundary.

Exercise 13. Prove the last statements about even and odd eigenfunctions of Δ on the fundamental domain for $SL(2, \mathbb{Z})$.

Computations of the eigenvalues of the non-Euclidean Laplacian for the Dirichlet and Neumann problems have been made by Cartier [1], Cartier and Hejhal [1], Haas [1], Hejhal [3], and Hejhal and Berg using a CRAY-1 (see Hejhal [1, Appendix C to Vol. II]). We give the last table mentioned as Table 3.5 which follows. This table has a shorter list of eigenvalues for cusp forms corresponding to the Neumann problem because the numerical method that was used produces "spurious eigenvalues" for the Neumann problem. These spurious eigenvalues would fill in the blanks for Neumann eigenvalues, except that they correspond to spurious cusp forms with a logarithmic singularity at the point $\exp(2\pi i/3)$ in the fundamental domain. Hejhal [3] proves this and shows that the s corresponding to the spurious eigenvalue $\lambda = s(s-1)$ are exactly the zeros of the Dedekind zeta function of the algebraic number field $\mathbb{Q}(\exp(2\pi i/3))$.

See §1.4 for a discussion of the Dedekind zeta function. The particular one that arises in Hejhal [3] is actually also an Epstein zeta function as well as the product of $\zeta(s)$ and $L(s, (-3/*))$, formed with the Kronecker symbol $(-3/*)$. See the discussion after formula (3.61) of §3.3 for references on the Kronecker symbol. The L-function appearing here is defined by

Table 3.5. Hejhal-Berg Table of Eigenvalues λ of Δ on $\mathcal{H}/SL(2,\mathbb{Z})$, for $\lambda = s(s-1)$, $s = \frac{1}{2} + it$ [a]

t Corresponding to Odd Cusp Forms Dirichlet Problem	t Corresponding to Even Cusp Forms Neumann Problem
9.53369 52613 536	13.77975 13518 907
12.17300 83246 797	17.73856 33811
14.35850 95182 59	19.42348 147
16.13807 31715 23	21.31579 6
16.64425 92018 8	
18.18091 78346	
19.48471 38	
20.10669 4	
21.47905 7	
22.19467	
24.41971 Incomplete	
25.05085 below the line	
26.0568	
26.4469	
27.28	

The last digit in each number is uncertain. In this and the rest of the tables of this chapter $\Delta u = y^2(u_{xx} + u_{yy})$.

[a] From Hejhal [1, Vol. II, pp. 653, 730].

Table 3.6. Zeros $s = \frac{1}{2} \pm i\gamma$ of $\zeta(s)L(s,(-3/*))$

γ for $\zeta(s)$	γ for $L(s,(-3/*))$
14.134725	8.039737
21.022040	11.249206
25.010858	15.704619
30.424876	18.261997
32.935062	20.455771

$$L(s,(-3/*)) = \sum_{n \geq 1} (-3/n)n^{-s} \quad \text{for Re } s > 1.$$

This is a very special example of a class of L-functions associated to Dirichlet characters. For the analytic continuation and functional equations of such L-functions, see for example Davenport [1]. The general theory is developed in Lang [3]. One can find tables of zeros of Dirichlet L-functions (see Spira [1]) and of Riemann's zeta function (see Haselgrove and Miller [1] and Brent [1]). We list a few of these zeros in Table 3.6.

Table 3.5 should be compared with Table 3.7 which was made by Cartier [1] and lists eigenvalues for the non-Euclidean Laplacian on a rectangle. It is also interesting to compare Table 3.5 with Table 3.8 of Haas. The main difference between the Haas table and that of Hejhal and Berg is the absence in the latter

3.5. Automorphic Forms—Not So Classical—Maass Waveforms

Table 3.7. Cartier's Table of Eigenvalues $\lambda = s(s-1)$, $s = \frac{1}{2} + it$, for Δ on the Rectangle $R = \{x + iy \mid 0 < x < \frac{1}{2}, 1 < y < 5\}$

t for Dirichlet's Problem	t for Neumann's Problem
9.790	8.906
12.421	11.631
14.590	13.863
16.522	15.837
17.019	16.108
18.291	17.644
19.948	19.330
20.309	19.465

Table 3.8. The Haas Table of Eigenvalues λ of Δ on $\mathcal{H}/SL(2,\mathbb{Z})$, $\lambda = s(s-1)$, $s = \frac{1}{2} + it$ as Corrected by Hejhal to Contain 18.261997 (which had been omitted by mistake)

t for Dirichlet's Problem	t for Neumann's Problem
9.533695	8.039738
12.17301	11.24921
14.35851	13.77975
16.13807	14.13473
16.64426	15.70462
18.18092	17.73856
19.48471	18.261997
20.10669	19.42348
	20.45578

This table includes spurious eigenvalues in the Neumann list as is explained in the text. The spurious eigenvalues correspond to numbers in Table 3.6.

table of the Neumann eigenvalues corresponding to the zeros of the Dedekind zeta function just mentioned. The "spurious eigenvalues" in the Haas table come from the numerical method used both by Haas and Hejhal and Berg. All start with the Fourier expansion of an even cusp form:

$$u(z) = \sum_{n>0} c_n y^{1/2} K_{it}(2\pi n y) \cos(2\pi n x)$$

and seek to find the unknown Fourier coefficients c_n and the number t, by choosing N point $z_1, \ldots, z_N$ in H and solving

$$\frac{u(z_j) - u(z_j^*)}{|z_j - z_j^*|} = 0, \quad j = 1, \ldots, N, \quad z^* = 1/\bar{z}.$$

The condition can be rewritten as a system of N linear equations in the N unknowns c_i, provided that one forgets the terms after c_N in the Fourier expansion of the cusp form u. Thus t must satisfy

$$\det(I_i(z_j, t))_{1 \le i,j \le N} = 0 \quad \text{for } I_j(z, t) \text{ involving } K\text{-Bessel functions.}$$

Hejhal proves in the above-mentioned paper that the eigenvalues which appear on the Haas list and not on the Hejhal list (the zeros of the Dedekind zeta function of $\mathbb{Q}(\exp(2\pi i/3))$ are spurious because they correspond to eigenfunctions that behave exactly like cusp forms except that they have a logarithmic singularity at the point $\exp(2\pi i/3)$ in the fundamental domain. So the Fourier expansion that is produced by this numerical method will converge only for $y > \sqrt{3}/2$ and not for $y \leq \sqrt{3}/2$. One can draw a very interesting moral on the care required in using computers (especially when dealing with Neumann boundary-value problems with continuous spectra).

Hejhal also notes that the Fourier coefficients c_n in the expansion

$$f(z) = \sum_{n \neq 0} c_n \exp(2\pi i n x) y^{1/2} K_{it}(2\pi |n| y)$$

for a spurious eigenform $f(z)$ do not satisfy the inequality

$$|c_n| \leq \sigma_0(n) n^{3/10}, \tag{3.71}$$

where $\sigma_0(n)$ is the number of positive divisors of n. However, as we shall see in Theorem 4 of §3.6, we can assume that the cusp form $f(z)$ is an eigenfunction for all the Hecke operators and we can normalize $f(z)$ so that the first Fourier coefficient $c_1 = 1$. For such forms, the inequality (3.71) has been proved (see Moreno [1] and Selberg [2]). This gives another indication that the spurious cusp forms are suspect.

The inequality (3.71) is a weak form of the *Ramanujan-Petersson conjecture for nonholomorphic cusp forms which are normalized eigenfunctions of the Hecke operators*. This conjecture says

$$|c_p| \leq 2 \quad \text{for all primes } p. \tag{3.72}$$

Such a result is not contained in Deligne's theorem stated in Theorem 4 of §3.3. Better bounds than that given in (3.71) have been obtained, but (3.72) is still unproved as we write this (see Marie-France Vignéras [5] for a discussion of the improvement of the exponent $\frac{3}{10}$ in (3.71) to $\frac{1}{4}$, as well as a discussion of the connections with group representations). Tables 3.9 and 3.10 provide a small amount of numerical evidence for the Ramanujan-Petersson conjecture (3.72), by listing the first few Fourier coefficients corresponding to various Maass waveforms.

The numerical methods used by Hejhal and most of the others are limited by the decreasing size of the K-Bessel functions in the Fourier expansions of cusp forms. Accurate evaluations of the K-Bessel functions are essential. Riho Terras [3] gives some very elegant procedures to accomplish this task (see also Purdy, Terras, Terras, and Williams [1]).

In producing Table 3.10 of Fourier coefficients for the first even nonholomorphic cusp form, Stark [10] used the fact that the cusp form is an eigenfunction for the Hecke operators to be considered in the next section. Thus we have the equation (on p. 223)

3.5. Automorphic Forms—Not So Classical—Maass Waveforms

Table 3.9. The Hejhal-Berg Table of Fourier Coefficients of Cusp Forms for the Eigenvalue λ of Δ on $\mathscr{H}/SL(2,\mathbb{Z})$, $\lambda = s(s-1)$, $s = \frac{1}{2} + it$ (c_n denoting the nth Fourier coefficient)[a]

t for Even Cusp Forms	c_2	c_3	c_5	c_7
13.77975 13518 907	1.54930447794	.24689977245	.7370604	−.2614
17.73856 33811	−.7654580566	−.9777789075	−1.0152735	1.1807
19.42348 147	−.69276198	1.5623543	−.038412	.313
21.31579 6	1.287529	1.251773	1.170	−.54

t for Odd Cusp Forms				
9.53369 52613 536	−1.0683335512	−.456197354	−.29067256	−.7449
12.17300 83246 797	.2892518714	−1.201858761	.03955272	.4481
14.35850 95182 59	−.2309151912	.6955949863	−1.2982845	−.4834
16.13807 31715 23	1.161855592	−1.281972561	−.7568063	−.2985
16.64425 92018 8	−1.540227825	.977492591	−.105242	−.693
18.18091 78346	.374063346	.101958698	.637331	−1.542
19.48471 38	−1.7001880	−.6145654	.8198	.063
20.10669 4	.858844	.187277	−1.395	.78
21.47905 7	−.656251	.226442	1.082	.42
22.19467	1.59685	−1.11648	−.637	−1.00

The last digit in each number is doubtful.

Hejhal appends this to his table: "Ausserdem ist in Erwägung zu ziehen, dass die oben angegebenen numerischen Werte von $v(x)$ mit Hilfe von Rechenmaschinen bestimmt wurden und daher ebenfalls im strengen Sinne unbewiesen sind" (from Siegel [5, Vol. III, p. 439]).*

[a] From Hejhal [1, Vol. II, pp. 653, 730].

Table 3.10. Stark's Table of Fourier Coefficients for the First Even Cusp Form[†]

P	C(P)	P	C(P)	P	C(P)
2	1.5493044779	269	−1.4815318734	617	−0.9413471724
3	0.2468997725	271	0.5040655357	619	−1.2022515538
5	0.7370603853	277	0.0799184237	631	1.2204182342
7	−0.2614200758	281	1.2661555488	641	−1.6267782096
11	−0.9535646526	283	−1.6137265420	643	−1.2717803924
13	0.2788270292	293	−0.9962645200	647	1.3158148270
17	1.3073417145	307	1.1244326539	653	−0.7308513965
19	0.0925585825	311	−0.0039967298	659	−1.2242624630
23	1.1380685214	313	0.4789645104	661	−0.0234437218
29	0.7521138455	317	−1.5835570331	673	1.3778855357
31	0.0248519535	331	1.1152062270	677	0.0265995820
37	0.1992656556	337	0.4390031691	683	0.5292669101
41	−0.3040329968	347	−1.4315675896	691	−1.3160485151

* "One also has to keep in mind that the numerical values of $v(x)$ were determined with the help of a calculator and have consequently not been proved in a strict sense."

Table 3.10. (cont.)

P	C(P)	P	C(P)	P	C(P)
43	0.7832393635	349	−0.3501353058	701	1.5888219871
47	0.3605684105	353	1.7625605824	709	−0.5251706918
53	1.3980657196	359	0.5325184967	719	−1.0362134271
59	−1.5877309619	367	−0.0033416821	727	−0.7502921970
61	1.1672759688	373	−0.9876462546	733	−0.9208587111
67	−0.0270938863	379	1.5933050198	739	0.8958405935
71	−1.6334238582	383	0.0502016847	743	1.6323865759
73	1.0678477369	389	−0.3273589916	751	−0.9959260442
79	−0.5311738889	397	−1.8931512743	757	1.3259008900
83	−0.9045323799	401	−0.8126034389	761	−0.5856192114
89	−1.0673531716	409	1.4753959508	769	−0.1033879313
97	−0.0032571155	419	0.7396796406	773	−1.4675416297
101	0.8641852969	421	0.8792514066	787	−0.4231228583
103	−1.2318448417	431	−1.2083667833	797	0.1724453075
107	−0.8126520455	433	−0.1571375484	809	−0.3987763947
109	−0.1537416810	439	−0.6020990403	811	−0.9581311790
113	0.8117725443	443	1.0132298238	821	−0.5238603585
127	1.1696610310	449	0.4461683165	823	0.1600228277
131	−0.6111258748	457	−0.7081760488	827	−0.1779829984
137	0.7718026731	461	0.3300809302	829	0.4892801545
139	0.0953561766	463	1.2726492305	839	1.5943055357
149	−0.1869675229	467	0.1481146844	853	−0.0944388002
151	−0.2836096280	479	1.1482791938	857	1.2009724917
157	1.1464542083	487	−1.1381929615	859	−0.9219874648
163	1.5036129830	491	−1.4887418682	863	1.1751819412
167	−0.7243411282	499	0.4359170070	877	−0.4805639483
173	0.0377358321	503	0.3463320737	881	−1.1622088814
179	−0.5924790734	509	0.7903047596	883	1.4610158527
181	1.8928188557	521	−0.8278509888	887	−1.9462665557
191	0.4549235996	523	−1.0508097534	907	−0.1627440731
193	−0.3510653631	541	0.6476703225	911	0.5225227593
197	−0.4934900086	547	0.0132913465	919	−1.0727170339
199	0.7615593277	557	−0.8512529528	929	−0.1305562742
211	1.6334955452	563	0.7157534392	937	−1.0585435062
223	−0.8898352549	569	−0.6677981282	941	−0.7542322227
227	−1.1170228371	571	−0.2379643425	947	1.3539019798
229	−1.1231294363	577	0.9500705634	953	0.4262552906
233	1.1318288732	587	1.0559415688	967	0.7553483561
239	−0.6112502497	593	0.1197484790	971	−0.0605613598
241	1.6168818500	599	0.8007081710	977	−0.3377517502
251	−0.4397565620	601	1.2754657449	983	−1.7315964075
257	0.3781399469	607	−1.5299526484	991	1.6842084188
263	1.0274020100	613	0.6464422791	997	−1.0060566454

† From Stark [10]. Reproduced with permission from R.A. Rankin (Ed.), *Modular Forms*, Ellis Horwood, Chichester, England, 1984.

Normalized eigenvalue = 13.7797513519.

3.5. Automorphic Forms—Not So Classical—Maass Waveforms

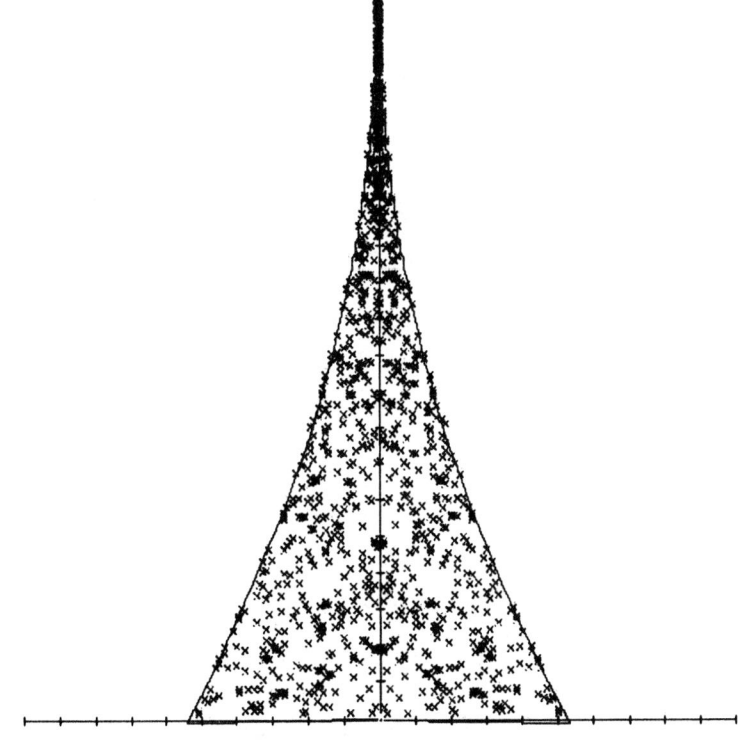

Figure 3.30. Images of points used by Stark in the computation of Table 6 for $p = 983$ and $z = 2i$ under the Cayley transform.

$$c_p f(z) = p^{-1/2} \sum_{j \bmod p} f\left(\frac{z+j}{p}\right) + p^{-1/2} f(pz)$$

(see Theorem 4 of §3.6). Stark then fixes a point $z \cong 1.4i$ and looks at points $z_j = (z+j)/p$. One can find a matrix $A_j \in SL(2, \mathbb{Z})$ such that $w_j = A_j z_j$ lies in the standard fundamental domain for $SL(2, \mathbb{Z})$. So our equation becomes

$$c_p f(z) = p^{-1/2} \sum_{j \bmod p} f(w_j) + p^{-1/2} f(pz). \quad (*)$$

One then uses $(*)$ to obtain the coefficients c_p recursively with more and more accuracy. You calculate $f(z)$ and the other values of f using the Fourier expansion of f, with a few Fourier coefficients; e.g., $c_1 = 1$. Then you get some digits of c_2. You plug those in and get more digits. This assumes that you know the eigenvalue, but you can also use the process to approximate eigenvalues by checking that the function f approximated is invariant under the substitution $z \to -1/z$. The method requires Riho Terras' recursive methods for computing the K-Bessel functions, applied to K_{it} (see R. Terras [3]).

Figure 3.30 shows a plot of the points w_j used by Stark in $(*)$ above for $p = 983$, after they have been moved to the unit disc by the Cayley transform

$z \to i(z - i)/(z + i)$. One can then ask whether these points w_j, j mod p, become uniformly distributed (with respect to the non-Euclidean area) in the fundamental domain as p approaches infinity. This would account for the fact that the sum $(*)$ of p terms times $p^{-1/2}$ seems to be giving a result which is smaller by a factor of $p^{-1/2}$ than one would expect. In fact, the Ramanujan-Petersson conjecture for Maass waveforms would follow from such uniform distribution of the points w_j in the fundamental domain. Indeed, Stark has suggested that perhaps one could use the Ramanujan-Petersson conjecture for holomorphic modular forms to show that the desired cancellation by $p^{-1/2}$ does occur.

Some History

In the spring of 1977 Riho Terras (referred to as *mon ex-mari* in Cartier & Hejhal [1]) and I spent a personally disastrous but academically exciting sabbatical at the University of Bonn. During that time we visited Paris and R.T. spoke to Cartier about computing K-Bessel functions to improve the eigenvalue programs. We were lucky enough to witness a general strike and R.T.'s talk had to be cancelled. Later that spring I visited H. Maass in Heidelberg and met H. Neuenhöffer, who showed me the calculations of Haas. No one mentioned $\zeta(s)$ and my thoughts were only on the eigenvalue problem (and my own troubles). Then Neuenhöffer attempted to send me the Haas manuscript, but only the outside envelope arrived. Was someone in the Post Office interested in the Riemann hypothesis? Unknown! But finally a letter from Neuenhöffer reached me which included the Haas tables. I put some of these tables in the first version of this book, on which I lectured at M.I.T. in Fall, 1978. Then H. Stark noticed the zeros of $\zeta(s)$ in the Haas table. But he overlooked the L-function zeros somehow. By that time probably 100 people had a preliminary version of this book with the Haas table in it. But very few people looked at it, perhaps. Anyway, ultimately D. Hejhal visited U.C.S.D. and I showed him the table. He immediately said that it gave him a headache. Evidently he went to the library and found the L-function zeros also, but he did not tell me that. Instead he wrote a very cryptic remark on my blackboard, which he does not want printed here. But I think that he did not believe the Haas table even then. He worked hard for several months, and the result was Hejhal [3], which was announced in summer 1979 at the Durham conference.

In the preliminary version of this book we had attempted to use the Courant minimax principle to compare eigenvalue problems such as those represented by Tables 3.7 and 3.8 (see Courant and Hilbert [1, Vol. I, p. 409]). However, the existence of a continuous spectrum makes these arguments go awry.

One might still wonder whether there is any chance to prove the Riemann hypothesis by interpreting the zeros $s = \frac{1}{2} + it$ of $\zeta(s)$ as giving eigenvalues $s(s - 1)$ of some self-adjoint operator. Pólya and Hilbert evidently suggested this independently around 1915, though they were probably not thinking of the eigenvalue problem discussed here. Hejhal's work appears to have laid to rest the possibility that Δ is the correct operator unless one can somehow deal

3.5. Automorphic Forms—Not So Classical—Maass Waveforms

with spurious eigenfunctions with logarithmic singularities. The fact that Epstein zeta functions do not satisfy the Riemann hypothesis in most cases leads one to become dubious about this approach. In fact, the Epstein zeta function for the matrix $\begin{pmatrix} 1 & 0 \\ 0 & 5 \end{pmatrix}$ has an infinite number of zeros in the region $\operatorname{Re} s > 1$ (see Titchmarsh [3, p. 244]). See Venkov [3, p. 159] for a unique published version of a conjectural meaning for the eigenvalues associated to Maass cusp forms and their analogues.

Exercise 14. Try to check Table 3.5.

Note. Hejhal says this is too hard. A class of beginning physics and engineering graduate students agreed when they tried the standard finite-element method on the problem. You need good K_{it} programs (e.g., those R. Terras [3]) if you use the Fourier expansion method that Hejhal used. We give an idea for a different method in §3.7 (see Theorem 3 of that section and the discussion following it). It would also be interesting to program Stark's method using Hecke operators from §3.6 to produce Table 3.10.

Maass [3] shows that there are cusp forms for congruence subgroups of $SL(2, \mathbb{Z})$ (but not for $SL(2, \mathbb{Z})$ itself) which arise from Hecke L-functions of real quadratic number fields. Marie-France Vignéras [1] derives this in another way by integrating a certain theta function. We will return to this topic in the next section. It does not appear that anyone has been able to use such constructions to obtain cusp forms for $SL(2, \mathbb{Z})$, however.

Theorem 2 (Estimates for Fourier Coefficients of Maass Waveforms).

(a) *Let* $E_s^*(z) = \pi^{-s}\Gamma(s)\zeta(2s)E_s(z)$ *and*

$$E_s^*(z) = ay^s + by^{1-s} + \sum_{n \neq 0} a_n y^{1/2} K_{s-1/2}(2\pi|n|y)\exp(2\pi i n x).$$

Then the Fourier coefficients a_n satisfy the inequality

$$|a_n| \leq C_\varepsilon |n|^{\operatorname{Re} s - 1/2 + \varepsilon},$$

where the positive constant C_ε exists for every $\varepsilon > 0$.

(b) *Suppose that $f \in \mathcal{SN}(SL(2, \mathbb{Z}), s(s-1))$ has the Fourier expansion*

$$f(z) = \sum_{n \neq 0} a_n y^{1/2} K_{s-1/2}(2\pi|n|y)\exp(2\pi i n x).$$

Then

$$|a_n| < C|n|^{1/2}.$$

PROOF.

(a) We know from Exercise 4 that, if $\operatorname{Re} s > 1$, then

$$|a_n| = 2|n|^{\operatorname{Re} s - 1/2}|\sigma_{1-2s}(n)| \leq 2|n|^{\operatorname{Re} s - 1/2}\zeta(2\operatorname{Re} s - 1).$$

However, if $\tfrac{1}{2} \leq \operatorname{Re} s \leq 1$, then $|a_n| < 2|n|^{\operatorname{Re} s - 1/2}\sigma_0(n)$. Then Hardy and Wright

tell us that given $\varepsilon > 0$, there is a positive constant C_ε such that $\sigma_0(n) < C_\varepsilon n^\varepsilon$ (see Hardy and Wright [1, pp. 260–262]).
(b) The proof proceeds as in Exercise 16 of §3.3. Use

$$a_n y^{1/2} K_{s-1/2}(2\pi|n|y) = \int_0^1 f(x+iy)\exp(-2\pi inx)\,dx,$$

and the fact that cusp forms are bounded on the upper half-plane. Then one sets $y = c/|n|$, where c is chosen so that $K_{s-1/2}(2\pi c) \neq 0$, to complete the proof. □

Exercise 15. Fill in the details in the proof of Theorem 2.

Note that in Theorem 2, when $\operatorname{Re} s = \frac{1}{2}$, one has a better estimate for the Fourier coefficients of Eisenstein series than one has for cusp forms. This is rather strange if we remember the holomorphic case. There are better estimates than that given in Theorem 2 when the cusp form is an eigenfunction for all the Hecke operators and the Ramanujan-Petersson conjecture (3.72) above would say that the same sort of estimate would then hold. See the earlier discussion of (3.71) and (3.72) for some references on Fourier coefficients of cusp forms. In addition, one can consult Bruggeman [1–3] and Kuznetsov [1].

Theorem 3. Suppose $\mathscr{SN}(SL(2,\mathbb{Z}),\lambda) \neq \{0\}$, then $\lambda \leq -3\pi^2/2$.

PROOF. (Roelcke [2, §7]). Suppose that

$$u(z) = \sum_{n \neq 0} a_n(y)\exp(2\pi inx) \in \mathscr{SN}(SL(2,\mathbb{Z}),\lambda).$$

Let D denote the usual fundamental domain for $SL(2,\mathbb{Z})$, given in Fig. 3.11. And let $D^* = \gamma(D)$, where $\gamma(z) = -1/z$. Then if $(u,u) = 1$, as in Exercise 9, part (a), we have

$$2|\lambda| = -2(\Delta u, u) = \int_{D \cup D^*} \left(\left|\frac{\partial u}{\partial x}\right|^2 + \left|\frac{\partial u}{\partial y}\right|^2 \right) dx\,dy \geq \int_{\substack{|x| \leq 1/2 \\ y \geq \sqrt{3}/2}} |u_x|^2 dx\,dy$$

$$= \int_{y \geq \sqrt{3}/2} \sum_{n \neq 0} 4\pi^2 n^2 |a_n(y)|^2 dy \geq 3\pi^2 \int_{y \geq \sqrt{3}/2} \sum_{n \neq 0} |a_n(y)|^2 y^{-2} dy$$

$$= 3\pi^2 \int_{y \geq \sqrt{3}/2} \int_{|x| \leq 1/2} |u(z)|^2 y^{-2} dx\,dy \geq 3\pi^2 \int_D |u|^2 y^{-2} dx\,dy = 3\pi^2.$$

This completes the proof. □

Note that Hejhal's Table 3.5 gives a much better estimate for the first eigenvalue corresponding to a cusp form. However, Theorem 3 is of interest, because the argument can be applied to a few congruence subgroups of small level.

3.5. Automorphic Forms—Not So Classical—Maass Waveforms

You might wonder whether there are discrete groups Γ in $SL(2, \mathbb{R})$ such that $0 \leq |\lambda| \leq \frac{1}{4}$ and $\mathscr{SN}(\Gamma, \lambda) \neq \{0\}$. Selberg [2] gives examples of such groups Γ. Randol [1] proves that such groups Γ exist with $\Gamma \backslash H$ compact and having as many eigenvalues λ in $(-a, 0)$ as you wish, for any given a. Moreover, Selberg [2, pp. 13–14] asserts that if Γ is or contains a congruence subgroup, then there is a lower bound of $\frac{3}{16}$ on $|\lambda|$ with $\mathscr{SN}(\Gamma, \lambda) \neq \{0\}$. And there are conjectures that then $|\lambda| \geq \frac{1}{4}$. Elstrodt [2], Hejhal [1, Vol. II], and Vignéras [5] give more history of this problem and many more references. In particular, Vignéras notes that this conjecture and that of Ramanujan and Petersson given in (3.72) above the "deux volets d'une même conjecture"* in representation theory (see also Satake's article in Borel and Mostow [1, pp. 261–262]). Piatetskii-Shapiro [3] notes that the statement in adelic representation theoretic language can fail, but not for $GL(n)$, using results of Jacquet and Shalika.

Theorem 4. *The vector space of cusp forms $\mathscr{SN}(SL(2, \mathbb{Z}), \lambda)$ is finite dimensional.*

PROOF (Maass [3, pp. 154–156], using an idea of Siegel [5, Vol. II, pp. 97–137]). Let $f \in \mathscr{SN}(SL(2, \mathbb{Z}), s(s - 1))$. Suppose that the Fourier coefficients in Exercise 3 are all zero, up to the coefficients a_n with $|n| = m$. Then we can show that if m is sufficiently large (depending on t, when $s = \frac{1}{2} + it$), it forces f to be identically zero. We know that f is bounded on the upper half-plane by Exercise 10. Thus $|f(z)|$ has a maximum at some point $z_0 \in H$. Set $z_0 = x_0 + iy_0$ and note that

$$a_n y_0^{1/2} K_{it}(2\pi |n| y_0) = \int_{x=-1/2}^{1/2} f(x + iy_0/2) \, dx \left(\frac{\sqrt{2} K_{it}(2\pi |n| y_0)}{K_{it}(\pi |n| y_0)} \right).$$

It follows that

$$M = |f(z_0)| < 2\sqrt{2} M \sum_{|n| > m} K_{it}(2\pi n y_0) / K_{it}(\pi n y_0).$$

The asymptotic formula for the K-Bessel function (see Exercise 2 of §3.2 or Lebedev [1, p. 123]) shows that if m is sufficiently large (depending on t), the preceding series of quotients of K-Bessel functions can be etimated by exponentials. The result is that

$$M \leq CM \exp(-\pi m y_0).$$

If m is larger than $(\log C)/\pi y_0$, it follows that M is zero and, thus, that f is identically zero, as was to be proved. □

Exercise 16. Use the error term in the asymptotic expansion for $K_{it}(y)$ to obtain a bound for the dimension of the space $\mathscr{SN}(SL(2, \mathbb{Z}), -(t^2 + \frac{1}{4}))$. This bound will be an increasing function of t.

---

* Two parts of the same conjecture.

Hint. See Lebedev [1, p. 123] for the requisite asymptotic expansion of the K-Bessel function.

Note. There are, perhaps doubtful, conjectures that all of the dimensions of the spaces of cusp forms in Theorem 4 are, in fact, equal to 1 (see Cartier [1]). Exercise 16 does nothing to support this conjecture, but Table 3.5 does give a small amount of evidence in favor of this suspicion. See Randol [2] for a proof that for sufficiently large eigenvalues, the dimensions of spaces of cusp forms for $\Gamma(p)$, p an odd prime, must be greater than 1, using knowledge of the degrees of irreducible representations of the finite simple group $PSL(2, \mathbb{Z}/p\mathbb{Z})$.

We can obtain an asymptotic formula for the sums of dimensions of spaces of cusp forms with eigenvalues less than or equal to x, as x approaches infinity, once we have proved the Selberg trace formula (see §3.7). The trace formula implies that if $N(x)$ denotes the number of eigenvalues λ_n of Δ corresponding to cusp forms, counted with multiplicity, such that $|\lambda_n| \leq x$, then

$$N(x) \sim \frac{\text{area}(SL(2,\mathbb{Z})\backslash H)}{4\pi} x = \frac{x}{12} \quad \text{as } x \to \infty. \tag{3.73}$$

The asymptotic formula (3.73) is the non-Euclidean analogue of the Weyl asymptotic law for the distribution of eigenvalues of the Euclidean Dirichlet problem in a compact domain in $\mathbb{R}^n$ (see Theorem 5 of §1.3 and Theorem 5 of §3.7).

In particular, formula (3.73) implies that there are infinitely many Maass cusp forms for $SL(2, \mathbb{Z})$. This application of the trace formula was pointed out by Selberg [6]. The argument only works for discrete subgroups Γ of $SL(2, \mathbb{Z})$ such that $\Gamma\backslash H$ is noncompact of finite volume, with the property that there are good bounds on the constant term coefficients in the Fourier expansion of the Eisenstein series on the line $\text{Re } s = \frac{1}{2}$. And Roelcke [2] gives more examples of groups Γ with infinitely many cusp forms. So there arose a conjecture, called the *Roelcke-Selberg conjecture*, to the effect that there are infinitely many cusp forms for very general discrete groups. Venkov has proved the infinite dimensionality of spaces of cusp forms with more general transformation properties for a wide class of discrete groups (see Venkov [1, 3]). But recently Phillips and Sarnak [1] have cast doubts on the conjecture.

It is natural to ask whether there are connections between nonanalytic and analytic cusp forms. We will find that there are even more analogies in the next section. There are certainly various constructions that allow one to go from forms of one type to those of another type. Often this is done using a theta function for indefinite quadratic forms as a kernel and integrating (see Kudla [1, 2] and Vignéras [1]). Theta functions of indefinite quadratic forms and related matters are also discussed in Siegel [5, Vol. III, in papers nos. 55, 58, 60]. A differential operator which maps nonholomorphic automorphic forms satisfying higher-order differential equations into holomorphic forms (in certain cases) is given by Schwandt [1]. See also Roelcke [1, Part I, §6] and H.

3.6. Automorphic Forms and Dirichlet Series. Hecke Theory and Generalizations

Neuenhöffer [1]. Neuenhöffer gives a construction of nonholomorphic cusp forms via residues of Poincaré series.

3.6. Automorphic Forms and Dirichlet Series. Hecke Theory and Generalizations

> ... Hecke took up the subject of modular functions and put it back into number theory where it had always belonged ....
>
> From Weil [5, Vol. III, p. 301].

We saw in the proof of Theorem 1 of §1.4 that the Mellin transform allows one to deduce the basic properties of the Epstein zeta function from those of the theta function. This is only a special case of a very general theory whose boundaries have not yet been seen.

In the 1930s Hecke looked at the following situation (see Hecke [1, pp. 591–626, 627–643, 644–671, 672–707, etc.] and [2]).

Definition. Suppose that f is a classical holomorphic modular form (see §3.4); i.e., $f \in \mathcal{M}(SL(2,\mathbb{Z}), k)$ with Fourier expansion

$$f(z) = \sum_{n \geq 0} a_n \exp(2\pi i n z), \qquad (3.74)$$

and consider

$$\Lambda_f(s) = \int_0^\infty (f(iy) - a_0) y^{s-1} \, dy. \qquad (3.75)$$

Then from Euler's formula for the gamma function, we know that for $\operatorname{Re} x > 0$ and $\operatorname{Re} s > 0$,

$$x^{-s} \Gamma(s) = \int_0^\infty y^{s-1} e^{-xy} \, dy$$

(see Lebedev [1, Ch. 1] for a discussion of $\Gamma(s)$). It follows that

$$\Lambda_f(s) = (2\pi)^{-s} \Gamma(s) L_f(s) \quad \text{with } L_f(s) = \sum_{n \geq 1} a_n n^{-s}. \qquad (3.76)$$

The L-series in (3.76) converges for $\operatorname{Re} s > k$, by Theorem 4 of §3.4.

Exercise 1 (Analytic Continuation and Functional Equation of L-Series Corresponding to Holomorphic Modular Forms).

(a) Show that $f \in \mathcal{M}(SL(2,\mathbb{Z}), k)$ implies that $\Lambda_f(s)$ defined by (3.75) can be continued to a meromorphic function of s in $\mathbb{C}$ so that

$$\Lambda_f(s) + a_0\left(\frac{1}{s} + \frac{i^k}{k-s}\right)$$

is entire and bounded in vertical strips (*EBV* for short).
(b) Show that, moreover, $\Lambda_f(s)$ satisfies the functional equation

$$\Lambda_f(s) = i^k \Lambda_f(k-s).$$

Hint. Imitate the proof of Theorem 1 of §1.4. That is, split up the Mellin transform representing $\Lambda_f(s)$ into

$$\int_0^1 + \int_1^\infty.$$

Then use the transformation formula $f(z) = z^{-k}f(-1/z)$ to rewrite the integral over $(0, 1)$. The result is that, as before, the *L*-series can be evaluated by an *incomplete gamma expansion*:

$$\Lambda_f(s) = a_0\left(\frac{i^k}{s-k} - \frac{1}{s}\right) + \sum_{n \geq 1} a_n\big(G(s, 2\pi n) + G(k-s, 2\pi n)i^k\big), \quad (3.77)$$

with the incomplete gamma function

$$G(s, x) = \int_1^\infty t^{s-1} \exp(-xt)\, dt \quad \text{for } \operatorname{Re} x > 0,$$

as in Theorem 1 of §1.4. The exercise follows from this expansion, using the exponential decay of $G(s, x)$, as x approaches infinity (see Exercise 7 of §1.4).

Hecke noticed that Mellin inversion (see §1.4) allows the argument of Exercise 1 to be turned around.

Exercise 2 (A Converse Result in Hecke Theory). Suppose that $f(z)$, with Fourier expansion in formula (3.74), is given and consider the function $\Lambda_f(s)$ defined by (3.75) and (3.76). Suppose that $\Lambda_f(s)$ has the functional equation

$$\Lambda_f(s) = i^k \Lambda_f(k-s)$$

and that $\Lambda_f(s) + a_0(1/s + i^k/(k-s))$ is entire and bounded in vertical strips (*EBV*). Prove that then $f(z)$ is a holomorphic modular form of weight k for $SL(2, \mathbb{Z})$; i.e., that f lies in $\mathscr{M}(SL(2, \mathbb{Z}), k)$.

Hints. Note that $z \mapsto z + 1$ and $z \mapsto -1/z$ generate $SL(2, \mathbb{Z})/\pm I = PSL(2, \mathbb{Z})$. Thus, it suffices for us to show that $f(z) = z^{-k}f(-1/z)$.

Use Mellin inversion to write

$$f(iy) - a_0 = \frac{1}{2\pi i} \int_{\operatorname{Re} s = c} y^{-s} \Lambda_f(s)\, ds,$$

where c is sufficiently large for the absolute convergence of Λ_f. Push the line of

3.6. Automorphic Forms and Dirichlet Series. Hecke Theory and Generalizations   231

integration to the left and pick up residues at $s = k$ and $s = 0$, to obtain

$$f(iy) - a_0 = \frac{1}{2\pi i} \int_{\operatorname{Re} s = k-c} y^{-s}\Lambda_f(s)\,ds + i^k y^{-k} a_0 - a_0.$$

Then use $\Lambda_f(k-s) = \Lambda_f(s)i^k$, to see that $f(iy) = (i/y)^k f(i/y)$ for all $y > 0$. To complete the exercise, note that if two holomorphic functions on H agree on a set with an accumulation point, they must agree on all of H.

Some references for these exercises are Hecke [1, 2], Lang [6], Ogg [1], and Shimura [1]. Note that more general groups than $SL(2, \mathbb{Z})$ can also be treated in this way. The following theorem follows from the exercises.

Theorem 1 (Hecke's Correspondence for Holomorphic Modular Forms for $SL(2, \mathbb{Z})$). *Suppose that $f(z)$ has the Fourier expansion given in formula (3.74). Suppose that the corresponding Dirichlet series is given by formulas (3.75) and (3.76). Then we have the equivalence*

$$\left(f \in \mathcal{M}(SL(2, \mathbb{Z}), k) \right) \Leftrightarrow \left(\begin{array}{c} \Lambda_f(s) + a_0\left(\dfrac{1}{s} + \dfrac{i^k}{k-s}\right) \text{ is EBV with} \\ \Lambda_f(s) = \Lambda_f(k-s)i^k. \end{array} \right)$$

Here EBV means entire and bounded in vertical strips.

From our point of view, the next logical question was answered in the 1940s by Maass [3]. This question is: What is the analogue of Theorem 1 for Maass waveforms? Another reference for the answer is Maass [1].

In Hecke's case we needed to know the Mellin transform

$$\int_0^\infty e^{-y} y^{s-1}\,dy = \Gamma(s) \qquad \text{for } \operatorname{Re} s > 0. \tag{3.78}$$

In Maass's case we need to know the Mellin transform

$$\int_0^\infty K_r(y) y^{s-1}\,dy = 2^{s-2} \Gamma\!\left(\frac{s+r}{2}\right) \Gamma\!\left(\frac{s-r}{2}\right), \qquad \text{if } \operatorname{Re} s > |\operatorname{Re} r|. \tag{3.79}$$

Exercise 3. Prove formula (3.79). (This was part (b) of Exercise 7 in §3.2).

Hint. Use the integral formula in Exercise 1 of §3.2 for the the K-Bessel function to see that

$$\int_{y>0} K_r(y) y^{s-1}\,dy = \frac{1}{2} \int_{y>0} \int_{t>0} y^{s-1} t^{r-1} \exp\left[-\tfrac{1}{2} y(t + t^{-1})\right] dt\,dy.$$

Then make the substitution $u = yt$, $v = y/t$, to complete the proof.

Definition. Let f be a Maass waveform; i.e., $f \in \mathcal{N}(SL(2, \mathbb{Z}), r(r-1))$, with Fourier expansion as given in Exercise 3 of §3.5:

$$f(z) = ay^r + by^{1-r} + \sum_{n \neq 0} a_n y^{1/2} K_{r-1/2}(2\pi|n|y) \exp(2\pi i n x). \tag{3.80}$$

Consider the Mellin transform:

$$W_f(s) = \int_0^\infty (f(iy) - ay^r - by^{1-r}) y^{s-1} \, dy. \tag{3.81}$$

Then, Exercise 3 shows that

$$W_f(s) = 2^{-5/2} \pi^{-(s+1/2)} \Gamma\left(\frac{s-r+1}{2}\right) \Gamma\left(\frac{s+r}{2}\right) L_f(s),$$

with $$\tag{3.82}$$

$$L_f(s) = \sum_{n \neq 0} a_n |n|^{-(s+1/2)}.$$

This converges by Theorem 2 of §3.5 when $\operatorname{Re} s > |\operatorname{Re} r|$, if f is not a cusp form, and for $\operatorname{Re} s > 1$, if f is a cusp form.

Exercise 4 (Analytic Continuation and Functional Equation of L-Series Corresponding to Nonholomorphic Modular Forms). Define the *higher-dimensional incomplete gamma functions* $G(s_1, s_2; a)$ by

$$G(s_1, s_2; a) = \iint_{uv \geq 1} u^{s_1 - 1} v^{s_2 - 1} \exp[-a(u+v)] \, du \, dv, \quad \text{for } \operatorname{Re} s > 0.$$

(a) Use the same trick as in Exercise 1 to show that $W_f(s)$ in (3.82) has the following incomplete gamma expansion, assuming that $f \in \mathcal{N}(SL(2, \mathbb{Z}), s(s-1))$:

$$W_f(s) = a\left(\frac{1}{s-r} + \frac{1}{-s-r}\right) + b\left(\frac{1}{s+r-1} + \frac{1}{-s+r-1}\right)$$

$$+ \sum_{n \neq 0} a_n \left(G\left(\frac{s+r}{2}, \frac{s-r+1}{2}; \pi|n|\right) + G\left(\frac{-s+r}{2}, \frac{1-s-r}{2}; \pi|n|\right) \right).$$

(b) Show that the incomplete gamma function $G(s_1, s_2; a)$ dies off exponentially as a approaches infinity.

(c) Deduce that $W_f(s)$ continues to a meromorphic function of s such that

$$W_f(s) - a\left(\frac{1}{s-r} + \frac{1}{-s-r}\right) - b\left(\frac{1}{s+r-1} + \frac{1}{-s+r-1}\right)$$

is EBV, with functional equation $W_f(s) = W_f(-s)$.

Hint. To prove (a) use the same substitution that appeared in Exercise 3.

3.6. Automorphic Forms and Dirichlet Series. Hecke Theory and Generalizations

Now we want to argue that Mellin inversion implies a converse to Exercise 4, analogous to Hecke's converse result in Theorem 1. However, in Exercise 2, we use a fact about holomorphic functions to show that $f(iy) = (i/y)^k f(i/y)$ for all $y > 0$ implies that $f(z) = z^{-k} f(-1/z)$ for all z in H. But real analytic functions $f(z)$ and $g(z)$ which coincide on Re $z = 0$ are not necessarily equal. For an example, let $f(z) = \exp(iy)$ and $g(z) = \exp(x + iy)$. Thus we need the following exercise from Maass [3].

Exercise 5 (Facts about Eigenfunctions of the Laplacian). Suppose that f is an eigenfunction of the non-Euclidean Laplacian; i.e., $\Delta f = \lambda f$, and suppose, in addition, that $f|_{x=0} = f_x|_{x=0} = 0$ for all $y > 0$. Here $f_x = \partial f/\partial x$. Show that f must be identically zero on H.

Hints. We know that $f(z) = \Sigma_{n \geq 0} c_n(y) x^n$. Thus $c_n(y)$ satisfies the recursion obtained from

$$0 = \Delta f - \lambda f = \sum_{n \geq 0} (n(n-1) y^2 c_n(y) x^{n-2} + y^2 c_n''(y) x^n - \lambda c_n(y) x^n).$$

The initial conditions then imply that the $c_n(y)$ are identically zero. Note that without the hypothesis that $\partial f/\partial x|_{x=0} = 0$, we could only prove that $c_{2n}(y) = 0$.

In order to obtain the second initial condition for the function $f(z) - f(-1/z)$, we need a *second Dirichlet series*:

$$L_{f_x}(s) = \sum_{n \neq 0} (2\pi i n) a_n |n|^{-(s+1/2)}. \tag{3.83}$$

This second Dirichlet series is also seen to be necessary because the first series L_f, from (3.82), will vanish identically if f is an odd function of x; i.e., if $f(x + iy) = -f(-x + iy)$. For then $a_n = -a_{-n}$. And Table 3.5 gives some meager evidence for the conjecture that the Dirichlet and Neumann problems have different eigenvalues. A cusp form f has the decomposition $f = u + v$, where u is a solution of the Dirichlet problem and v is a solution of the Neumann problem. So, probably, either $f = u$ or $f = v$; i.e., f is either even or odd, as a function of x. Thus, if $f(z)$ denotes a Maass waveform solving the Dirichlet problem, we see that the L-series $L_f(s) = 0$, and we must consider $L_{f_x}(s)$ to obtain a nonzero series.

Using the differentiated modular form is an old trick in number theory. One needs it to obtain the analytic continuation of Dirichlet L-series:

$$\sum_{n \geq 1} \chi(n) n^{-s}$$

when $\chi(-1) = -\chi(1)$ and χ is a multiplicative character on the group of units in $\mathbb{Z}/m\mathbb{Z}$, viewed as a function on $\mathbb{Z}$ by writing $\chi(a) = 0$ if $(a, m) \neq 1$. A reference for the analytic continuation of these L-series is Davenport [1]. Hecke generalized this to grossencharacters for algebraic number fields (see Hecke [1, pp. 215–234, 249–289]).

Exercise 6 (Analytic Continuation and Functional Equation for the Second L-Series Corresponding to a Nonholomorphic Modular Form). Note that $f(-1/z) = f(z)$ implies that $f_x(-1/z) = z^2 f_x(z)$; so f_x has weight 2. Imitate the argument in Exercise 4 to obtain the analytic continuation and functional equation of L_{f_x}.

Theorem 2 (The Maass Correspondence between Nonholomorphic Modular Forms and Dirichlet Series). *Suppose that $f(z)$ has the Fourier expansion given by formula* (3.80). *Then $f \in \mathcal{N}(SL(2, \mathbb{Z}), r(r-1))$ is equivalent to the following assertions about the Dirichlet series defined in formulas* (3.82)–(3.83):

(1) $W_f(s) - a\left(\dfrac{1}{s-r} + \dfrac{1}{-s-r}\right) - b\left(\dfrac{1}{s+r-1} + \dfrac{1}{-s+r-1}\right)$

  is EBV with functional equation $W_f(s) = W_f(-s)$, and
(2) W_{f_x} *has the analogous properties (see Exercise 6).*

Exercise 7. Complete the proof of Theorem 2.

Maass actually proved a result like Theorem 1 for functions $f(z)$ satisfying more general differential equations and invariant under more general discrete subgroups of $SL(2, \mathbb{R})$.

Expansions of Dirichlet series in series of higher-dimensional incomplete gamma functions have been studied by many people. They arise whenever there are several gamma factors in the functional equation of the Dirichlet series (see Goldfeld and Viola [1], Lavrik [2], and Terras [1, 2]).

Examples. The Dirichlet series corresponding to the classical Eisenstein series G_k from §3.4 is

$$\sum_{n\geq 1} \sigma_{k-1}(n) n^{-s} = \zeta(s)\zeta(s+1-k). \tag{3.84}$$

The functional equation can actually be deduced from that of the Riemann zeta function, using the duplication formula for the gamma function and the functional equation of the gamma function.

The Dirichlet series corresponding to the nonholomorphic Eisenstein series $E_r^*(z) = \pi^{-r}\Gamma(r)\zeta(2r)E_r(z)$ is

$$4\sum_{n\geq 1} \sigma_{1-2r}(n) n^{r-1/2} n^{-s-1/2} = 4\zeta(s-r+1)\zeta(s+r). \tag{3.85}$$

Here again the functional equation can be deduced from that of the Riemann zeta function.

Exercise 8. Prove all the assertions made in the preceding examples.

One can use slight extensions of Hecke's theorem to derive the product expansion of Δ and θ in §3.4 (see Ogg [1, Ch. I, pp. 43–46], [2]). However, many important applications require a major extension of the theory, in particular, to congruence subgroups such as $\Gamma(N)$, which is defined after formula (3.46). Hecke and Maass both considered extensions of the theory to congruence subgroups. A simple example to explain why Hecke was interested in congruence subgroups can be found in Hecke's lectures, which were given at the Institute for Advanced Study at Princeton in 1938 (see Hecke [2, p. 43]). There it is noted that the zeta function for the imaginary quadratic field $\mathbb{Q}(\sqrt{-7})$ has as its inverse Mellin transform a theta function which is a modular form for $\Gamma_0(7)$ = the congruence subgroup of $SL(2, \mathbb{Z})$ consisting of matrices whose lower left entry is divisible by 7. Hecke also made connections with the representations of the finite group $SL(2, \mathbb{Z}/n\mathbb{Z})$ (see also Eichler [2]).

In 1967, Weil extended Hecke's correspondence between modular forms and Dirichlet series to congruence subgroups in a different way (see Weil [5, Vol. 3, pp. 165–172]). It would be very interesting to connect the two points of view. In fact, Weil extended Hecke's theory to the subgroup $\Gamma_0(N)$ consisting of all elements of $SL(2, \mathbb{Z})$ whose lower left entry is divisible by N. Here Weil discovered that the converse theorem requires more than one functional equation. Because the groups involved do not even contain the inversion $z \mapsto -1/z$, one has to add the matrix

$$\begin{pmatrix} 0 & -1 \\ N & 1 \end{pmatrix},$$

forming a larger group than $\Gamma_0(N)$. The resulting group will have more than two generators, in general. Weil finds that to make $f(z)$, with Fourier coefficients a_n, a modular form for $\Gamma_0(N)$, one requires the functional equations of the L-functions

$$\sum_{n \geq 1} \chi(n) a_n n^{-s}$$

for sufficiently many primitive characters χ of the unit group $(\mathbb{Z}/m\mathbb{Z})^* = \{b \bmod m | (b, m) = 1\}$. Here one considers only m which are relatively prime to N. The functional equations of these L-functions involve Gauss sums. Other references for Weil's result are Ogg [1] and Razar [2].

Weil's result involves explicit formulas for elements of the congruence subgroup. From the point of view of harmonic analysis, one might expect that harmonic analysis on $\Gamma(N)\backslash H$ should be a combination of analysis on $\Gamma(1)\backslash H$ and that on $\Gamma(1)/\Gamma(N)$. The latter quotient is isomorphic to $SL(2, \mathbb{Z}/N\mathbb{Z})$. The principal series representations of the finite special linear group are related to characters of the group $(\mathbb{Z}/N\mathbb{Z})^*$ (see Kirillov [1, p. 267]). This sort of thinking would lead one to expect a result similar to Weil's but involving L-functions formed with characters mod N itself, rather than mod m, with $(m, N) = 1$ (see Razar [2]).

Maass [3] actually obtains the correspondence between nonholomorphic modular forms for congruence subgroups and Dirichlet series with functional equations involving two gamma factors. This allows Maass to prove the existence of nonholomorphic cusp forms for congruence subgroups, distinct from $SL(2,\mathbb{Z})$ itself, by using the fact that the functional equations of Hecke L-functions with grossencharacters for real quadratic fields involve two gamma factors.

Jacquet and Langlands [1] give an adelic view of Hecke theory. See also Gelbart [1] and Weil [4], as well as Borel and Casselman [1]. Recently Winnie Li [1,2,3] has obtained an adelic version of Hecke theory closer to that envisioned above from the point of view of harmonic analysis.

Hecke theory has implications for certain conjectures of number theory and geometry. Number theorists have been interested in Weil's result for weight 1 modular forms because of the *Artin conjecture* that Artin L-functions associated to representations of Galois groups of number fields are entire provided the representation is nontrivial and irreducible. A reference for Artin L-functions is Heilbronn's article in Cassels and Fröhlich [1, pp. 204–230]. This conjecture has implications for the Dedekind zeta functions of number fields. For example, it implies that whenever a number field F is contained in a larger number field K, the quotient of Dedekind zeta functions given by $\zeta_K(s)/\zeta_F(s)$ is an entire function of s. This result is known if K/F is Galois or normal. Putting results of Weil, Langlands, Deligne, and Serre together, one sees that (modulo the Artin conjecture) certain forms of weight 1 for $\Gamma_0(N)$ correspond to Artin L-functions for irreducible two-dimensional representations of the Galois group of the algebraic closure of the rationals. Because Artin L-functions have functional equations involving many gamma factors, there is a general feeling that these L-functions must somehow correspond to automorphic forms for $GL(n)$. This sort of feeling is part of the "Langlands philosophy." References for some of these things are Borel and Casselman [1], Serre [2], Tate [2], and Weil [5, Vol. III, notes especially]. Recently Langlands [2] and Tunnell [1] have proved some new cases of the Artin conjecture for two-dimensional representations of Galois groups using base change and twisted versions of the trace formula.

There is also a geometric conjecture related to Hecke theory. This is the *conjecture of Taniyama and Weil* which says that any elliptic curve over the rationals has a zeta function coming from a modular form of weight 2. It is not hard to prove that modular forms of weight 2 give rise to L-series which are zeta functions of elliptic curves or abelian varieties (with "real multiplication"). See Weil [5, Vol. III, notes] for some discussion of the Taniyama-Weil conjecture. There are also many open questions on the Hasse-Weil zeta functions of algebraic varieties (even elliptic curves). References include Katz [1, pp. 300–301] and Swinnerton-Dyer [1].

Exercise 9 (Operators Which Raise the Level).

(a) Suppose that χ is a multiplicative character mod m; i.e.,

$$\chi:(\mathbb{Z}/m\mathbb{Z})^* \to \mathbb{T}, \quad \text{a homormorphism of multiplicative groups,}$$

where $(\mathbb{Z}/m\mathbb{Z})^*$ is the multiplicative group of integers a mod m, with a

relatively prime to m, and $\mathbb{T}$ is the multiplicative group of complex numbers of norm one. We extend χ to a function on $\mathbb{Z}^+$ by setting $\chi(n) = 0$ if n and m are not relatively prime. Suppose that $f \in \mathcal{M}(SL(2, Z), k)$. Define

$$L_\chi f(z) = \frac{1}{m} \sum_{v,\, v \bmod m} \chi(u) \exp(-2\pi i u v/m) f(z + v/m).$$

Show that $L_\chi f$ is a modular form for the congruence group $\Gamma_0(m^2)$.

(b) Suppose that a is relatively prime to m and that a' is an integer such that $aa' \equiv 1 \pmod{m}$. Let R_a be a matrix in $SL(2, \mathbb{Z})$ such that

$$R_a \equiv \begin{pmatrix} a' & 0 \\ 0 & a \end{pmatrix} \pmod{N}.$$

R_a is not well defined but the coset of R_a in $\Gamma(N)\backslash SL(2, \mathbb{Z})$ is well defined. Set $N = m^2$ and show that

$$L_\chi f|_{R_a} = \chi^2(a) L_\chi f \qquad \text{for } f \text{ as in part (a).}$$

Here the slash operator is defined by

$$g|_L(z) = (cz + d)^{-k} g((az + b)/(cz + d)) \qquad \text{if}$$

$$L = \begin{pmatrix} a & b \\ c & d \end{pmatrix} \in SL(2, \mathbb{Z}),\ g \text{ of weight } k.$$

(c) Suppose that $f(z)$ is as in part (a) and $f(z) = \sum_{n \geq 0} a_n \exp(2\pi i n z)$. Show that

$$L_\chi f(z) = \sum_{n \geq 0} a_n \chi(n) \exp(2\pi i n z).$$

This leads to the twisted L-function studied by Weil.

(d) Suppose that $f(z)$ is as in part (a). Let χ be a character mod m. Define

$$f_\chi(z) = \sum_{\substack{a \bmod m \\ (a, m) = 1}} \chi(a) f(z + a/m).$$

Show that f_χ is a modular form of weight k for $\Gamma_0(m^2)$ and that, using the notation of part (b), $f_\chi|_{R_a} = \chi^{-2}(a) f_\chi$.

(e) Can you connect the operators in parts (a) and (d)?

Hints. Parts (a)–(c) can be found in Ogg [1, Ch. IV, pp. 37–38]. Part (d) comes from Eichler [2, pp. 147–149]. See also Razar [2].

It will help to note that there is a chain of subgroups of $\Gamma = SL(2, \mathbb{Z})$ given by

$$\Gamma(N) \subset \Gamma_1(N) \subset \Gamma_0(n) \subset \Gamma,$$

where $\Gamma_1(N) = \{\gamma \in \Gamma \mid \gamma \equiv \left(\begin{smallmatrix} 1 & * \\ 0 & 1 \end{smallmatrix}\right) \bmod N\}$. The quotients

$$\Gamma_1(N)/\Gamma(N) \quad \text{and} \quad \Gamma_0(N)/\Gamma_1(N)$$

are abelian; the first isomorphic to the additive group $\mathbb{Z}/N\mathbb{Z}$ and the second isomorphic to the multiplicative group $(\mathbb{Z}/N\mathbb{Z})^*$.

Note also that (as in part (a) of Exercise 13 which follows), if L is a matrix of determinant N, then transforming $f(z)$ by L using the slash operator

$$f|_L(z) = N^{k/2}(cz+d)^{-k}f((az+b)/(cz+d)) \qquad \text{for } L = \begin{pmatrix} a & b \\ c & d \end{pmatrix}$$

sends a form for $SL(2, \mathbb{Z})$ of weight k to a form for $\Gamma(N)$ of weight k.

Note. Exercise 9, and Exercise 13 which follows, give a large number of examples of modular forms for congruence subgroups. Classically, examples of the form $f(mz)$ arose in the theory of equations for $f(mz)$ in terms of $f(z)$, for example, when $f(z)$ is a modular function, such as $j(z)$. See Shimura [1, pp. 109–110] for a discussion of the modular polynomial, also Weber [1, Vol. III].

Hecke also developed a theory which will determine what modular forms correspond to Dirichlet series with Euler products. The first Euler product to arise in number theory was that for the Riemann zeta function:

$$\zeta(s) = \prod_{p \text{ prime}} (1 - p^{-s})^{-1}, \qquad \text{Re } s > 1 \tag{3.86}$$

(see Exercise 4 of §1.4). It is this result which governs the application of the Riemann zeta function to the study of the distribution of primes.

Around 1935, Hecke used Hecke operators to answer the question, Which modular forms correspond to L-series with Euler products? (see Hecke [1, pp. 577–707] and [2]). Such operators had already been considered by Hurwitz [1, pp. 163–188] and Mordell [1]. It turns out that Hecke operators are related to the analogue for p-adic groups of the spherical functions considered in §3.2 (see Gelbart [1, pp. 47ff]). As usual, we shall restrict ourselves to $SL(2, \mathbb{Z})$, although Hecke did not. For each positive integer n, let M_n denote the set of all 2×2 matrices of determinant n. Set $\Gamma = SL(2, \mathbb{Z})$. It is easy to see that M_n is a disjoint union:

$$M_n = \bigcup_{\substack{ad=n,\, d>0, \\ b \bmod d}} \Gamma \begin{pmatrix} a & b \\ c & d \end{pmatrix}. \tag{3.87}$$

Exercise 10. Prove formula (3.87).

Hint. It is clear that given a matrix $L \in M_n$, there exists a matrix $\gamma \in SL(2, \mathbb{Z})$ such that γL has zero for its lower left entry. Here we use the fact that a row vector of integers is the second row of a matrix in $SL(2, \mathbb{Z})$ if and only if the entries of the vector are relatively prime.

Definition. The *Hecke operator* T_n is defined by

$$T_n f(z) = n^{k-1} \sum_{\substack{ad=n,\, d>0 \\ b \bmod d}} d^{-k} f\left(\frac{az+b}{d}\right). \tag{3.88}$$

3.6. Automorphic Forms and Dirichlet Series. Hecke Theory and Generalizations

Note that the Hecke operator T_n also depends on k, which will be the weight of the modular form $f(z)$ on which T_n acts. If $f(z)$ is a modular form of weight k, then the Hecke operator is actually independent of the choice of representatives for $SL(2, \mathbb{Z}) \backslash M_n$. Thus we can consider the sum in (3.88) to be a sum over $SL(2, \mathbb{Z}) \backslash M_n$. We will see that analogous sums arise in the theory of automorphic forms for $SL(n, \mathbb{Z})$ in Chapter 4. Other references for Hecke operators are Apostol [3, pp. 120–138], Lang [6, Chs. 2, 3], Ogg [1], Serre [1], and Shimura [1]. The last reference gives a very general treatment.

Theorem 3 (Properties of Hecke Operators on Spaces of Holomorphic Modular Forms).

(1) *The Hecke operator T_n defined by formula* (3.88) *has the property that if $f \in \mathcal{M}(SL(2, \mathbb{Z}), k)$, then $T_n f \in \mathcal{M}(SL(2, \mathbb{Z}), k)$. Moreover, cusp forms go to cusp forms.*

(2) *Suppose that $f \in \mathcal{M}(SL(2, \mathbb{Z}), k)$ has Fourier expansion*
$$f(z) = \sum_{m \geq 0} a_m \exp(2\pi i m z).$$
Then $T_n f$ has Fourier expansion
$$T_n f(z) = \sum_{m \geq 0} b_m \exp(2\pi i m z) \quad \text{with } b_m = \sum_{d \mid (m, n)} d^{k-1} a_{mn/d^2}.$$

(3) *We have the multiplication formula*
$$T_n T_m = \sum_{d \mid (m, n)} d^{k-1} T_{mn/d^2}.$$
And the ring of Hecke operators $\{T_n, n \geq 1\}$ is commutative.

(4) *Define the Petersson inner product of two cusp forms $f, g \in \mathcal{S}(SL(2, \mathbb{Z}), k)$ by*
$$[f, g] = \int_{SL(2, \mathbb{Z}) \backslash H} f(z) \overline{g(z)} y^{k-2} \, dx \, dy.$$
Then $[T_n f, g] = [f, T_n g]$. It follows that the Hecke operators can be simultaneously diagonalized on $\mathcal{M}(SL(2, \mathbb{Z}), k)$.

(5) *Suppose that $f \in \mathcal{M}(SL(2, \mathbb{Z}), k)$ with*
$$f(z) = \sum_{m \geq 0} a_m \exp(2\pi i m z),$$
and $a_1 = 1$, satisfies $T_n f = c_n f$, for $n = 1, 2, 3, \ldots$. Then $a_n = c_n$, for all $n = 1, 2, 3, \ldots$. Moreover, the Dirichlet series associated to f has an Euler product of the form
$$L_f(s) = \sum_{n \geq 1} a_n n^{-s} = \prod_{p \text{ prime}} (1 - a_p p^{-s} + p^{k-1-2s})^{-1}.$$
Conversely, if $L_f(s) = \sum_{n \geq 0} a_n n^{-s}$ has such an Euler product and is known to correspond to a modular form f in $\mathcal{M}(SL(2, \mathbb{Z}), k)$, then f is an eigenfunction for all of the Hecke operators T_n.

(6) *The eigenvalues of the Hecke operators T_n are totally real algebraic numbers.*

Exercise 11. Prove the preceding theorem using the same sorts of arguments that we give in the proof of Theorem 4, which concerns the analogue for nonholomorphic modular forms. Details can also be found in the references preceding Theorem 3. The only part of Theorem 3 that does not have an analogue in Theorem 4 is part (6). To see this, use the fact that the space of modular forms of weight k has a basis consisting of elements all of whose Fourier coefficients are rational (see Ogg [1, III–12]).

Examples. The Eisenstein series G_k gives an example of an eigenfunction of the Hecke operators, since the corresponding Dirichlet series has an Euler product:
$$L_{G_k}(s) = \zeta(s)\zeta(s + 1 - k).$$
The cusp form Δ of weight 12 must also be an eigenfunction of the Hecke operators, since it is unique, up to a constant multiple. Thus the corresponding Dirichlet series
$$L_\Delta(s) = \sum_{n \geq 1} \tau(n) n^{-s}$$
has an Euler product, a result first proved by Mordell [1] in 1917. Ramanujan had stated this result some years before Mordell's proof, as well as the Ramanujan conjecture (see §3.4). Hardy [1, Ch. X] discusses these matters in his book on Ramanujan, beginning with the remark: "We may seem to be straying into one of the backwaters of mathematics, but the genesis of $\tau(n)$ as a coefficient in so fundamental a function compels us to treat it with respect." Weil considers this remark in his 1972 Columbia lectures (see Weil [5, Vol. 3, pp. 280–281]) as "a typical example of the deep gulf that separates number-theorists from an analyst like Hardy." This sort of mathematical name-calling is fun, but history clearly has a way of refuting all such assertions, including many made by the present author in this and earlier versions of these notes.

More examples can be found in Hecke [1, 2]. In particular paper no. 37 in Hecke's collected works [1] gives a list of Dirichlet series L_f, showing that the presence of an Euler product is neither necessary nor sufficient for the existence of infinitely many zeros on the line of symmetry on $\operatorname{Re} s = k/2$.

Goldstein [3] has considered analogues of the Riemann hypothesis and Mertens conjecture for the Dirichlet series attached to cusp forms which are eigenfunctions for all the Hecke operators. R.J. Anderson [1] disproved the Mertens conjecture for these Dirichlet series corresponding to such cusp forms when the weight is sufficiently large. Recently Odlyzko disproved the original Mertens conjecture.

Bounds for the degree of the number fields containing the eigenvalues of Hecke operators are obtained in Rankin and Rushforth [1]. Naomi Jochnowitz [1] studied the ring $T^{(k)}$ of Hecke operators acting on holomorphic cusp forms of weight k for $SL(2, \mathbb{Z})$. Then $T^{(k)} \otimes \mathbb{Q}$ is a direct product of totally real number fields whose dimension over $\mathbb{Q}$ equals the dimension of the space

$\mathscr{S}(SL(2,\mathbb{Z}),k)$ over $\mathbb{C}$. Moreover, in all known cases the direct product is actually a single number field. For example, Hecke [1, p. 671] shows that $T^{(24)} \otimes \mathbb{Q} \simeq \mathbb{Q}(\sqrt{144169})$ and that $T^{(24)}$ is the unique suborder of index 24 in the ring of integers of this field. Jochnowitz shows that if N is any integer and k is sufficiently large then N divides the index of $T^{(k)}$ in the maximal order of $T^{(k)} \otimes \mathbb{Q}$.

Leveque's reviews in number theory [1] contain references to many papers on Hecke operators, etc. For example, there are reviews of some of Hecke's papers (see LeVeque [1, Vol. 2, pp. 448ff) as well as a formula of Petersson for $\zeta(3)$, involving his inner product and theta functions. And Wohlfahrt [1] generalizes the theory of Hecke operators to automorphic forms of real (rather than integral) weight. Rademacher [2] shows that T_{p^r} is essentially a Tchebychef polynomial of the second kind in T_p. Atkin and Lehner [1] complete the theory of Hecke operators for the congruence group $\Gamma_0(N)$. The problem was to find a satisfactory theory of operators for primes dividing the level of the congruence subgroup. Ihara [1] uses the Eichler-Selberg trace formula to connect Ramanujan's conjecture with the Weil conjectures on algebraic varieties—a project ultimately completed by Deligne, as we noted in §3.4.

The Hecke operators in the nonholomorphic case were developed by Maass [1, 3].

Definition. For a Maass waveform $f \in \mathcal{N}(SL(2,\mathbb{Z}), r(r-1))$, and $n = 1, 2, 3, \ldots,$ define the *Hecke operator* T_n by

$$T_n f(z) = n^{-1/2} \sum_{\substack{ad=n,\, d>0,\\ b \bmod d}} f\left(\frac{az+b}{d}\right). \tag{3.89}$$

The normalizing factor $n^{-1/2}$ is not consistent with formula (3.88), since our Maass waveforms have weight 0, but it has become standard in the Maass waveform literature.

Theorem 4 (Properties of Hecke Operators on Maass Waveforms).

(1) *The Hecke operator T_n, defined by formula (3.89), has the property that if $f \in \mathcal{N}(SL(2,\mathbb{Z}), r(r-1))$, then $T_n f \in \mathcal{N}(SL(2,\mathbb{Z}), r(r-1))$. Moreover, if f is a cusp form, then so is $T_n f$. And the map $f \mapsto T_n f$ is linear.*

(2) *Suppose that $f \in \mathcal{N}(SL(2,\mathbb{Z}), r(r-1))$ has the Fourier expansion*

$$f(z) = cy^r + c'y^{1-r} + \sum_{m\neq 0} c_m y^{1/2} K_{r-1/2}(2\pi|m|y)\exp(2\pi imx),$$

and

$$T_n f(z) = by^r + b'y^{1-r} + \sum_{m\neq 0} b_m y^{1/2} K_{r-1/2}(2\pi|m|y)\exp(2\pi imx).$$

Then $b = n^{1/2-r}\sigma_{2r-1}(n)c$, $\quad b' = n^{r-1/2}\sigma_{1-2r}(n)c'$, and

$$b_m = \sum_{d|(m,n)} c_{mn/d^2}.$$

(3) The algebra generated by the Hecke operators is commutative and
$$T_n T_m = \sum_{d|(m,n)} T_{mn/d^2}.$$

(4) Define the Petersson inner product for nonholomorphic cusp forms $f, g \in \mathscr{SN}(SL(2, \mathbb{Z}), r(r-1))$ by
$$(f, g) = \int_{SL(2,\mathbb{Z})\backslash H} f(z)\overline{g(z)}\, y^{-2}\, dx\, dy.$$

Then $(T_n f, g) = (f, T_n g)$ and thus the Hecke operators can be simultaneously diagonalized on $\mathscr{N}(SL(2, \mathbb{Z}), r(r-1))$. Note that in this case the Petersson inner product is the same as the usual L^2-inner product.

(5) Suppose that $f \in \mathscr{N}(SL(2,\mathbb{Z}), r(r-1))$ has the Fourier expansion
$$f(z) = cy^r + c'y^{1-r} + \sum_{m \neq 0} c_m y^{1/2} K_{r-1/2}(2\pi|m|y) \exp(2\pi i m x)$$

and that f is an eigenfunction of all the Hecke operators; i.e.,
$$T_n f = u_n f \quad \text{for } n = 1, 2, 3, \ldots.$$

Then
$$c_n = u_{|n|} c_{n/|n|} \quad \text{for } n \neq 0;$$

and the two associated Dirichlet series have Euler products
$$L_f(s - 1/2) = \sum_{m \neq 0} c_m |m|^{-s} = (c_1 + c_{-1}) \prod_{p \text{ prime}} (1 - u_p p^{-s} + p^{-2s})^{-1},$$

$$L_{f_x}(s + 1/2) = \sum_{m \neq 0} (2\pi i m) c_m |m|^{-s-1}$$
$$= 2\pi i (c_1 - c_{-1}) \prod_{p \text{ prime}} (1 - u_p p^{-s} + p^{-2s})^{-1}.$$

Conversely, if $f \in \mathscr{N}(SL(2, \mathbb{Z}), r(r-1))$ corresponds to Dirichlet series L_f and L_{f_x} with Euler products such as those above, then f is an eigenfunction for all the Hecke operators.

PROOF.

(1) First note that the Hecke operators clearly commute with the non-Euclidean Laplace operator. It is easy to see that $T_n f$ is invariant under $SL(2, \mathbb{Z})$ if f is invariant. For if M_n denotes the 2×2 integer matrices of determinant n and L runs through a full set of representatives for $SL(2, \mathbb{Z})\backslash M_n$, and $\gamma \in SL(2, \mathbb{Z})$, then $L\gamma$ also runs through a full set of representatives for $SL(2, \mathbb{Z})\backslash M_n$. Finally, it is clear that if $f(z)$ has most polynomial growth in y, then so does $T_n f$, as y approaches infinity.

(2) Suppose that T_n is given by formula (3.89). First note that $\operatorname{Im}\left(\dfrac{az+b}{d}\right)$

3.6. Automorphic Forms and Dirichlet Series. Hecke Theory and Generalizations 243

$= nyd^{-2}$ and $\operatorname{Re}\left(\dfrac{az+b}{d}\right) = (ax+b)/d$. It follows that

$$T_n f(z) = n^{-1/2} \sum_{\substack{n=ad,\, d>0 \\ b \bmod d}} \left(c(nyd^{-2})^r + c'(nyd^{-2})^{1-r}\right)$$

$$+ n^{-1/2} \sum_{\substack{n=ad,\, d>0 \\ b \bmod d}} \sum_{m \neq 0} c_m (nyd^{-2})^{1/2} K_{r-1/2}(2\pi |m| nyd^{-2})$$

$$\times \exp[2\pi i m(ax+b)/d].$$

The first sum is

$$y^r c n^{r-1/2} \sum_{0<d|n} d^{1-2r} + y^{1-r} c' n^{1/2-r} \sum_{0<d|n} d^{2r-1}$$

$$= c y^r n^{r-1/2} \sigma_{1-2r}(n) + c' y^{1-r} n^{1/2-r} \sigma_{2r-1}(n).$$

The second sum simplifies since the sum over b is zero unless d divides m. So the second sum is

$$y^{1/2} \sum_{d|(m,n)} c_m K_{r-1/2}(2\pi |m| nyd^{-2}) \exp(2\pi i m n x\, d^{-2}).$$

Sum over $a = n/d$ instead of d and set $v = ma/d$ (so that $m = nva^{-2}$) to obtain

$$\sum_{\substack{0<a|(n,v) \\ v \in \mathbb{Z}}} c_{nva^{-2}} y^{1/2} K_{r-1/2}(2\pi |v| y) \exp(2\pi i v x).$$

This gives the stated result.

(3) The proof is easy if $(m, n) = 1$. For suppose the matrices

$$\begin{pmatrix} d_1 & d_{12} \\ 0 & d_2 \end{pmatrix}$$

run through a complete set of representatives for M_n modulo $SL(2, \mathbb{Z})$, as in formula (3.87) and that the matrices

$$\begin{pmatrix} c_1 & c_{12} \\ 0 & c_2 \end{pmatrix}$$

run through a complete set of representatives for M_m modulo $SL(2, \mathbb{Z})$. The product is

$$\begin{pmatrix} d_1 c_1 & d_1 c_{12} + c_2 d_{12} \\ 0 & d_2 c_2 \end{pmatrix}.$$

If d_{12} runs through a complete set of representatives mod d_2 and c_{12} runs through a complete set of representatives mod c_2, then $d_1 c_{12} + c_2 d_{12}$ runs through a complete set of representatives mod $d_2 c_2$. Why?

To complete the proof of (3), we want to show that

$$T_{p^k} T_p = T_{p^{k+1}} + T_{p^{k-1}} \qquad \text{for } k \geq 1. \tag{3.90}$$

Let us set up the following notation. We shall call two matrices *equivalent* and write $A \sim B$ if $f(Az) = f(Bz)$, for all $f \in \mathcal{N}(SL(2,\mathbb{Z}), r(r-1))$.*

In order to prove (3.90), we shall multiply matrices in the set of representatives for $SL(2,\mathbb{Z})\backslash M_n$ which are summed over in the definition (3.89) of the Hecke operator. We have, for $e + f = k$,

$$\begin{pmatrix} 1 & a \bmod p \\ 0 & p \end{pmatrix}\begin{pmatrix} p^e & b \bmod p^f \\ 0 & p^f \end{pmatrix} = \begin{pmatrix} p^e & b \bmod p^f + p^f(a \bmod p) \\ 0 & p^{f+1} \end{pmatrix}$$
$$= \begin{pmatrix} p^e & b \bmod p^{f+1} \\ 0 & p^{f+1} \end{pmatrix}, \tag{3.91}$$

$$\begin{pmatrix} p & 0 \\ 0 & 1 \end{pmatrix}\begin{pmatrix} p^e & b \bmod p^f \\ 0 & p^f \end{pmatrix} = \begin{pmatrix} p^{e+1} & p(b \bmod p^f) \\ 0 & p^f \end{pmatrix} \sim \begin{pmatrix} p^e & b \bmod p^f \\ 0 & p^{f-1} \end{pmatrix}$$
$$= \begin{pmatrix} p^e & b_1 \bmod p^{f-1} + p^{f-1}(c_1 \bmod p) \\ 0 & p^{f-1} \end{pmatrix}$$
$$= \begin{pmatrix} 1 & c_1 \bmod p \\ 0 & 1 \end{pmatrix}\begin{pmatrix} p^e & b_1 \bmod p^{f-1} \\ 0 & p^{f-1} \end{pmatrix} \tag{3.92}$$
$$\sim \begin{pmatrix} p^e & b_1 \bmod p^{f-1} \\ 0 & p^{f-1} \end{pmatrix}.$$

Equation (3.91) leads to operators in $T_{p^{k+1}}$, except that the $f + 1 = 0$ term is missing. Equation (3.92) gives that term when $f = 1$. And what remains in equation (3.92) gives operators from $T_{p^{k-1}}$, each of them taken p times, when we take account of the $c_1 \bmod p$. This completes the proof of formula (3.90). Why?

Mathematical induction and (3.90) imply that

$$T_{p^e} T_{p^f} = \sum_{t=0}^{\min(e,f)} T_{p^{e+f-2t}}. \tag{3.93}$$

From this result, part (3) follows easily.

Note also that (3.90) implies the following formula for the formal power series in the indeterminate X (for prime p):

$$\sum_{e \geq 0} T_{p^e} X^e = (I - T_p X + X^2)^{-1}. \tag{3.94}$$

To prove this you must simply multiply both sides by $(I - T_p X + X^2)$ and make use of (3.90). Formula (3.94) will lead to Euler products for L-functions corresponding to eigenfunctions of all the Hecke operators.

Exercise 12. Fill in the details that were omitted in the proof of part (3); e.g., in the proofs of (3.93) and (3.94).

---

* This amounts to saying that $c\gamma A = B$ for some $\gamma \in SL(2,\mathbb{Z})$, $c \in \mathbb{R} - 0$.

(4) This argument goes back to Petersson [1]. Recall that M_n is the set of 2×2 integral matrices of determinant n and $\Gamma = SL(2, \mathbb{Z})$. Thus (3.89) says that the Hecke operator T_n is a sum over $\Gamma \backslash M_n$ and thus

$$n^{1/2}(T_n f, g) = \sum_{A \in \Gamma \backslash M_n} \int_{\Gamma \backslash H} f(Az) \overline{g(z)} y^{-2} \, dx \, dy.$$

For $A \in M_n$, set $h(z) = f(Az)$. Note that $h(Bz) = h(z)$, for all B in the congruence subgroup

$$\Gamma(n) - \{B \in SL(2, \mathbb{Z}) = \Gamma | B \equiv I (\text{mod } n)\}$$

(see Exercise 13).

Since a fundamental domain $\Gamma(n) \backslash H$ can be taken to be $[\Gamma : \Gamma(n)]$ copies of $\Gamma \backslash H$ (by Exercise 8 of §3.3), we have

$$n^{1/2}(T_n f, g) = \sum_{A \in \Gamma \backslash M_n} [\Gamma : \Gamma(n)]^{-1} \int_{\Gamma(n) \backslash H} f(Az) \overline{g(z)} y^{-2} \, dx \, dy$$

$$= \sum_{A \in \Gamma \backslash M_n} [\Gamma : A\Gamma(n) A^{-1}]^{-1} \int_{A\Gamma(n) A^{-1} \backslash H} f(w) \overline{g(A^{-1} w)} \, v^{-2} \, du \, dv,$$

setting $w = u + iv = Az$ and noting that $[\Gamma : A\Gamma(n) A^{-1}] = [\Gamma : \Gamma(n)]$. Part (b) of Exercise 13 tells us that the last quantity is none other than $(f, T_n g) n^{1/2}$ and part (4) of Theorem 4 is proved.

Exercise 13.

(a) Suppose that A is a 2×2 integral matrix of determinant n and that f is a modular form for $\Gamma = SL(2, \mathbb{Z})$. Define $h(z) = f(Az)$. Prove that $h(z)$ is a modular form for the congruence subgroup $\Gamma(n)$.
(b) If A runs through a complete set of representatives for $\Gamma \backslash M_n$, where M_n is the set of integral 2×2 matrices of determinant n, $\Gamma = SL(2, \mathbb{Z})$, show that nA^{-1} also runs through a complete set of representatives for M_n/Γ. Then show that there is a common set of representatives for M_n/Γ and $\Gamma \backslash M_n$.

Hints.

(a) Note that $nA^{-1} \in \mathbb{Z}^{2 \times 2}$ and that $B \in \Gamma(n)$ implies that $nABA^{-1} \equiv 0 \,(\text{mod } n)$. But then $ABA^{-1} \in SL(2, \mathbb{Z})$.
(b) For the last part of this Exercise see Shimura [1, pp. 53–54]. You need to know that the number of left Γ-cosets of $\Gamma A \Gamma$ is the same as the number of right Γ cosets of $\Gamma A \Gamma$. It is easy to prove that these numbers are the same in the case at hand.

(5) First suppose that $T_n f = u_n f$. From part (2) of the theorem, it follows that

$$c_n = u_n c_1 \quad \text{and} \quad c_{-n} = u_n c_{-1}.$$

Thus

$$L_f(s - \tfrac{1}{2}) = \sum_{n \neq 0} c_n |n|^{-s} = (c_1 + c_{-1}) \sum_{n \geq 1} u_n n^{-s}.$$

and there is an analogous formula for L_{f_x}. The Euler product follows easily from formula (3.94).

Conversely, suppose that L_f and L_{f_x} have Euler products of the indicated type; e.g.,

$$L_f(s - \tfrac{1}{2}) = \sum_{m \neq 0} c_m |m|^{-s} = (c_1 + c_{-1}) \prod_{p \text{ prime}} (1 - u_p p^{-s} + p^{-2s})^{-1}.$$

Then the following Dirichlet series has no terms $|m|^{-s}$ such that p divides m:

$$(1 - u_p p^{-s} + p^{-2s}) \sum_{m \neq 0} c_m |m|^{-s} = \sum_{m \neq 0} (c_m - u_p c_{m/p} + c_{m/p^2}) |m|^{-s}.$$

Plug in $m = np$ to see that $c_{np} - u_p c_n + c_{n/p} = 0$. Thus

$$c_{np} + c_{n/p} = u_p c_n.$$

The left-hand side of this last equality is the nth Fourier coefficient of $T_p f$ and the right-hand side is the nth Fourier coefficient of $u_p f$. Thus

$$T_p f = u_p f \qquad \text{for all primes } p.$$

By formula (3.90), we know that $T_{p^e} = T_p T_{p^{e-1}} - T_{p^{e-2}}$, if $e \geq 2$. Thus we can show that f is an eigenfunction for T_{p^e} by mathematical induction. Then $T_m T_n = T_{mn}$ for $(m, n) = 1$ completes the proof of the converse in part (5), and the proof of the entire theorem. □

Exercise 14. Show directly that both the holomorphic and the nonholomorphic Eisenstein series $G_k(z)$ and $E_s(z)$ are eigenfunctions of the appropriate Hecke operators T_n.

Hint. Since the Hecke operator is a sum over $\Gamma \backslash M_n$ and the Eisenstein series is a sum over $\Gamma_\infty \backslash \Gamma$, the problem involves the determination of representatives of the quotient $\Gamma_\infty \backslash M_n$. That is, you must sum over a, c with $(a, c) = t$, for t dividing n. Then you need to notice that the solutions b, d of $ad - bc = n$ are not unique. Elementary number theory says there are t of them modulo Γ_∞. For set $a = a_1 t, c = c_1 t, b = b_0 + a_1 u, d = d_0 + c_1 u, u \in \mathbb{Z}$. Then

$$\begin{pmatrix} a & b \\ c & d \end{pmatrix} = \begin{pmatrix} a & b_0 \\ c & d_0 \end{pmatrix} \begin{pmatrix} 1 & u/t \\ 0 & 1 \end{pmatrix}.$$

You can check your calculation by noting that the eigenvalues are essentially the coefficients of the Fourier expansions of $G_k(z)$ and $E_s(z)$.

Example (The Rankin-Selberg Method). There is another sort of Dirichlet series that can be associated to a modular form. We used this idea already in Exercise 8 of §3.5. The method goes back to Rankin [3] and Selberg [5], who used the method to obtain good estimates (though not so good as the

Ramanujan-Petersson conjecture) for the Fourier coefficients a_n of holomorphic cusp forms $f(z)$ of weight k,

$$f(z) = \sum_{n \geq 1} a_n \exp(2\pi i n z),$$

by considering

$$L_{f^*}(s) = \sum_{n \geq 1} |a_n|^2 n^{1-k-s}. \tag{3.95}$$

The trick is to note that

$$(4\pi)^{1-s-k} \Gamma(s+k-1) L_{f^*}(s) = \int_{0 < y, |x| \leq 1/2} y^{k+s} |f(z)|^2 y^{-2} \, dy \, dx. \tag{3.96}$$

The integral here is over a fundamental domain for $\Gamma_\infty \backslash H$, where Γ_∞ is the subgroup of $\Gamma = SL(2, \mathbb{Z})$ fixing $i\infty$, defined by formula (3.67).

Next note that we can write the right-hand side of (3.96) as

$$\int_{\Gamma \backslash H} \sum_{\gamma \in \Gamma_\infty \backslash \Gamma} \text{Im}(\gamma z)^s y^k |f(z)|^2 y^{-2} \, dx \, dy = \int_{\Gamma \backslash H} E_s(z) y^k |f(z)|^2 y^{-2} \, dx \, dy. \tag{3.97}$$

Exercise 15. Prove formulas (3.96) and (3.97).

Then the analytic continuation of the Eisenstein series leads to the desired estimates for the coefficients a_n, using methods from analytic number theory. This method works for congruence subgroups. However, Selberg [2] gives examples of discrete groups Γ' such that the Eisenstein series has poles in $(1 - \delta, 1)$. In this same paper, Selberg shows how to get around this problem for Γ' of finite index in Γ, by going from scalar modular forms for Γ' to vector-valued modular forms for Γ.

Part of Deligne's proof of the Weil conjectures was motivated by the Rankin-Selberg argument (see Katz [1]). Many other applications of the Rankin-Selberg method have appeared recently. Some examples are Andrianov [1], Moreno [1], Novodvorsky and Piatetskii-Shapiro [1], Piatetskii-Shapiro [1], Shimura [1–3], Stark [3, Part II], and Zagier [1].

There are other classical methods of estimating Fourier coefficients of modular forms using Poincaré series and Kloosterman sums (see Petersson [2] and Selberg [2]).

R.A. Smith [1] showed that if f is a cuspidal Maass waveform in $\mathcal{S}\mathcal{N}(SL(2, \mathbb{Z}), r(r-1))$, $r = \frac{1}{2} + it$, $t > 0$, and f is a common eigenfunction for all the Hecke operators, normalized so that its first Fourier coefficient c_1 has absolute value 1, then the L^2-norm of f on the fundamental domain satisfies the inequality

$$\tfrac{1}{6} |\Gamma(it)| \leq (f, f)^{1/2} \leq 2 |\Gamma(1 + it)| \quad \text{for } t \geq 200.$$

Thus the norm of f decreases exponentially as $t \to \infty$.

Exercise 16 (Uniform Distribution of Horocycles in the Fundamental Domain as the Horocycle Approaches the Real Axis). Let $C_y \subset \Gamma\backslash H$ be the horocycle $\Gamma_\infty\backslash\mathbb{R} + iy$, which is a closed curve in the fundamental domain such that the length of C_y is $1/y$. Show that C_y fills up $\Gamma\backslash H$ in a very uniform way as $y \to 0^+$, in the sense that, for any open set U in $\Gamma\backslash H$, we have

$$\frac{\text{length}(C_y \cap U)}{\text{length}(C_y)} \sim \frac{\text{area}(U)}{\text{area}(\Gamma\backslash H)} \qquad \text{as } y \to 0+. \tag{3.98}$$

Here all lengths and areas are non-Euclidean!

Hints (see Zagier [1]). Use $k = 0$ and $f(z) = \chi_U(z) =$ the characteristic function of U, which is 1 on U and 0 outside U, in formulas (3.96) and (3.97). Then

$$C(U, y) = \int_0^1 \chi_U(x + iy)\, dx = \frac{\text{length}(C_y \cap U)}{\text{length}(C_y)}$$

and

$$I(U, s) = \int_{y=0}^{\infty} C(U, y) y^{s-2}\, dy = \int_{\Gamma\backslash H} \chi_U(z) E_s(z) y^{-2}\, dx\, dy.$$

Use a Tauberian theorem for Mellin transforms, after noting that

$$\operatorname{Res}_{s=1} I(U, s) = \frac{\text{area}(U)}{\text{area}(\Gamma\backslash H)}.$$

Note. Zagier [1] goes on to show that if the error term in (3.98) is $O(y^{3/4-\varepsilon})$ the Riemann hypothesis is true.*

One can view the integral

$$\int_0^{i\infty} f(z) z^s\, dz$$

as a special case of the period integral (for $s \in \mathbb{Z}$)

$$\int_a^b f(z) z^s\, dz$$

with $a, b \in P_\Gamma = \{\text{cusps of the discrete group } \Gamma\}$. This observation allows Manin [1, 2, 3, 4] to develop a systematic way of calculating these integrals or modular symbols, making use of facts about cohomology, Farey fractions, and continued fractions. Write $\{a, b\}_\Gamma$ for the modular symbol or the homology class determined by the geodesic joining the cusps a and b in P_Γ. Then one has

$$\{\gamma a, \gamma b\}_\Gamma = \{a, b\}_\Gamma \qquad \text{for all } \gamma \in \Gamma,$$

and

* The notation "$O(y^P)$" stands for a function which when divided by y^P is bounded as $y \to \infty$.

$\{a_1, a_2\}_\Gamma + \cdots + \{a_k, a_1\}_\Gamma = 0$ for all sequences of cusps $a_1, \ldots, a_k$.

The second sort of relation brings in the Farey fractions. One can also transfer the action of the Hecke operators to the homology. As a result (after some modification of the theory when the weight of the cusp form is not 2), one can prove, for example, the following fact about ratios of L-series corresponding to the cusp form $\Delta(z)$ of weight 12 (the discriminant defined in Exercise 3 of §3.4).

$$\left(\int_0^{i\infty} \Delta(z)\, dz : \int_0^{i\infty} \Delta(z) z^2\, dz : \int_0^{i\infty} \Delta(z) z^4\, dz \right) = \left(1 : \frac{-691}{2^2 3^4 5} : \frac{-691}{2^3 3^2 5 \cdot 7} \right).$$

Such work was begun by Eichler [4] and Shimura [4]. A reference is Lang [6]. It is proved, for example, that ratios of certain L-values in the critical strip lie in the field generated by the Fourier coefficients of the corresponding modular form. This fits into a general philosophy of Deligne (see Deligne's article in Borel and Casselman [1, Vol. 2, pp. 313–346] and Zagier [3, pp. 118–120]). Another paper on this subject is Razar [1].

Voronin [1] considers a similar sort of Dirichlet series to that in formula (3.76). Suppose that $f(z) = \Delta(z)$, as in formula (3.55). Let C be the vertical line in H starting at the point a on the real axis. Thus C is a geodesic. Consider the integral $I(s) = \int_C (-i(z-a))^{s-1} f(z)\, dz$. It is easy to see that for $\operatorname{Re} s > 12$, $I(s) = i(2\pi)^{-s} \Gamma(s) \Phi(s)$, where

$$\Phi(s) = \sum_{n \geq 1} \tau(n) \exp(2\pi i n a) n^{-s}, \qquad \operatorname{Re} s > 12.$$

Here $\tau(n)$ denotes the Ramanujan tau-function defined by formula (3.53). Voronin manages to obtain an analytic continuation of $\Phi(s)$ as a meromorphic function in the whole complex s-plane.

Exercise 17 (Fourier Expansions of Other Eisenstein Series). Compute the Fourier expansions for some other Eisenstein series; e.g., that for the leaky torus in Exercise 13 of §3.3 or for some congruence subgroup.

If you have a home computer attached to a television set, the following exercise should be possible. It illuminates what is happening in Exercise 16 above, as well as in Fig. 3.30.

Exercise 18 (H. Stark) (Computer-Generated Movie of a Non-Euclidean Shock Wave). Consider the following points on a horocycle: $z_j(y) = iy + j/N$, $j = 1, 2, \ldots, N$, holding N fixed. Find $\gamma \in SL(2, \mathbb{Z})$ such that $\gamma z_j(y) = w_j(y)$ lies in the standard fundamental domain for $SL(2, \mathbb{Z}) \backslash H$ (as in Fig. 3.11). Start with $y = 2$. Let y approach zero from above and watch what happens to the points $w_1(y), \ldots, w_N(y)$. At first, you see points on a horizontal line segment of length 1 and height 2. As the segment moves down ($y \to 0+$), the individual points $z_j(y)$ move along geodesics. You can consider the points as forming a non-Euclidean shock wave. The wave reflects from the boundaries of the fundamental domain as y passes through $y = 1$. Then for smaller values of y, more reflections occur and the picture begins to look quite chaotic. The maximum amount of chaos appears to occur around $y = 1/N$. After that the picture begins to become less random. Ultimately the points form various horizontal line segments which move up to the cusp at infinity. For example, if $N = 51$, there are four lines corresponding to points with reduced denominators 1, 3, 17, 15. This is related to the rational cusps approached by the $z_j(y)$, as $y \to 0$. This movie can be generated on a home computer attached to a TV set (e.g., an Apple II).

3.7. Harmonic Analysis on the Fundamental Domain. The Roelcke-Selberg Spectral Resolution of the Laplacian, and the Selberg Trace Formula

The Selberg theory allows us to actually perform calculations on noncocompact but finite volume universes which are of interest to those working in general relativity.

From N. Hurt [1, p. xiv].

In this section we seek to describe the non-Euclidean analogue of Fourier series for $L^2(SL(2, \mathbb{Z})\backslash H)$ as well as two related analogues of the Poisson summation formula. See §1.3 and §1.4 for the Euclidean versions and compare them with the non-Euclidean results given in Theorems 1, 3, and 4 which follow. We shall discuss various applications in number theory, analysis, and geometry. Number-theoretic applications include a non-Euclidean analogue of the circle problem on the asymptotics of the number of lattice points inside a circle of radius x, as x approaches infinity (cf. Theorem 5 of §1.3 and formula (3.104) that follows). We will also consider the asymptotics of units in real quadratic fields (in Theorem 6). There are other number-theoretic applications that will not be discussed here; e.g., the Eichler-Selberg trace formula for the trace of a Hecke operator, the Riemann hypothesis for Selberg's zeta function, formulas for dimensions of spaces of holomorphic modular forms, and Langlands work on the Artin conjecture for Artin L-functions using twisted trace formulas. More analytic applications include the asymptotics of the eigenvalues of the Laplacian on $L^2(SL(2, \mathbb{Z})\backslash H)$, which is another non-Euclidean version of Theorem 5 of §1.3 (see Theorem 5 which follows), and the solution of the non-Euclidean heat equation on $SL(2, \mathbb{Z})\backslash H$ (Exercise 16). The Selberg trace formula (Theorem 4) provides a duality between the eigenvalues of the Laplacian on $M = \Gamma\backslash H$ and closed geodesics in M (if the contribution from the elliptic and parabolic elements of Γ is small). Thus one obtains geometric information about M.

3.7. Harmonic Analysis on the Fundamental Domain

The last remarks are related to the question: "Can one hear the shape of a drum?" discussed by M. Kac [1, pp. 474–496]. Kac describes the problem for a plane membrane M, held fixed along its boundary C. If the membrane is set in motion and $u(z, t)$ denotes the vertical displacement above $z \in M$ at time t, then u satisfies the wave equation

$$\Delta_z u(z, t) = u_{tt}(z, t) \quad \text{for } z \text{ in } M,$$

$$u(z, t) = 0 \quad \text{for } z \text{ on the boundary } C.$$

The solutions will be superpositions of normal modes $\exp(i\omega_n t)v_n(z)$. Here $v_n(z)$, $n = 1, 2, 3, \ldots$, denotes a complete orthonormal set of eigenfunctions of the Laplacian on M which vanish on the boundary C, with eigenvalue $-\omega_n^2 = \lambda_n$. This comes from Theorem 7 of §1.3 on the spectra of compact self-adjoint operators. One hears the normal modes or pure tones ω_n corresponding to the eigenvalues of the Laplacian. The question is then, "What can you say about the geometry of M if you know the spectrum λ_n, $n = 1, 2, 3, \ldots$?" It is of course possible to pose the analogous question for any Riemannian manifold. Milnor showed that there are two noncongruent lattices L in $\mathbb{R}^{16}$ such that the corresponding tori $\mathbb{R}^{16}/L$ have the same spectra. Thus you cannot hear everything about a torus. Marie-France Vignéras [3] found analogous examples for $\Gamma\backslash H$, with Γ's coming from quaternion algebras.

Although you cannot expect to hear everything about M from the eigenvalues of the Laplacian on M, there are still many geometric quantities that can be determined from the spectrum. For example, Weyl showed (in the Euclidean case) that you can hear the area of M. A non-Euclidean analogue of this result is contained in Theorem 5 which follows.

One can also consider the problem of material diffusing through a plane region M with boundary C, starting at a position z, in such a way that it is absorbed upon hitting the boundary curve C. The density of matter $p_M(z, w; t)$ at position w will satisfy the heat or diffusion equation

$$\frac{\partial}{\partial t} p_M(z, w; t) = \Delta_w p_M(z, w; t),$$

$$p_M(z, w; t) \to 0 \quad \text{as } w \text{ approaches the boundary,}$$

$$p_M(z, w; t) \to \delta(z - w) \quad \text{as } t \to 0+.$$

Here $\delta(x)$ denotes the Dirac delta distribution centered at the origin in the plane. If v_n and λ_n are as above, then one can express p_M as follows:

$$p_M(z, w; t) = \sum_{n \geq 1} \exp(\lambda_n t)\overline{v_n(z)}v_n(w).$$

As $t \to 0+$, it is reasonable to expect that "particles of the diffusing stuff will not have had enough time to have felt the influence of the boundary C. As particles begin to diffuse they may not be aware, so to speak, of the disaster that awaits them when they reach the boundary," according to M. Kac [1, p. 481]. Thus one expects that

$$p_M(z,w;t) \sim \frac{1}{4\pi t}\exp\left[-\frac{\|z-w\|^2}{4t}\right] \quad \text{as } t \to 0+$$

(see Exercise 13 of §1.2). It will follow upon integration that

$$\sum_{n\geq 1} \exp(-\lambda_n t) \sim \text{area}(M)/4\pi t \quad \text{as } t \to 0+.$$

This is the Laplace transform of the result of Weyl mentioned earlier, which allows one to hear the area of the drum. We consider a non-Euclidean analogue of this result in the proof of Theorem 5 below. There has been a large amount of work on further terms in the asymptotic expansion given above (see Berger [1] and Molchanov [1] for references). Kac [loc. cit.] provides connections between the asymptotic result above and an asymptotic formula relating quantum statistical mechanics and classical mechanics, as well as connections with the Wiener integral. This latter integral interprets diffusion in terms of a measure defined on the space of all continuous curves emanating from the origin. The measure is defined to agree with the measure coming from the Einstein-Smoluchowski theory of Brownian motion.

These considerations have many connections with physics, as Molchanov [1, p. 7] describes in the following quotation: "In the physics literature the idea of 'integrating along trajectories' as a method of studying spectra has found broad application in quantum mechanics, mainly in the work of the Feynman school; the monograph [Feynman [1]] is devoted to the theory of Feynman integrals. It concerns 'measures' in a space of trajectories, which are constructed from a fundamental solution of the Schrödinger wave equation. A mathematically rigorous foundation for the Feynman theory, at least in certain of its facets, is even now still wanting (see the discussion in [Yu. L. Daletskii [1].)" Another reference on Feynman integrals is the paper of Cecile M. DeWitt [1]. Hurt [1,2], Gutzwiller [1-3], Dowker [1-2] give some insight into the connections between the Selberg trace formula and the Feynman integral picture, and more.

Most of the results on harmonic analysis on $\Gamma\backslash H$, for Fuchsian groups Γ of the first kind, are due to A. Selberg, who lectured on them at Göttingen in 1954. The manuscript of the part of the lectures concerning compact fundamental domains seems to have been lost (thanks to the absence of Xerox machines). However, the second part of the lectures concerning noncompact fundamental domains such as that for $SL(2,\mathbb{Z})$ can be found in the Göttingen library (this is Selberg [6]) and may soon be reprinted. W. Roelcke was working independently on the subject around the same time (see Roelcke [2]). Previously Delsarte has worked out some of the theory for the compact fundamental domains (see Delsarte [1, Tome II, pp. 599–601, 829–845]), motivated by non-Euclidean lattice point problems.

In this section (and indeed this chapter), we have chosen to consider only the example of $\Gamma = SL(2,\mathbb{Z})$. In one respect, this is a nice example, for one can make everything in the trace formula very concrete. However, in another respect, this example, is a difficult one. For it produces parabolic terms in the trace formula—parabolic terms that are rather complicated. Even the elliptic

3.7. Harmonic Analysis on the Fundamental Domain

terms are not trivial. So the reader might want to replace Γ by some nice arithmetic group with compact fundamental domain. Examples of such groups are discussed by Fricke and Klein [1, Vol. I, pp. 502–634] and more briefly by Magnus [1, pp. 123–133]. These groups arise as groups of units of ternary quadratic and binary hermitian forms. Other references are Vignéras [1], Gelfand, Graev, and Piatetskii-Shapiro [1].

General references for this section include Borel and Casselman [1], Borel and Mostow [1], Duistermaat, Kolk, and Varadarajan [1], Elstrodt [1, 2], Elstrodt, Grunewald, and Mennicke [1], Faddeev [1], Fay [1], Gangolli [1], Gangolli and Warner [1, 2], Gelfand, Graev, and Piatetskii-Shapiro [1], Godement [1–4], Goldfeld and Husemoller [1], Hejhal [1–10], Hurt [1], Kubota [1], Lang [4], Lax and Phillips [1], McKean [1], Roelcke [1–3], Sarnak [1–3], Selberg [1, 3, 6], Subia [1], Tamagawa [1], Venkov [1, 3], Venkov, Kalinin, and Faddeev [1], Vignéras [2], Wallace [1], and Warner [2].

We shall begin by discussing the spectral resolution of the non-Euclidean Laplacian on the fundamental domain. This consists of an expansion of an "arbitrary" function $f: SL(2,\mathbb{Z})\backslash H \to \mathbb{C}$ in eigenfunctions of Δ. Such an expansion is a non-Euclidean analogue of Fourier series, but it involves a mixture of series and integrals. Because the fundamental domain is not compact, the eigenvalue problem for Δ on $SL(2,\mathbb{Z})\backslash H$ is said to be *singular*. There may be continuous or discrete spectra (or both) in such cases. For example, recall the eigenvalue problem arising from the quantum mechanics of the hydrogen atom (see formula (2.13)). See Stakgold [1, 2] and Titchmarsh [1] for other examples of eigenvalue problems with mixtures of continuous and discrete spectra.

The reader should recall from §3.5 the basic facts about the spectrum of the non-Euclidean Laplacian on $SL(2,\mathbb{Z})\backslash H$. Let us now collect *the various parts of the spectrum of the non-Euclidean Laplacian on $SL(2,\mathbb{Z})\backslash H$*:

(a) the *discrete spectrum* = {constants, cusp forms} with orthonormal basis $\{v_n\}_{n\geq 0}$, $v_0 = \sqrt{3/\pi}$, v_n = cusp form, $n \geq 1$. $\Delta v_n = s_n(s_n - 1)v_n$, $s_n \in \frac{1}{2} + i\mathbb{R}$, $v_n(z) \sim 0$, as $z \to \infty$, for $n \geq 1$.

(b) the *continuous spectrum* = $\{E_s | s = \frac{1}{2} + it, t \in \mathbb{R}\}$, E_s the Eisenstein series defined in formula (3.66), $\Delta E_s = s(s-1)E_s$,

$$E_s(z) \sim y^s + \frac{\Lambda(1-s)}{\Lambda(s)} y^{1-s}, \quad \text{as } z \to \infty, \qquad (3.99)$$

for $\Lambda(s) = \pi^{-s}\Gamma(s)\zeta(2s)$.

Exercise 1. Show that the three spaces of Eisenstein series, constants, and cusp forms, respectively, are mutually orthonormal.

Note. Throughout this section the inner product on $L^2(SL(2,\mathbb{Z})\backslash H)$ is

$$(f, g) = \int_{SL(2,\mathbb{Z})\backslash H} f(z)\overline{g(z)} y^{-2} \, dx \, dy. \qquad (3.100)$$

The Eisenstein series are not in $L^2(SL(2,\mathbb{Z})\backslash H)$ by Exercise 11 of §3.5, but they are in $L^1(SL(2,\mathbb{Z})\backslash H)$, when s is the critical strip. See also Exercise 12 of §3.5 and Selberg [3, pp. 183–184] for a way of truncating Eisenstein series in order to make sense of inner product formulas for them.

Our first goal is to prove the following.

Theorem 1 (The Roelcke-Selberg Spectral Decomposition of Δ on $L^2(SL(2,\mathbb{Z})\backslash H)$). *Any f in $L^2(SL(2,\mathbb{Z})\backslash H)$ has the non-Euclidean Fourier expansion*

$$f(z) = \sum_{n \geq 0} (f, v_n) v_n(z) + \frac{1}{4\pi i} \int_{\operatorname{Re} s = 1/2} (f, E_s) E_s(z) \, ds.$$

Here we use the notation of (3.99) *and* (3.100).

Before proving Theorem 1, note that the spectral measure $(4\pi i)^{-1} ds$ for the continuous part of the decomposition is computed from the asymptotics and functional equation of the Eisenstein series in accordance with the principle that worked in Theorems 1 and 2 of §3.2. For we know from Exercise 2 of §3.5 that

$$E_s(z) \sim y^s, \text{ as } y \to \infty, \text{ for } s \text{ fixed with } \operatorname{Re} s > 1.$$

And formula (3.68) shows that $E_s(z)$ has a functional equation relating s to $1 - s$. Thus we find that the spectral measure should be

$$\frac{\text{spectral measure for usual Mellin inversion in §1.4}}{\text{number of functional equations}} = \frac{(2\pi i)^{-1}}{2}.$$

Lemma 1 which follows gives a rigorous derivation of this spectral measure. Since $E_s(z)$ has a pole at $s = 1$, there is more happening in Theorem 1 than we saw in Theorems 1 and 2 of §3.2. And we will find that Cauchy's residue theorem throws $\operatorname{Res}_{s=1} E_s(z)$ into the discrete part of the spectrum of Δ on $SL(2,\mathbb{Z})\backslash H$.

The proof that follows is decomposed into a sequence of lemmas that will only end with Theorem 2 concerning the discrete part of the spectrum. The method is that of R. Godement (see Borel and Mostow [1, pp. 211–234]). Other references for this discussion are Kubota [1, Ch. 5, including footnotes and a remark after Theorem 7.5.5] and Lang [4, Ch. 13].

The main idea of the proof is to reduce to the ordinary Mellin inversion formula by making use of series which we shall call "incomplete theta series." Suppose that $\psi : \mathbb{R}^+ \to \mathbb{C}$ decreases sufficiently rapidly at 0 and ∞; e.g., if there is a number $k > 1$ such that $|\psi(y)| \leq y^k$, for all $y > 0$. Then make the following definition.

Definition. The *incomplete theta series* (or Poincaré series) $T\psi$ is given by

$$T\psi(z) = \sum_{\gamma \in \Gamma_\infty \backslash \Gamma} \psi(\operatorname{Im}(\gamma z)) \quad \text{for } z \in H. \tag{3.101}$$

Here $\Gamma_\infty = \{\pm \begin{pmatrix} 1 & n \\ 0 & 1 \end{pmatrix} | n \in \mathbb{Z}\}$ as in formula (3.67).

3.7. Harmonic Analysis on the Fundamental Domain

Exercise 2.

(a) Show that if ψ has compact support, then the series (3.100) has only finitely many terms.
(b) Show that if there exists a $k > 1$ such that $|\psi(y)| \leq y^k$, for all $y > 0$, then series (3.101) converges.
(c) Give conditions on ψ that suffice to put $T\psi$ in $L^2(SL(2, \mathbb{Z}) \backslash H)$.

Exercise 3 (An Adjoint for the Operator T).

(a) Define the constant coefficient operator on $f \in L^2(SL(2, \mathbb{Z}) \backslash H)$ by

$$Cf(y) = \int_{-1/2}^{1/2} f(x + iy) \, dx.$$

Show that this operator pulls out the constant term in the Fourier expansion

$$f(z) = \sum_{n=-\infty}^{+\infty} c_n(y) e^{2\pi i n x}.$$

That is, show that $(Cf)(y) = c_0(y)$.

(b) Show that if T denotes the incomplete theta series (3.101),

$$(T\psi, f) = \int_{y>0} \psi(y)(Cf)(y) y^{-2} \, dy,$$

assuming f is real-valued.

Hint. Note that

$$\int_{\Gamma \backslash H} \sum_{\Gamma_\infty \backslash \Gamma} = \int_{\Gamma_\infty \backslash H}.$$

This same idea was used in formulas (3.96) and (3.97) during the discussion of the Rankin-Selberg method.

Lang [4, pp. 241–243] notes that the adjointness relation in the last exercise happens for any pair of closed subgroups Γ, N of a Lie group (or just a locally compact topological group) G. Here

$$N = \left\{ \pm \begin{pmatrix} 1 & x \\ 0 & 1 \end{pmatrix} \bigg| x \in \mathbb{R} \right\}, \qquad \Gamma = SL(2, \mathbb{Z}), \; G = SL(2, \mathbb{R}).$$

And $N \backslash G \cong \mathbb{R}^2 - 0$ via the identification $Ng \to (0, 1)g$, $g \in G$. The G-invariant measure on $N \backslash G$ is Lebesgue measure on $\mathbb{R}^2$. Lang shows that the same calculations that give part (b) of Exercise 4 say that

$$(T\psi, f)_{\Gamma \backslash G} = \int_{\Gamma \backslash G} T\psi(y) f(y) \, dy = \int_{N \backslash G} \psi(y) \int_{\Gamma \cap N \backslash N} f(ny) \, dn \, dy$$

$$= (\psi, Cf)_{N \backslash G},$$

defining

$$T\psi(y) = \int_{\gamma \in N \cap \Gamma \backslash \Gamma} \psi(\gamma y) \, d\gamma \quad \text{and} \quad Cf(y) = \int_{\Gamma \cap N \backslash N} f(ny) \, dn.$$

Lemma 1 (Properties of Incomplete Theta Series). *Let θ_0 denote the closed subspace of $L^2(SL(2,\mathbb{Z}) \backslash H)$ generated by $T\psi$ such that $M\psi(-1) = (T\psi, 1) = 0$, where $T\psi$ is the incomplete theta series (3.101) and ψ is smooth with compact support on $\mathbb{R}^+$ and Mellin transform*

$$M\psi(s) = \int_0^\infty \psi(y) y^{s-1} \, dy.$$

Define

$$L_0^2(SL(2,\mathbb{Z}) \backslash H) = \{f \in L^2(SL(2,\mathbb{Z}) \backslash H) | (Cf)(y) = 0 \text{ for almost all } y > 0\}.$$

Here $C(y)$ is the constant term operator in Exercise 3(a). Then we have the following list of assertions.

(1) *There is an orthogonal decomposition*

$$L^2(SL(2,\mathbb{Z}) \backslash H) = L_0^2(SL(2,\mathbb{Z}) \backslash H) \oplus \mathbb{C} \oplus \theta_0.$$

(2) *For $c > 1$ and $E_s(z)$ the Eisenstein series defined by formula (3.66),*

$$T\psi(z) = \frac{1}{2\pi i} \int_{\text{Re } s = c} M\psi(-s) E_s(z) \, ds.$$

(3) *If $\varphi(s) = \Lambda(1-s)/\Lambda(s)$ and $\Lambda(s) = \pi^{-s} \Gamma(s) \zeta(2s)$, then*

$$(T\psi, E_s) = M\psi(\bar{s} - 1) + \varphi(\bar{s}) M\psi(-\bar{s}).$$

(4) *Set $v_0^2 = (\text{area}(SL(2,\mathbb{Z}) \backslash H))^{-1} = \text{Res}_{s=1} E_s(z)$. Then v_0 enters into the discrete part of the spectral decomposition of Δ on $SL(2,\mathbb{Z}) \backslash H$, and*

$$T\psi(z) = (T\psi, v_0) v_0 + \frac{1}{4\pi i} \int_{\text{Re } s = 1/2} (T\psi, E_s) E_s(z) \, ds.$$

PROOF.

(1) From Exercise 3, we have

$$(T\psi, g) = \int_{y > 0} \psi(y)(Cg)(y) y^{-2} \, dy,$$

where $C(g)$ is the constant term in the Fourier expansion of g. It follows that the orthogonal complement of the L^2-space spanned by the $T\psi$'s is the space $L_0^2(SL(2,\mathbb{Z}) \backslash H)$. Next use Exercise 3 to see that

$$(T\psi, 1) = M\psi(-1).$$

This means that we can decompose the L^2-span of the incomplete theta series into an orthogonal direct sum of the constants and θ_0. To see why the

3.7. Harmonic Analysis on the Fundamental Domain

constants come out discretely in the spectral decomposition, see the argument in the proof of part (4).

(2) From the Mellin inversion formula of §1.4 it follows that

$$\psi(y) = \frac{1}{2\pi i} \int_{\mathrm{Re}\, s = c} M\psi(-s) y^s\, ds.$$

In order to prove (2), use the definitions of $T\psi$ in (3.100) and the Eisenstein series E_s in formula (3.66) and sum this Mellin inversion formula over $\Gamma_\infty \backslash \Gamma$ for $\mathrm{Re}\, s = c > 1$. This last condition is necessary for absolute convergence.

(3) This follows easily from Exercise 3 and the Fourier expansion of E_s given in Exercise 4 of §3.5.

(4) For $c > 1$, we have from part (2),

$$2T\psi(z) = \frac{1}{2\pi i} \int_{\mathrm{Re}\, s = c} 2M\psi(-s) E_s(z)\, ds.$$

To use the residue theorem and move the integral over to the line $\mathrm{Re}\, s = \frac{1}{2}$, we must find the poles of the integrand in $\frac{1}{2} \leq \mathrm{Re}\, s \leq c$. Now $M\psi(-s)$ is entire, since ψ has compact support. Thus the only pole of the integrand in the region is at $s = 1$ with residue

$$2M\psi(-1) \mathrm{Res}_{s=1}\, E_s(z) = 2(T\psi, 1)(\mathrm{vol}(\Gamma \backslash H))^{-1}.$$

So we have

$$2T\psi(z) = 2(T\psi, v_0) v_0 + \frac{1}{2\pi i} \int_{\mathrm{Re}\, s = 1/2} 2M\psi(-s) E_s(z)\, ds.$$

Now we use the functional equation of E_s (noting that $|\varphi(s)| = 1$, for $\mathrm{Re}\, s = \frac{1}{2}$) to write

$$\int_{\mathrm{Re}\, s = 1/2} 2M\psi(-s) E_s(z)\, ds$$

$$= \int_{\mathrm{Re}\, s = 1/2} (M\psi(-s) E_s(z) + \varphi(s) M\psi(-s) E_{1-s}(z))\, ds$$

$$= \int_{\mathrm{Re}\, s = 1/2} (M\psi(-s) E_s(z) + \varphi(1-s) M\psi(-(1-s)) E_s(z))\, ds.$$

The last equality follows from replacing s by $1 - s$ in the second part of the integrand. Now note that when $\mathrm{Re}\, s = \frac{1}{2}$, we have $-s = \bar{s} - 1$ and $1 - s = \bar{s}$. Thus, it follows from part (3) that

$$2T\psi(s) = 2(T\psi, v_0) v_0 + \frac{1}{2\pi i} \int_{\mathrm{Re}\, s = 1/2} (T\psi, E_s) E_s(z)\, ds,$$

as was to be proved. □

Exercise 4. Fill in the details in the proofs of (2)–(4) above. In (4), note that you need bounds on the integrand to move the lines of integration. The argument is similar to that used for ordinary Mellin transforms.

Exercise 5 (The Size of $E_s(z)$ as t Approaches Infinity and the Residue at $s = 1$).

(a) Show that $E_s(z) = O(t^{1/2+\varepsilon})$, as t approaches infinity (for fixed z), if $s = \sigma + it$ and $\sigma \geq \frac{1}{2}$.

Hint. Use the Phragmen-Lindelof theorem (see Lang [3, pp. 262–267] or Titchmarsh [3]). Note that $E_s(z) = O(\exp(t^a))$ for some $a > 0$, using the incomplete gamma expansion (Theorem 1 of §1.4) or the Fourier series (Exercise 4 of §3.5). Deduce that for Re $s > 1$, $E_s(z) = O(1)$, using the definition of $E_s(z)$ as a Dirichlet series. Obtain the estimate for $E_s(z)$ when Re $s > 0$, using the functional equation

$$E_s(z) = \frac{\Lambda(1-s)}{\Lambda(s)} E_{1-s}(z) = \frac{\Lambda(1-s)}{\Lambda(\frac{1}{2}-s)} E_{1-s}(z), \quad \text{with } \Lambda(s) = \pi^{-s}\Gamma(s)\zeta(2s),$$

and using Stirling's formula for $\Gamma(s)$. One could actually improve the estimate, making use of Hadamard's three lines theorem, and one could do even better if one knew the Riemann hypothesis.

(b) Show that $\mathrm{Res}_{s=1} E_s(z) = (\mathrm{vol}(SL(2,\mathbb{Z})\backslash H))^{-1}$. You could use the Fourier expansion of E_s to do this, for example, plus the fact that $\zeta(2) = \pi^2/6$. Or you could use the incomplete gamma expansion of Epstein's zeta function (Theorem 1 of §1.4). This is a general property of Eisenstein series as we shall see in later chapters (see also Sarnak [4] and Siegel [5, Vol. I, pp. 459–468; Vol. III, pp. 328–333]).

Lemma 1 completes the portion of the proof of Theorem 1 coming from the continuous spectrum and the one-dimensional space of constants in the discrete spectrum. This part of the discrete spectrum is often called the residual spectrum since it comes from a residue of the Eisenstein series. There are groups Γ such that the Eisenstein series have arbitrarily many poles on the real line (see Selberg [3, p. 184]). Next we must deal with the rest of the discrete spectrum of Δ. We will show that $L_0^2(SL(2,\mathbb{Z})\backslash H)$ lies in the discrete spectrum of Δ (Theorem 2).

At this point one might wonder why $L_0^2(SL(2,\mathbb{Z})\backslash H)$ is not the zero space. This is equivalent to showing that nonzero cusp forms exist, as we will show. One can deduce this from the Selberg trace formula itself. It is fairly easy to show that there are odd cusp forms for $SL(2,\mathbb{Z})$, as the following exercise shows. See our discussion of the Roelcke-Selberg conjecture at the end of §3.5.

Exercise 6.

(a) Show that $L_0^2(SL(2,\mathbb{Z})\backslash H)$ contains a nonzero element.

3.7. Harmonic Analysis on the Fundamental Domain

Hint (H. Stark). Take an odd function $g(x)$ of period 1; e.g., $\sin(2\pi nx)$, $x \in \mathbb{R}$, $n \in \mathbb{Z}$. Then form a function $f(z)$ on the standard fundamental domain D for $SL(2, \mathbb{Z})$ pictured in Fig. 3.11 by setting

$$f(z) = \begin{cases} g(x) & \text{for } z = x + iy, \text{ with } y \geq 2, \\ 0 & \text{for } z = x + iy, \text{ with } y < 2. \end{cases}$$

We could actually do this smoothly. Then extend $f(z)$ to a function $F(z)$ by writing $F(z) = f(\gamma z)$, if $\gamma \in SL(2, \mathbb{Z})$ is such that $\gamma z \in D$. Note that the lack of uniqueness of γ for z on the boundary of D does not matter since f is the same at equivalent boundary points. You need to show that the function $F(z)$ is an odd function of x. This comes from the fact that $f(z)$ was an odd function of x and the identity

$$\gamma^*(-\bar{z}) = -\overline{\gamma z} \quad \text{for } \gamma = \begin{pmatrix} a & b \\ c & d \end{pmatrix} \in SL(2, \mathbb{Z}) \text{ and } \gamma^* = \begin{pmatrix} a & -b \\ -c & d \end{pmatrix}.$$

(b) What goes wrong with trying to find an even element of $L_0^2(SL(2, \mathbb{Z})\backslash H)$ by starting with a classical holomorphic cusp form of weight k and forming $g(z) = y^{k/2}|f(z)|$?

In order to study the discrete spectrum of Δ it is natural to attempt to replace Δ with integral operators. Here we shall use convolution operators. It is also possible to study the *resolvent*:

$$(\Delta - \lambda)^{-1} f(z) = \int_{w \in \Gamma \backslash H, w = u + iv} G_\lambda(z, w) f(w) v^{-2} \, du \, dv. \tag{3.102}$$

The kernel $G_\lambda(z, w)$ in formula (3.102) is called the Green's function for $\Delta - \lambda$. We studied the Euclidean analogue of this Green's function in formulas (1.15)–(1.17). References for the non-Euclidean Green's function are Elstrodt [1, 2], Faddeev [1], Fay [1], Hejhal [1], Lang [4], Neuenhöffer [1], and Roelcke [1, 2]. The poles in λ of the Green's function G_λ are the eigenvalues of Δ on $L^2(SL(2, \mathbb{Z})\backslash H)$; i.e., the discrete spectrum. Branching in λ for G_λ gives rise to the continuous spectrum and the Kodaira-Titchmarsh formula allows the computation of the spectral measure from the jump across the real axis (see formula (2.24)). See Example 2 and Exercises 10–15 of this section for more information on Green's functions.

In order to define convolution as in §3.2 we must think of functions on $H \cong G/K$, with $G = SL(2, \mathbb{R})$ and $K = SO(2)$, as functions on G by writing $f(a) = f(ai)$. Here ai means the point in H to which $a \in SL(2, \mathbb{R})$ maps $i = \sqrt{-1}$ by fractional linear transformation (see Exercise 8 of §3.1). Suppose then that we have two integrable functions $f, g : H \to \mathbb{C}$. Then define the *convolution* of f and g by

$$L_g f(a) = (f * g)(a) = \int_G f(b) g(b^{-1} a) \, db. \tag{3.103}$$

Here db denotes left = right Haar measure on G (see §2.1 and Exercise 19 of §3.2). The measure db is unique up to a positive constant and is characterized by its invariance under the group action. General results on Haar measure can be found in Helgason [2] and Lang [2]. We do not need any explicit formula for db, just its existence.

Note. There is a slightly confusing aspect of the identification $f(ai) = f(a)$, for $a \in SL(2, \mathbb{R})$. For if we adjust the Haar measure da so that

$$\int_G f(ai)\,da = \int_H f(x+iy)y^{-2}\,dx\,dy,$$

we discover quickly that the integral on the left is invariant under the substitution of a^{-1} for a, but not the integral on the right! For

$$x + iy = \begin{pmatrix} 1 & x \\ 0 & 1 \end{pmatrix}\begin{pmatrix} y^{1/2} & 0 \\ 0 & y^{-1/2} \end{pmatrix};$$

and

$$\left(\begin{pmatrix} 1 & x \\ 0 & 1 \end{pmatrix}\begin{pmatrix} y^{1/2} & 0 \\ 0 & y^{-1/2} \end{pmatrix}\right)^{-1} i = \begin{pmatrix} y^{-1/2} & 0 \\ 0 & y^{1/2} \end{pmatrix}\begin{pmatrix} 1 & -x \\ 0 & 1 \end{pmatrix} i = \frac{i-x}{y}.$$

Let $u + iv = (i - x)/y$. The Jacobian of this change of variables is

$$\frac{\partial(u,v)}{\partial(x,y)} = y^{-3}.$$

Therefore

$$\int_H f\left(\left(\begin{pmatrix} 1 & x \\ 0 & 1 \end{pmatrix}\begin{pmatrix} y^{1/2} & 0 \\ 0 & y^{-1/2} \end{pmatrix}\right)^{-1}\right) y^{-2}\,dx\,dy = \int_H f(u+iv)v^{-1}\,du\,dv.$$

This is a reflection of the non-unimodularity of the group of upper triangular matrices with determinant 1 and positive diagonal. The moral is that we should be somewhat careful about our identifications.

As an example, let us compute $L_g f(i)$, for $f(z) = p_s(z) = (\operatorname{Im} z)^s$. The result is

$$L_g p_s(i) = \int_G p_s(bi) g(b^{-1})\,db = \int_H y^s g\left(\frac{i-x}{y}\right) y^{-2}\,dx\,dy$$

$$= \int_H y^{-s} g(z) y^{-1}\,dx\,dy = \hat{g}(1-\bar{s}),$$

where $\hat{g}(s)$ is the Helgason transform of g defined in (3.27) assuming that g is in fact rotation-invariant.

Lemma 2 (Properties of Convolution Operators). *Suppose throughout that $g: H \to \mathbb{C}$ is infinitely differentiable with compact support.*

(1) *The operator L_g defined by (3.103) commutes with the left action of x in G on*

functions. Thus L_g is a G-invariant integral operator such that

$$L_g : L^2(SL(2,\mathbb{Z})\backslash H) \to C^\infty(SL(2,\mathbb{Z})\backslash H).$$

(2) *If $g(a) = g(a^{-1})$ for all a in G and if g is real-valued, then L_g is a self-adjoint operator on $L^2(SL(2,\mathbb{Z})\backslash H)$; i.e., $(L_g f, h) = (f, L_g h)$, using the inner product* (3.100).

(3) $L_{g*h} f = L_g L_h f$.

(4) *The operators L_g commute if g is K-bi-invariant, $K = SO(2)$; i.e., if $g : K\backslash G/K \to \mathbb{C}$.*

(5) *If $g(a) = g(a^{-1})$ for all $a \in G$, then $\Delta(L_g) = L_g \Delta$ and $\Delta L_g = L_{\Delta g}$.*

PROOF.

(1) For x in G, set $f^x(y) = f(xy)$. Then, using the left invariance of Haar measure db, we obtain

$$L_g(f^x)(a) = \int_G f(xb) g(b^{-1}a)\, db = \int_G f(b) g(b^{-1}xa)\, db = (L_g f)^x(a).$$

(2) To see this, write

$$\begin{aligned}(L_g f, h) &= \int_{\Gamma\backslash G/K} \int_{G/K} f(b) g(b^{-1}a)\, db\, \overline{h(a)}\, da \\ &= \int_{\Gamma\backslash G/K} \int_{\Gamma\backslash G/K} f(b) \sum_{\gamma\in\Gamma} g(b^{-1}\gamma a)\, db\, \overline{h(a)}\, da \\ &= \int_{G/K} \int_{G/K} g(b^{-1}a) \overline{h(a)}\, da\, f(b)\, db \\ &= (f, L_g h) \qquad \text{since } g(a) = g(a^{-1}) \text{ and } g \text{ is real-valued.}\end{aligned}$$

(3) This follows from the associative law for the convolution of L^1 functions.

(4) First note that if g is K-bi-invariant, then if ${}^t x$ denotes the transpose of x, we have $g(x) = g({}^t x)$ for all x in G. For we know from §3.1 that any matrix in G has the form $k_1 a k_2$, with a = diagonal and $k_j \in K$. And clearly $g(a) = g({}^t a)$ for diagonal matrices a. Thus if h and g are both K-bi-invariant, we have

$$\begin{aligned}(h * g)(a) &= \int g(b^{-1}a) h(b)\, db \\ &= \int h(ac^{-1}) g(c)\, dc = (g * h)({}^t a) = (g * h)(a).\end{aligned}$$

(5) See Exercise 7. □

Exercise 7. Prove part (5) of Lemma 2. Do both formulas require $g(a) = g(a^{-1})$? If g is K-bi-invariant, does it follow that $g(a) = g(a^{-1})$?

Lemma 3 (The Connection between the Eigenvalues of Convolution Operators and Eigenvalues of the Laplacian. The Selberg Transform and the Helgason Transform).

(a) *Suppose that $L_g f = \lambda_g f$, $\lambda_g \in \mathbb{C}$, for all $g \in C_c^\infty(K \backslash G / K)$. Then $\Delta f = \mu f$, and the transform $\mu \to \lambda_g$, which is often called the Selberg transform, is essentially the Helgason transform defined by formula (3.27) for rotation-invariant functions g.*

(b) *More precisely suppose that $g : K \backslash G / K \to \mathbb{C}$ is infinitely differentiable with compact support. Let ϕ be any eigenfunction of the non-Euclidean Laplacian with $\Delta \phi = s(s-1)\phi$. For example, ϕ could be a cusp form, or y^s, or the spherical function*

$$h_s(z) = \int_{k \in K} (\operatorname{Im}(kz))^s \, dk = P_{s-1}(\cosh r) \qquad \text{if } z - ke^{-r}i \in H,$$

with P_{s-1} = the Legendre function of Exercises 9 and 10 of §3.2. Then $\Delta \phi = s(s-1)\phi$ implies that

$$\frac{L_g \phi}{\phi}(x) = \frac{\phi * g}{\phi}(x) = \hat{g}(1 - \bar{s}) = \int_H g(z) y^{-s-1} \, dx \, dy$$

$$= \text{the Helgason transform;}$$

i.e., if $\mu = s(s-1)$, then $\lambda_g = \hat{g}(1 - \bar{s})$.

PROOF.

(a) Suppose that $f * g = \lambda_g f$. Then

$$\lambda_g \Delta f = \Delta(f * g) = f * (\Delta g) = \lambda_{\Delta g} f \quad \text{implies} \quad \Delta f = \mu f,$$

with $\mu = \lambda_{\Delta g}/\lambda_g$. Note that if g runs through a Dirac family at the identity, we can assume that λ_g is nonzero.

Question. Does it suffice to know that $L_g f = \lambda_g f$ for all g in a Dirac family at the identity?

(b) Let $\mathcal{M}$ be the operator that averages over the compact group $K = SO(2)$. Then

$$\mathcal{M}\phi(x) = \int_K \phi(kx) \, dk,$$

where dk is left (which equals right) Haar measure on $K = SO(2)$, normalized so that

$$\int_K dk = 1.$$

By the uniqueness of the spherical function $h_s(z) = P_{s-1}(\cosh r)$, for $z = ke^{-r}i$, with $k \in K$, $r > 0$ (see formula (3.21) with $a = 0$ and Exercise 9 of

3.7. Harmonic Analysis on the Fundamental Domain

that section), we have $\mathcal{M}\phi = ch_s$ with $c = \phi(i)$, for $i = \sqrt{-1}$. It follows that
$$\mathcal{M}(\phi * g)(a) = ((\mathcal{M}\phi) * g)(a) = ((ch_s) * g)(a) = c(h_s * g)(a).$$

Evaluate this at $a = $ the identity, I, to obtain:
$$L_g\phi(I) = (\phi * g)(I) = c(h_s * g)(I) \quad \text{where } c = \phi(I) = \phi(i),$$

by our identification. Therefore we can set $f(z) = y^s$ and obtain
$$\lambda_g = \frac{L_g f}{f}(i) = \frac{L_g \phi}{\phi}(i) = \int_H g(z) y^{-s-1}\, dx\, dy = \hat{g}(1 - \bar{s}),$$

as we saw in the note after Definition (3.103).

Note that if $\Delta f = \mu f$, then $L_g f = \lambda_g f$. For if $a \in G$, set $f^a(x) = f(ax)$, when $x \in G$. Then $\Delta(f^a) = (\Delta f)^a = \mu f^a$. So, using the preceding results, we find that:
$$L_g(f^a)(I) = (L_g f)^a(I) = \hat{g}(1 - \bar{s})(f^a)(I),$$

which implies that $(L_g f)(a) = \hat{g}(1 - \bar{s}) f(a)$. $\square$

We are now in a position to prove the remaining part of Theorem 1.

Theorem 2. *Suppose $g: H \to \mathbb{C}$ is infinitely differentiable with compact support. Let L_g be the convolution operator in (3.103). Then L_g is a compact operator on the space $L_0^2(\Gamma \backslash H)$, $\Gamma = SL(2, \mathbb{Z})$, defined in Lemma 1. Thus the spectrum of Δ on $L_0^2(\Gamma \backslash H)$ is discrete and the spectral theorem for compact, self-adjoint operators (see Theorem 7 of §1.3) assures the existence of a complete orthonormal set of cusp forms spanning $L_0^2(\Gamma \backslash H)$.*

PROOF. Let $g: K \backslash G / K \to \mathbb{C}$ be C^∞ with compact support for $G = SL(2, \mathbb{R})$, $K = SO(2)$. We want to show that L_g is a compact operator on $L_0^2(\Gamma \backslash H)$. Suppose that $\| \ \|$ denotes the sup norm for functions on G; i.e.,
$$\|f\| = \sup \{|f(x)|, x \in G\}.$$

And let $\| \ \|_2$ denote the L^2 norm for functions on $\Gamma \backslash H$; i.e.,
$$\|f\|_2 = (f, f)^{1/2} \quad \text{with the inner product in (3.100).}$$

We want to show that, for any f in $L_0^2(\Gamma \backslash H)$,
$$\|L_g f\| \le C \|f\|_2 \quad \text{for some positive constant } C, \tag{*}$$

This will imply that L_g is a compact operator on $L_0^2(\Gamma \backslash H)$. To see this, note that the image of L_0^2 under L_g is an equicontinuous family of continuous functions. If we can show the family to be uniformly bounded, then the theorem of Arzelà-Ascoli (see Kolmogorov and Fomin [1]) implies that sequences $\{L_g f_n\}$ must have subsequences converging on compacta in $\Gamma \backslash H$ to continuous functions. Lang [4] gives another proof of this.

Now, in order to bound the $L_g f$'s by the L_2 norms of the f's, we argue as follows, using the Poisson summation formula (§1.3). Assuming that $g(a) = g(a^{-1})$, we can write

$$L_g f(a) = \int_H f(z) g(a^{-1} z) y^{-2} \, dx \, dy.$$

Suppose next that

$$a^{-1} = k \begin{pmatrix} d^{1/2} & 0 \\ 0 & d^{-1/2} \end{pmatrix} \begin{pmatrix} 1 & v \\ 0 & 1 \end{pmatrix} \quad \text{with } k \in K, d > 0, v \in \mathbb{R}.$$

It is possible to make this decomposition by Exercise 9 of §3.1. In the general case of Chapter 5, this is called the Iwasawa decomposition of a in G. We are trying to show that $L_g f(a)$ is bounded as $ai = -v + i/d$ approaches infinity in H; i.e., as v approaches $+\infty$ or $-\infty$, or d approaches 0 or infinity. We will assume that $g(x + iy) = h(x)k(y)$ to simplify the calculations, since g can be approximated by such functions. Now set

$$\Gamma_\infty = \left\{ \begin{pmatrix} 1 & n \\ 0 & 1 \end{pmatrix} \middle| n \in \mathbb{Z} \right\}.$$

Then

$$L_g f(a) = \int_{\Gamma_\infty \backslash H} \sum_{n \in \mathbb{Z}} g(d(z + n + v)) f(z) y^{-2} \, dx \, dy$$

$$= \int_{\Gamma_\infty \backslash H} \sum_{n \in \mathbb{Z}} h(d(x + n + v)) k(dy) f(z) y^{-2} \, dx \, dy.$$

Poisson tells us (using the dilation and translation properties of the Fourier transform on $\mathbb{R}$) that

$$\sum_{n \in \mathbb{Z}} h(d(x + n + v)) = d^{-1} \sum_{n \in \mathbb{Z}} \hat{h}(n/d) \exp[2\pi i n(x + v)],$$

where $\hat{h}$ denotes the Fourier transform of h over $\mathbb{R}$. And the fact that h is C^∞ with compact support says that $|\hat{h}(x)| \leq c|x|^{-p}$ for any power p. So if $n \neq 0$, we have a bound. If $n = 0$, since f is in L_0^2, we find that

$$\int_{\Gamma_\infty \backslash H} \hat{h}(0) k(dy) f(z) y^{-2} \, dx \, dy = 0.$$

Thus we have

$$|L_g f(a)| \leq c d^{p-1} \int_{\Gamma_\infty \backslash H} \sum_{n \neq 0} |n|^{-p} |k(dy)| |f(z)| y^{-2} \, dx \, dy.$$

We have assumed that the function k has compact support. Thus for d small enough, the preceding integral will be bounded by a constant times

$$d^{p-1} \int_{|x| \leq 1/2, y > 2} |f(z)| y^{-2} \, dx \, dy.$$

3.7. Harmonic Analysis on the Fundamental Domain

The Cauchy-Schwarz inequality completes the proof of the desired inequality (∗). Finally, the proofs of Theorems 1 and 2 are complete. □

Question. Suppose we replace $SL(2,\mathbb{Z})$ in Theorem 1 by some congruence subgroup $\Gamma(N)$. Can you formulate harmonic analysis on $\Gamma(N)\backslash H$ as some sort of product of that on $SL(2,\mathbb{Z})\backslash H$ with that on the finite group $SL(2,\mathbb{Z})/\Gamma(N)$?

It is now possible to develop a non-Euclidean analogue of the Poisson summation formula, using the same argument that proved Theorem 2 of §1.3.

Theorem 3 (A Non-Euclidean Poisson Summation Formula). *Let $f: H \to \mathbb{C}$ be $K = SO(2)$-invariant; i.e., $f(kz) = f(z)$ for all $k \in K$, $z \in H$ and suppose also that for $\Gamma = SL(2,\mathbb{Z})$, the series*

$$g(z) = \sum_{\gamma \in \Gamma/\pm I} f(\gamma z)$$

converges absolutely and uniformly on compacta in H to a function $g(z)$ such that $\Delta g \in L^2(\Gamma\backslash H)$. Then, using the notation (3.99),

$$g(z) = \sum_{n \geq 0} \hat{f}(s_n)\overline{v_n(i)}\, v_n(z) + \frac{1}{4\pi i} \cdot \int_{\operatorname{Re} s = 1/2} \hat{f}(s)\overline{E_s(i)}\, E_s(z)\, ds,$$

with $\hat{f}(s) =$ the non-Euclidean Fourier (or Helgason transform) given by

$$\hat{f}(s) = \int_H f(z) \overline{y^s}\, y^{-2}\, dx\, dy.$$

PROOF. By Theorem 1, we can write

$$g(z) = \sum_{n \geq 0} (g, v_n) v_n(z) + \frac{1}{4\pi i} \int_{\operatorname{Re} s = 1/2} (g, E_s) E_s(z)\, ds.$$

Convergence is uniform and absolute by Exercise 17 that follows. Using the fact that H is a disjoint union of translates by $\gamma \in \Gamma/\pm I$ of the fundamental domain $\Gamma\backslash H$, we see that

$$(g, v_n) = \int_{\Gamma\backslash H} \sum_{\gamma \in \Gamma/\pm I} f(\gamma z)\overline{v_n(z)} y^{-2}\, dx\, dy = \int_H f(z)\overline{v_n(z)} y^{-2}\, dx\, dy.$$

Then Lemma 3 implies that $(g, v_n) = \hat{f}(s_n)\overline{v_n(i)}$. The formula with v_n replaced by E_s is proved similarly, to complete the proof of Theorem 3. □

Later, while discussing Selberg's trace formula, we shall find another derivation of Theorem 3 (from the non-Euclidean analogue of Mercer's theorem which was Theorem 8 of §1.3).

The analogue of Theorem 3, for Γ such that $\Gamma\backslash H$ is compact, goes back to Delsarte in 1942. Delsarte used this result to study non-Euclidean lattice point problems such as the asymptotic result given in formula (3.104) below. Others have used Theorem 3 to study Green's functions for the Laplacian on $\Gamma\backslash H$ (see Exercise 14 below). It is also possible to use the non-Euclidean Poisson summation formula to study the Poincaré series that arise in work on the Ramanujan-Petersson conjecture for Fourier coefficients of modular forms (see Bruggeman [1], Deshouillers and Iwaniec [1], Goldfeld and Sarnak [1], Kuznetsov [1], and Selberg [2]). Kudla and Millson [1] use non-Euclidean Poisson summation to give an explicit construction of the harmonic one-form dual to an oriented closed geodesic on an oriented Riemann surface of genus greater than 1. Takhtajan and Vinogradov [2] use such results to obtain relations between fundamental units of real quadratic fields and the existence of cusp forms with eigenvalue $\frac{1}{4}$ for $\Gamma_0(d)$.

Example 1 (Non-Euclidean Lattice Point Problems). References include Delsarte [1, Tome II, pp. 599–601, 829–845], Huber [1], Mennicke [1], and Patterson [1].

Recall first that in Theorem 5 of §1.3 we found an asymptotic result for the number of lattice points; i.e., elements of $\mathbb{Z}^2$, in a circle, as the radius of the circle blows up. One non-Euclidean analogue of this is

$$N_\Gamma(x) = \#\{\gamma \in \Gamma/\pm I | \cosh d(i, \gamma i) \leq x\} \sim \frac{2\pi x}{\text{area}(\Gamma\backslash H)}, \quad x \to \infty, \quad (3.104)$$

for $\Gamma = SL(2,\mathbb{Z})$, where $d(z, w)$ denotes the non-Euclidean distance between points z in w in H. Note that $2\pi/\text{area}(\Gamma\backslash H) = 6$ in this case. Now, you may worry about the hyperbolic cosine in the definition of $N_\Gamma(x)$. Hopefully the following exercise will allay these worries by indicating that there is a generalization to $SL(n,\mathbb{Z})$ or $GL(n,\mathbb{Z})$ (see Chapter 4).

Exercise 8.

(a) Recall Exercises 9 and 10 of §3.1. These exercises show that if $z = ke^{-r}i$, for $k \in K = SO(2)$ and $r > 0$, and W_z is the corresponding 2×2 matrix in $\mathscr{SP}_2$, then

$$W_z = \begin{pmatrix} 1 & 0 \\ 0 & 1 \end{pmatrix} \left[\begin{pmatrix} \exp(-r/2) & 0 \\ 0 & \exp(r/2) \end{pmatrix} k \right],$$

Here we use the notation $Y[A] = {}^t A Y A$, with ${}^t A$ = transpose of A. Show that $\cosh d(i, z) = \frac{1}{2}(e^r + e^{-r}) = \frac{1}{2}\text{Tr}(W_z)$. Then show that if

$$\gamma = \begin{pmatrix} a & b \\ c & d \end{pmatrix} \in \Gamma,$$

$$2\cosh d(\gamma i, i) = \text{Tr}({}^t\gamma\gamma) = a^2 + b^2 + c^2 + d^2.$$

(b) Prove that

$$2\cosh d(z, w) + 2 = \frac{|z - \bar{w}|^2}{\text{Im } z \, \text{Im } w}.$$

Hint. See Exercise 25 of §3.2.

From Exercise 8, we see that formula (3.104) is

$$\#\{\gamma \in \Gamma/\pm I \mid \tfrac{1}{2}\operatorname{Tr}({}^t\gamma\gamma) \le x\} \sim 6x, \qquad x \to \infty,$$

which means that

$$\#\{a, b, c, d \in \mathbb{Z} \mid a^2 + b^2 + c^2 + d^2 \le x, ad - bc = 1\} \sim 6x, \qquad x \to \infty. \tag{3.105}$$

We know from Theorem 5 of §1.3 that

$$\#\{a, b, c, d \in \mathbb{Z} \mid a^2 + b^2 + c^2 + d^2 \le x\} \sim \pi^2 x^2, \qquad x \to \infty. \tag{3.106}$$

A comparison of these two asymptotic results gives one a good feeling for the relative densities of these two sets.

In order to prove (3.104), one may imitate the proof of Theorem 5 of §1.3. To this end we form the *non-Euclidean theta function*† for $a > 0$:

$$\theta_\Gamma(a) = \sum_{\gamma \in \Gamma/\pm I} \exp[-\tfrac{1}{2}a\operatorname{Tr}({}^t\gamma\gamma)] = \sum_{\gamma \in \Gamma/\pm I} \exp[-a\cosh d(i, \gamma i)]. \tag{3.107}$$

Set

$$f_a(z) = \exp[-a\cosh d(i, z)] = \exp[-\tfrac{1}{2}a\operatorname{Tr}(W_z)] \qquad \text{for } z \in H \text{ and } W_z$$

the corresponding matrix in $\mathcal{SP}_2$. The Helgason transform is

$$\hat{f}_a(s) = \int_{W_z \in \mathcal{SP}_2} \exp[-\tfrac{1}{2}a\operatorname{Tr}(W_z)]y^{\bar{s}-2}\,dx\,dy = 2(2\pi/a)^{1/2}K_{\bar{s}-1/2}(a). \tag{3.108}$$

To prove (3.108), note that Exercise 9 of §3.1 implies that for $z = x + iy$,

$$W_z = \begin{pmatrix} 1/y & 0 \\ 0 & y \end{pmatrix}\begin{bmatrix} 1 & x \\ 0 & 1 \end{bmatrix} = \begin{pmatrix} 1/y & * \\ * & (x^2 + y^2)/y \end{pmatrix}.$$

Therefore $\operatorname{Tr}(W_z) = 1/y + (x^2 + y^2)/y$, and

$$\hat{f}_a(s) = \int_{y>0}\int_{x \in \mathbb{R}} \exp[-\tfrac{1}{2}a(1/y + (x^2 + y^2)/y)]y^{\bar{s}-2}\,dx\,dy.$$

Perform the integral over x and obtain

$$\int_{x \in \mathbb{R}} \exp(-ax^2/2y)\,dx = (2\pi y/a)^{1/2}.$$

It follows that

$$\hat{f}_a(s) = (2\pi/a)^{1/2}\int_{y>0} y^{\bar{s}-1/2}\exp[-\tfrac{1}{2}a(y + 1/y)]y^{-1}\,dy.$$

A formula from Exercise 1 of §3.2 finishes the proof of (3.108).

† Our non-Euclidean theta function in (3.107) is not the same as that considered by differential geometers (see Molchanov [1, p. 46] and the proof of Theorem 5).

Applying Theorem 3 to $f_a(z) = \exp[-\frac{1}{2}a\operatorname{Tr}(W_z)]$, we have

$$\theta_\Gamma(a) = \sum_{\gamma \in \Gamma/\pm I} \exp[-\tfrac{1}{2}a\operatorname{Tr}({}^t\gamma\gamma)] = 2(2\pi/a)^{1/2} \sum_{n \geq 0} K_{\bar{s}_n - 1/2}(a)|v_n(i)|^2$$

$$+ 2(2\pi/a)^{1/2} \frac{1}{4\pi i} \int_{\operatorname{Re} s = 1/2} K_{\bar{s} - 1/2}(a)|E_s(i)|^2 \, ds. \tag{3.109}$$

In order to obtain the asymptotics of $\theta_\Gamma(a)$, as $a \to 0+$, we need the following results:

$$K_{iu}(a) = (2\pi/u)^{1/2} e^{-u\pi/2}\left(\cos\left[u\log\frac{a}{2} - u\log u + u + \frac{\pi}{4}\right] + o(1)\right), \tag{3.110}$$

$u \to \infty$, $a \to 0$;

$$E_{1/2 + iu}(z) = O(|u|^{1/2 + \varepsilon}) \qquad \text{as } u \to \infty. \tag{3.111}$$

Formula (3.111) was Exercise 5 and formula (3.110) comes from Exercise 9 below.

One finds that the main term on the right-hand side of (3.109) is the term corresponding to v_0, and the rest is $O(1)$; i.e., bounded. It follows that

$$\theta_\Gamma(a) \sim \frac{2\pi}{\operatorname{area}(\Gamma \backslash H)a} \qquad \text{as } a \to 0+. \tag{3.112}$$

One finishes the proof of (3.104) with the same Tauberian theorem that sufficed to prove Theorem 5 of §1.3 (namely, Theorem 5 of §1.2).

Exercise 9. Prove formula (3.110) using the following facts about I and K Bessel functions (see Lebedev [1, p. 140]):

$$K_s(z) = \frac{\pi}{2} \frac{I_{-s}(z) - I_s(z)}{\sin(s\pi)} \qquad \text{(taking limits if } s \in \mathbb{Z}\text{)};$$

$$I_s(z) = \left(\frac{z}{2}\right)^s \sum_{k=0}^\infty \frac{(z^2/4)^k}{k!\Gamma(s + k + 1)}.$$

Example 2 (Green's Functions for $\Delta - \lambda$ on $\Gamma \backslash H$ by the Method of Images). References for this subject are Elstrodt [1, 2], Faddeev [1], Fay [1], Hejhal [1, Vol. 2; 4], Lang [4, Ch. 14], and Neuenhöffer [1]. We discussed the method of images briefly in §1.3. Here we seek a non-Euclidean analogue of the Green's function in Exercise 20 of §1.3.

The *Green's function* $G_\lambda(z, w)$ *for the resolvent* $R_\lambda = (\Delta - \lambda I)^{-1}$ *on the fundamental domain* $\Gamma \backslash H$, $\Gamma = SL(2, \mathbb{Z})$, *is given by*

$$R_\lambda f(z) = \int_{w = u + iv \in \Gamma \backslash H} G_\lambda(z, w) f(w) v^{-2} \, du \, dv. \tag{3.113}$$

The *method of images* says that if $g_\lambda(z, w)$ is the kernel for the resolvent R_λ on all of H, then

$$G_\lambda(z, w) = \sum_{\gamma \in \Gamma/\pm I} g_\lambda(z, \gamma w). \tag{3.114}$$

So let us find $g_\lambda(z, w)$. In order to do this, one should recall the basic facts about Green's functions (see Courant and Hilbert [1, pp. 363–388], Garabedian [1] or Stakgold [1, 2]). In particular, $g_\lambda(z, w)$ should have the same sort of singularity as that for the Euclidean Green's function for the Euclidean Laplacian $(\partial^2/\partial x^2 + \partial^2/\partial y^2)$ on $\mathbb{R}^2$. The reason for this is that the y^2 in the non-Euclidean Laplacian is cancelled out by the y^{-2} in the G-invariant area element on H. This always happens when calculating Green's functions for differential operators with weights. So the *singularity* of $g_\lambda(z, w)$ is the same as that of $(1/2\pi)\log|z - w|$. Also, $g_\lambda(z, w)$ must be $G = SL(2, \mathbb{R})$-*invariant*; i.e., $g_\lambda(az, aw) = g_\lambda(z, w)$ for all z, w in H and a in G. And $g_\lambda(z, w)$ should be as small as possible as $z \to \infty$. Of course, g_λ should also satisfy the *differential equation*

$$y^2\left(\frac{\partial^2}{\partial x^2} + \frac{\partial^2}{\partial y^2}\right) g_\lambda(z, w) = \lambda g_\lambda(z, w) \qquad \text{for } z \neq w.$$

Exercise 10.

(a) Show that the Green's function for $L = \partial^2/\partial x^2 + \partial^2/\partial y^2$ in $\mathbb{R}^2$ is $k(z, w) = (1/2\pi)\log|z - w|$; i.e.,

$$L^{-1}f(z) = \int_{w = u + iv \in \mathbb{R}^2} k(z, w) f(w)\, du\, dv.$$

Hint. Use Green's theorem or Fourier transforms (see Courant and Hilbert [1], Garabedian [1], or Vladimirov [1]). Compare with Exercise 6 of §1.1 and Example 2 of §1.2.

(b) Explain why the non-Euclidean Green's function $g_\lambda(z, w)$ for $(\Delta - \lambda)$ on H should have the same singularity as $k(z, w)$ as z approaches w.

Hint. See Lang [4, pp. 276–280].

Because $g_\lambda(z, w)$ is G-invariant, we can move w to i. Also, $g_\lambda(z, i)$ is a function of $d(z, i) = r =$ the geodesic radial coordinate of $z = ke^{-r}i$, for $k \in K = SO(2)$ and $r > 0$. Then, by separation of variables on

$$y^2\left(\frac{\partial^2}{\partial x^2} + \frac{\partial^2}{\partial y^2}\right) g_\lambda(z, w) = \lambda g_\lambda(z, w) \qquad \text{for } z = x + iy \neq w = u + iv,$$

we find as in Exercise 11 of §3.2, that solutions which are functions of the radial variable alone have to be of the form

$$f(r) = AP_{-s}(\cosh r) + BQ_{-s}(\cosh r), \qquad \lambda = s(s - 1).$$

Here $P_{-s}(u)$ and $Q_{-s}(u)$ are the *Legendre functions* of the first and second kinds, respectively (see Lebedev [1, Ch. 7]. In order to make $f(r)$ behave as $r \to \infty$, we choose $A = 0$. In order to obtain a singularity $(1/2\pi)\log|z - i|$ as $z \to i$ or $r \to 0$, we need the following exercise.

Exercise 11. Show that if $z = ke^{-r}i$, with $k \in K = SO(2)$ and $r > 0$, then
$$\log|z - i| \sim \log(\cosh r - 1) \qquad \text{as } r \to 0+.$$

Hint. From Exercise 8, we have
$$2\cosh d(z, i) + 2 = \frac{|z + i|^2}{\operatorname{Im} z}.$$

Thus,
$$2\cosh d(z, i) - 2 = \frac{|z - i|^2}{\operatorname{Im} z}.$$

Next note (see Erdélyi et al. [1, Vol. I, p. 163]) that
$$Q_{s-1}(\cosh r) \sim -\tfrac{1}{2}\log(\cosh r - 1) \qquad \text{as } r \to 0+, \qquad (3.115)$$
if $s \neq -1, -2, \ldots$. Thus, in order to obtain the correct singularity, the (*free-space*) *Green's function* for $\Delta - \lambda$ on H must be
$$g_\lambda(z, w) = \frac{-1}{\pi} Q_{s-1}(\cosh d(z, w)), \qquad \lambda = s(s - 1). \qquad (3.116)$$

We can check this against Elstrodt [2, p. 67], where g_λ is given as follows, once we note that Elstrodt sums over Γ and not $\Gamma/\pm I$ and replaces Δ by $-\Delta$:
$$g_\lambda(z, w) = \frac{-\Gamma(s)^2}{2\pi\Gamma(2s)} \left(\frac{u + 1}{2}\right)^{-s} F\left(s, s, 2s; \frac{2}{u+1}\right), \qquad (3.117)$$
if $u = \cosh d(z, w)$, $\lambda = s(s - 1)$, and $F(a, b, c; z)$ is the Gauss hypergeometric function (see Lebedev [1, Ch. 7]). For from Lebedev [1, p. 3 and p. 200], we have
$$\frac{-1}{\pi} Q_{s-1}(u) = \frac{-1}{\sqrt{\pi} 2^s} \frac{\Gamma(s)}{\Gamma(s + \tfrac{1}{2})}(u+1)^{-s} F\left(s, s, 2s; \frac{2}{u+1}\right)$$
$$= \frac{-1}{2\pi} \frac{\Gamma(s)^2}{\Gamma(2s)} \left(\frac{u+1}{2}\right)^{-s} F\left(s, s, 2s; \frac{2}{u+1}\right).$$

Next we use formula (3.114) to build up the Green's function for $\Delta - \lambda I$ on the fundamental domain $\Gamma \backslash H$.

Exercise 12. Show that the series (3.114) converges for (z, w, s) in compact sets inside $H \times H \times \{s | \operatorname{Re} s > 1\}$.

Hint. You need to know (see Lebedev [1, p. 176]) that
$$Q_s(z) \sim \frac{\sqrt{\pi}\,\Gamma(s+1)}{\Gamma(s+\tfrac{3}{2})(2z)^{s+1}} \qquad \text{as } |z| \to \infty, \; s \neq -1, -2, \ldots.$$

3.7. Harmonic Analysis on the Fundamental Domain

This allows you to bound the series (3.114) by a constant (dependent on s) times the *non-Euclidean Eisenstein series* or *Epstein zeta function*

$$Z_\Gamma(z, w; s) = \sum_{\gamma \in \Gamma/\pm I} (\cosh d(z, \gamma w))^{-s}. \tag{3.118}$$

Note that (3.118) is the Mellin transform of the theta function defined in (3.107); i.e.,

$$\Gamma(s) Z_\Gamma(i, i, s) = \int_{a=0}^{\infty} a^{s-1} \theta_\Gamma(a) \, da. \tag{3.119}$$

You might find it hard to believe that the series (3.118) converges for Re $s > 1$, since Epstein's zeta function in §1.4 given by

$$Z(I_4, s) = \sum_{a,b,c,d \neq 0} (a^2 + b^2 + c^2 + d^2)^{-s}$$

only converges for Re $s > 2$ (see Exercise 5 and Theorem 1 of §1.4). However, there are many fewer terms in the sum for $Z_\Gamma(z, w; s)$, as we saw in (3.105) and (3.106). In fact, one could use (3.105) to do Exercise 11 and the fact $Z_\Gamma(i, i; s)$ has a pole at $s = 1$ to prove (3.105) via a Tauberian theorem for Dirichlet series.

One can study the convergence of (3.118) using an integral test. For one has

$$\int_{w=u+iv \in \Gamma \backslash H} \sum_{\substack{\gamma \in \Gamma/\pm I \\ \cosh d(i, \gamma w) \geq 1}} (\cosh d(i, \gamma w))^{-s} v^{-2} \, du \, dv = 2\pi \int_{\cosh r \geq 1} (\cosh r)^{-s} \sinh r \, dr,$$

which is finite if Re $s > 1$.

Exercise 13. Apply the non-Euclidean Poisson summation formula of Theorem 3 to $Z_\Gamma(i, i; s)$, or Mellin transform formula (3.109).

It follows from Exercise 12 that the analytic continuation of the Green's function $G_\lambda(z, w)$ in $\lambda = s(s-1)$ to Re $s = \frac{1}{2}$ is necessary in order to reach the spectrum of the non-Euclidean Laplacian. There are many ways to approach the analytic continuation of G_λ. We shall not go into this here, beyond giving a few more exercises. One wants this analytic continuation because it provides another proof of Theorem 1 (see formulas (2.22)–(2.24), since one can deduce the spectral theory of Δ on $\Gamma \backslash H$ form the behavior of G_λ. The poles of G_λ correspond to the discrete spectrum of Δ, and the jump discontinuities as λ crosses the real axis correspond to the continuous spectrum. Recall here that real $\lambda = s(s-1)$ correspond to s with Re $s = \frac{1}{2}$ or $s \in [0, 1]$.

Exercise 14.

(a) Show that $G_\lambda(z, w)$ given by formula (3.114) is an L^2-function and thus has an L^2-expansion

$$-G_\lambda(z, w) = \sum_{n \geq 0} \frac{v_n(z) \overline{v_n(w)}}{\lambda - \lambda_n} + \frac{1}{4\pi i} \int_{\text{Re } s = 1/2} \frac{E_s(z) \overline{E_s(w)}}{\lambda - s(s-1)} \, ds.$$

Note. This gives a non-Euclidean Poisson sum formula for $G_\lambda(z,w)$.
(b) For absolute convergence, show that one should look at $G_\lambda - G_\mu$, since $(\lambda - \mu)R_\lambda R_\mu = R_\lambda - R_\mu$, if $R_\lambda = (\Delta - \lambda I)^{-1}$ = the resolvent.

Exercise 15. Since G_λ is Γ-invariant in z, it has a Fourier expansion as a periodic function of $x = \operatorname{Re} z$. Show that this expansion is

$$-G_\lambda(z,w) = E_s(w)\frac{y^{1-s}}{2s-1} + \sum_{n\neq 0} F_n(w;s) y^{1/2} K_{s-1/2}(2\pi|n|y) \exp(2\pi inx),$$

with

$$F_n(w;s) = \sum_{\gamma \in \Gamma_\infty\backslash\Gamma} (\operatorname{Im}(\gamma w))^{1/2} I_{s-1/2}(2\pi|n|\operatorname{Im}(\gamma w)) \exp(2\pi in \operatorname{Re}(\gamma w)).$$

Here $I_s(z)$ denotes the I-Bessel function (see Lebedev [1, §5.7]). One must keep $\operatorname{Im} z$ large compared with $\operatorname{Im}(\gamma w)$, $\gamma \in \Gamma$, in this expansion.

Exercise 15 helps to explain Hejhal's work [3] which was discussed briefly in §3.5, showing that spurious eigenvalues $\lambda = s(s-1)$ of Δ on $\Gamma\backslash H$ correspond to zeros of Epstein's zeta function:

$$Z(W,s) = \zeta(2s)E_s(\rho) = \zeta(s)L(s,(-3/*)),$$

with

$$\rho = \exp(2\pi i/3) = (-1 + i\sqrt{3})/2, \qquad W = 3^{-1/2}\begin{pmatrix} 2 & -1 \\ -1 & 2 \end{pmatrix}.$$

For at zeros s of $E_s(\rho)$, the Green's function $G_{s(s-1)}(z,\rho)$ looks like a cusp form, by Exercise 15, except that it has a logarithmic singularity at $z = \rho$.

Hejhal [4] uses Fourier-type expansions of $G_\lambda(z,w)$ coming from certain hyperbolic matrices to show that the solutions of the congruence

$$y^2 \equiv 5 \pmod{L}$$

are such that $y/L \pmod 1$ is uniformly distributed as L approaches infinity, and that the finer properties of this distribution are ruled by the eigenvalues of Δ of $\Gamma\backslash H$. The result on distribution of solutions of quadratic congruences was first proved by Hooley (see Hooley [1]). The generalization to higher-degree congruences remains open. However, it is possible that extensions of Hejhal's Green's function identities to $SL(n,\mathbb{Z})$ might allow one to attack such questions, especially if one recalls the work of Dorothy Wallace [1] connecting units in higher-degree number fields and hyperbolic elements of $SL(n,\mathbb{Z})$. See Hejhal [6–10] and Fay [1] for related work.

Exercise 16 (The Heat Equation on $\Gamma\backslash H$). Given some initial heat distribution function f on $\Gamma\backslash H$, find $u = u(z,t)$ satisfying $u_t = \Delta_z u$, $\Delta_z = y^2(\partial^2/\partial x^2 + \partial^2/\partial y^2)$, $z = x + iy$, and $u(z,0) = f(z)$, for $z \in \Gamma\backslash H$, $t > 0$.

3.7. Harmonic Analysis on the Fundamental Domain

Answer. Using the notation (1), we obtain

$$u(z,t) = \sum_{n \geq 0} A_n \exp(\lambda_n t) v_n(z) + \frac{1}{4\pi i} \int_{\operatorname{Re} s=1/2} A_s \exp\{s(s-1)t\} E_s(z)\, ds,$$

where $A_n = (f, v_n)$, and $A_s = (f, E_s)$.

One way to see this is to use the method of images to see that the fundamental solution for the heat equation on $\Gamma \backslash H$ is

$$\theta^\#(z,t) = \sum_{\gamma \in \Gamma/\pm I} g_t(\gamma z),$$

$g_t =$ the fundamental solution for the heat equation on H itself given by formulas (3.32) and (3.33) of §3.2. Apply the non-Euclidean Poisson sum formula to see that

$$\theta^\#(z,t) = \sum_{n \geq 0} \exp(\lambda_n t) \overline{v_n(i)} v_n(z) + \frac{1}{4\pi i} \int_{\operatorname{Re} s=1/2} \exp(s(s-1)t) \overline{E_s(i)} E_s(z)\, ds$$

where $\lambda_n = s_n(s_n - 1)$.

Exercise 17 (Remarks on Convergence of Non-Euclidean Fourier Series). Suppose that $\Delta f \in L^2(\Gamma \backslash H)$. Show that the non-Euclidean Fourier "series" for f (see Theorem 1) converges uniformly and absolutely.

Answer. Imitate the proof of part (3) of Theorem 1 of §1.3. Look at

$$\sum_{n \geq 0} |(f, v_n) v_n(z)| + \frac{1}{4\pi i} \int_{\operatorname{Re} s=1/2} |(f, E_s) E_s(z)|\, ds$$

$$= \sum_{n \geq 0} \left| \frac{(\Delta f, v_n) v_n(z)}{s_n(s_n - 1)} \right| + \frac{1}{4\pi i} \int_{\operatorname{Re} s=1/2} \left| \frac{(\Delta f, E_s) E_s(z)}{s(s-1)} \right| ds$$

$$\leq \left(\sum_{n \geq 0} \left| \frac{v_n(z)}{s_n(s_n - 1)} \right|^2 \right)^{1/2} \left(\sum_{n \geq 0} |(\Delta f, v_n)|^2 \right)^{1/2}$$

$$+ \left(\frac{1}{4\pi i} \int_{\operatorname{Re} s=1/2} \left| \frac{E_s(z)}{s(s-1)} \right|^2 ds \right)^{1/2} \left(\int_{\operatorname{Re} s=1/2} |(\Delta f, E_s)|^2 ds \right)^{1/2}.$$

Then use the fact that Δf and $G_\lambda =$ the Green's function for $\Delta - \lambda$ are both square-integrable.

There is another application of Theorem 1, which was given by Kaori Imai [1] showing that there is a higher-dimensional analogue of Hecke's Theorem 1 of §3.6. We shall discuss this in more detail in Chapter 5. Briefly, one wants to Mellin transform Siegel modular forms for the symplectic group $Sp(n, \mathbb{Z})$. When $n = 2$, Imai shows that the converse theorem in Hecke theory can be proved using the Roelcke-Selberg spectral decomposition of the Laplacian on $L^2(SL(2, \mathbb{Z}) \backslash H)$. When n is larger than 2, this converse theorem requires a workable generalization of Theorem 1 for $SL(n, \mathbb{Z})$—a subject to be discussed in Chapter 4.

Elstrodt, Grunewald, and Mennicke [1–3] and Mennicke [1, 2] have used the non-Euclidean Poisson sum formula for groups such as $SL(2, \mathbb{Z}[i])$ (to be discussed in Chapter 5) in order to derive many sorts of alebraic and analytic results. One very interesting idea announced by Mennicke [2] is to use Theorem 3 in order to study the eigenvalues of the Laplacian corresponding to Maass-type cusp forms, as well as to study the cusp forms themselves.

Selberg's Trace Formula for $SL(2, \mathbb{Z})$. Asymptotics of the Eigenvalues of the Laplacian on the Fundamental Domain and of Units in Real Quadratic Fields (the Length Spectrum)*

The discussion of Selberg's trace formula should begin with a review of the part of §1.3 which gave an interpretation of Poisson's sum formula as a trace formula. However, the Selberg analogue of Poisson summation is complicated by the lack of commutativity of the groups G, Γ involved, as well as the lack of compactness of the fundamental domain.

We wish to find traces of the "compact part" of the convolution operators in (3.103) and Lemma 2, defined by

$$L_g f(a) = \int_{G/K} f(b) g(b^{-1}a) \, db \qquad (3.120)$$

viewed as integral operators on functions f in $L^2(\Gamma \backslash H)$, for K-invariant functions $g \in C_c^\infty(H)$. We can rewrite (3.120) as an integral operator on the fundamental domain as follows:

$$\begin{aligned} L_g f(a) &= \int_{\Gamma \backslash G/K} f(b) K_g(a, b) \, db, \\ K_g(a, b) &= \sum_{\gamma \in \bar{\Gamma}} g(b^{-1} \gamma a), \qquad \bar{\Gamma} = \Gamma/\{\pm I\} = PSL(2, \mathbb{Z}). \end{aligned} \qquad (3.121)$$

Note that we must use $\bar{\Gamma} = \Gamma/\{\pm I\} = $ *the projective linear group*, or the formula will be off by a factor of 2. For $\bar{\Gamma} \backslash H = \Gamma \backslash H$, since the fractional linear transformation corresponding to $-\gamma$ is the same as that for $+\gamma$, and

$$H = \bigcup_{\gamma \in \bar{\Gamma}} \gamma(D), \qquad D = \text{a fundamental domain for } \Gamma \backslash H,$$

is a disjoint union.

We will write

$$\begin{aligned} K_g(a, b) &= K_g(a(i), b(i)), \\ k_g(a, b) &= k_g(a(i), b(i)) = g(b^{-1}a), \end{aligned} \qquad (3.122)$$

---

* This section has been greatly influenced by the work of Dorothy Wallace [1, 2].

for a, b in $G = SL(2, \mathbb{R})$. Here, as usual, $a(i)$ is the point in H to which the fractional linear transformation corresponding to $a \in G$ sends $i = \sqrt{-1}$.

In this setting, using the notation in (3.99), involving v_n, s_n, and E_s, *Mercer's theorem* (see Theorem 8 of §1.3) becomes

$$\left.\begin{aligned} K_g(a,b) &= \hat{K}_g(a,b) + E_g(a,b), \\ \text{with} \\ \hat{K}_g(a,b) &= \sum_{n \geq 0} \hat{g}(s_n) v_n(a(i)) \overline{v_n(b(i))} \\ E_g(a,b) &= \frac{1}{4\pi i} \int_{\operatorname{Re} s = 1/2} \hat{g}(s) E_s(a(i)) \overline{E_s(b(i))}\, ds. \end{aligned}\right\} \tag{3.123}$$

Here $\hat{g}$ denotes the *Helgason transform* of §3.2 (which is also called the Selberg transform in this context):

$$\hat{g}(s) = \int_H g(z) y^{\bar{s}-2}\, dx\, dy, \tag{3.124}$$

for K-invariant functions g. Formula (3.123) is equivalent to the non-Euclidean Poisson summation formula in Theorem 3 above.

Next define the *integral operator* $\hat{L}_g$ corresponding to $\hat{K}_g$ by

$$\hat{L}_g f(a) = \int_{\Gamma \backslash G/K} f(b) \hat{K}_g(a,b)\, db. \tag{3.125}$$

Our discussion of the Selberg trace formula begins by writing

$$\operatorname{Trace} \hat{L}_g = \sum_{n \geq 0} \hat{g}(s_n) = \int_{\Gamma \backslash G/K} \hat{K}_g(a,a)\, d\bar{a}. \tag{3.126}$$

In order to proceed further, we must decompose $SL(2, \mathbb{Z})$ into conjugacy classes of various sorts, according to the Jordan forms of these matrices.

Definition. *Classification of Elements* of $SL(2, \mathbb{Z}) = \Gamma$ (cf. §3.3).

(a) *Central Elements*: $+I$, $-I$;

(b) *Parabolic Elements*: those having Jordan form

$$\pm \begin{pmatrix} 1 & a \\ 0 & 1 \end{pmatrix} \qquad \text{with } a \neq 0;$$

(c) *Elliptic Elements*: those with Jordan form

$$\begin{pmatrix} a & 0 \\ 0 & 1/a \end{pmatrix}, \qquad \text{with } a \notin \mathbb{R},\ |a| = 1;$$

(d) *Hyperbolic Elements*: those with Jordan form

$$\begin{pmatrix} a & 0 \\ 0 & 1/a \end{pmatrix}, \qquad \text{with } a \text{ real, } a \neq +1, -1, 0.$$

In the case of hyperbolic elements γ, we define $a^2 = N\gamma$, the *norm* of the element, where a is chosen so that $|a| > 1$.

Definition. For $\tau \in \Gamma$, the *conjugacy class* of τ in Γ is

$$\{\tau\} = \{\gamma\tau\gamma^{-1} | \gamma \in \Gamma\}. \tag{3.127}$$

Definition. For $\tau \in \Gamma$, the *centralizer* of τ in Γ is

$$\Gamma_\tau = \{\gamma \in \Gamma | \gamma\tau = \tau\gamma\}. \tag{3.128}$$

Exercise 18 (Explicit Characterizations of the Conjugacy Classes in $SL(2, \mathbb{Z})$).

(a) Show that the conjugacy classes of parabolic elements of $SL(2, \mathbb{Z})$ are represented by

$$\pm \begin{pmatrix} 1 & a \\ 0 & 1 \end{pmatrix}, \quad a \in \mathbb{Z}.$$

(b) Show that the conjugacy classes of elliptic elements of $SL(2, \mathbb{Z})$ are represented by

$$\pm \begin{pmatrix} 0 & 1 \\ -1 & 0 \end{pmatrix}, \pm \begin{pmatrix} 1 & -1 \\ 1 & 0 \end{pmatrix}.$$

(c) Show that the conjugacy classes of hyperbolic elements of $SL(2, \mathbb{Z})$ are represented by units ε_d in orders $\mathcal{O}$ in real quadratic fields $\mathbb{Q}(\sqrt{d})$, with multiplicity h_d = the narrow class number of the order $\mathcal{O}$. The narrow class number is defined similarly to the class number in §1.4, except that the equivalence relation between two ideals $\mathfrak{a}$, $\mathfrak{b}$ in $\mathbb{Z}[\varepsilon_d]$ is

$$\mathfrak{a} \sim \mathfrak{b} \Leftrightarrow \mathfrak{a} = c\mathfrak{b} \quad \text{with } c \in \mathbb{Q}(\sqrt{d}) \text{ with norm} = +1.$$

The order $\mathcal{O}$ need not be the whole ring of integers in $\mathbb{Q}(\sqrt{d})$.

Hints. See Olga Taussky [1, 2], Shimura [1], Schoeneberg [1], and Dorothy Wallace [1, 2]. It helps to look at the points in $H \cup \mathbb{R}$ which are fixed by $\gamma \in \Gamma$. Part (c) is the most complicated. Following Taussky [1, 2] and Wallace [1, 2], you should note that if γ is a hyperbolic element of $SL(2, \mathbb{Z})$, the diagonal elements ε in the Jordan form of γ must be units in a real quadratic field K, since ε is a root of the characteristic polynomial of γ. The eigenvectors ${}^t(w_1, w_2)$ in K^2 generate an ideal $\mathbb{Z}w_1 \oplus \mathbb{Z}w_2$ in an order $\mathcal{O}$ of K. Note that this order need not be the maximal order; i.e., the whole ring of integers of K defined in §1.4. However, one can still form the narrow ideal classes (which do not necessarily form a group). And one can relate the number of these narrow ideal classes to the ordinary class number of K (*see* Lang [5]). Sarnak [1] gives a description of part (c) of this exercise in the language of quadratic forms which goes back to Gauss.

3.7. Harmonic Analysis on the Fundamental Domain

The following exercise shows that the concept of primitive hyperbolic element of $SL(2, \mathbb{Z})$ is not the same thing as that of fundamental unit in the relevant real quadratic number field.

Exercise 19.

(a) Let γ be a hyperbolic element of $\Gamma = SL(2, \mathbb{Z})$. Show that the centralizer Γ_γ of γ in Γ is an infinite cyclic group generated by γ_0 which is called a *primitive hyperbolic element of* Γ.

(b) Show that $\begin{pmatrix} 2 & 9 \\ 1 & 5 \end{pmatrix}$ is a primitive hyperbolic matrix in Γ with eigenvalue $z^4 = (7 + 3\sqrt{5})/2$, where $z = (1 + \sqrt{5})/2$.

Hint (see Hejhal [1, Ch. 1]). Use the fact that Γ is discrete as well as the possibility of simultaneously diagonalizing commuting elements of Γ.

Exercise 20 (Hyperbolic Conjugacy Classes and the Length Spectrum of Closed Geodesics on the Fundamental Domain).

(a) Let $C(z, w)$ be a geodesic line or circle in H connecting two points z, w on $\mathbb{R} \cup \{\infty\}$. Consider the image $\overline{C(z, w)}$ in the standard fundamental domain for $SL(2, \mathbb{Z}) \backslash H$ which is given in Fig. 3.11. We say that $\overline{C(z, w)}$ is a closed geodesic if it is a closed curve in the fundamental domain. Here we mean "closed" in the sense that, once correctly parameterized, the beginning of the curve is the same point as the end of the curve. We do not refer to the topological notion of closed (i.e., the set of points on the curve has open complement). Show that $\overline{C(z, w)}$ is a closed geodesic in $SL(2, \mathbb{Z}) \backslash H$ if and only if there is an element γ of $SL(2, \mathbb{Z})$ such that $\gamma C(z, w) \subset C(z, w)$ (as sets of points). Show then that $\overline{C(z, w)}$ is a closed geodesic if and only if z and w are the fixed points of a hyperbolic element γ of $SL(2, \mathbb{Z})$.
(b) Suppose that $\overline{C(z, w)}$ is a closed geodesic in $SL(2, \mathbb{Z}) \backslash H$, with z, w the fixed points of a hyperbolic element γ of $SL(2, \mathbb{Z})$. Show that if a point q lies on $C(z, w)$, then so does γq and the hyperbolic distance between q and γq is $\log N\gamma$, where $N\gamma$ is the norm of γ, defined before formula (3.127).
(c) Using a computer, graph $\overline{C(z, w)}$ for various choices of z, w. We did this in Fig. 3.31–3.34 (and then mapped everything into the unit disc via the Cayley transform $z \to i(z - i)/(z + i)$.

It follows from Exercises 18–20 that the primitive hyperbolic conjugacy classes in $SL(2, \mathbb{Z})$ have both a number-theoretic and a geometric interpretation. The number-theoretic interpretation is that they correspond to fundamental units in orders in real quadratic number fields. The geometric interpretation is that they correspond to closed geodesics on the Riemann surface $SL(2, \mathbb{Z}) \backslash H$—the *length spectrum* of $SL(2, \mathbb{Z}) \backslash H$. The number of hyperbolic conjugacy classes with a given trace can thus be viewed either as a class

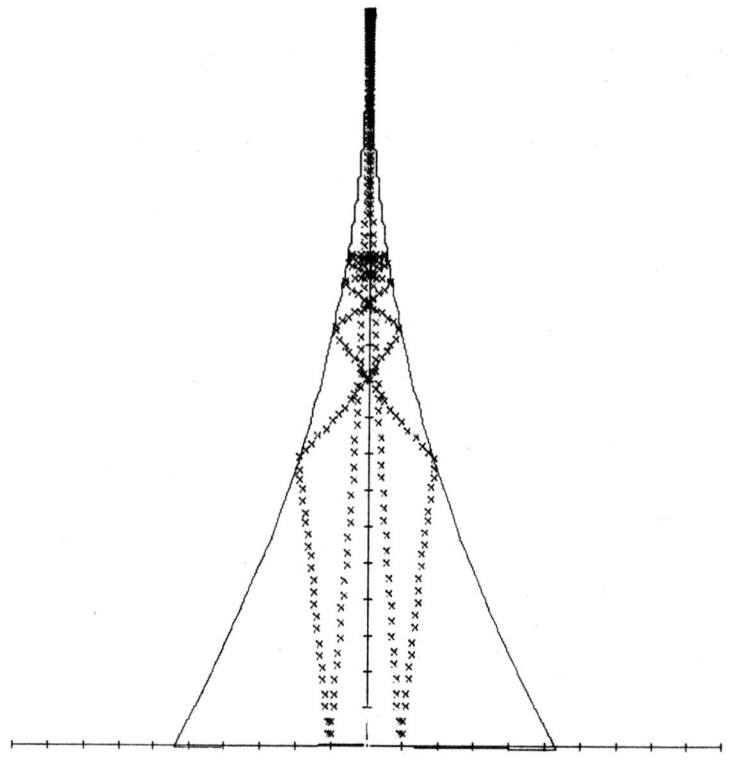

Figure 3.31. Images of points on geodesic circle of center 0 and radius 5 after mapping by $SL(2, \mathbb{Z})$ into the fundamental domain and then Cayley transform into the unit disc.

number or as the number of closed geodesics with a given length. Selberg's trace formula (Theorem 4) gives relations between the length spectrum and the eigenvalue spectrum of Δ on $SL(2,\mathbb{Z})\backslash H$. This results in an analogue of the Prime Number Theorem (see Theorem 6), which can thus be given either a number-theoretic or a geometric interpretation.

In Exercise 3 of §1.3, we saw an equidistribution property of numbers on a circle. There is also an analogue for the torus $\mathbb{R}^2/\mathbb{Z}^2$, showing that a line in $\mathbb{R}^2$ which makes an irrational angle with the x-axis will correspond to a densely wound line in the torus $\mathbb{R}^2/\mathbb{Z}^2$. In Exercise 16 of §3.6, we saw that the image of a horocycle at height y in $SL(2,\mathbb{Z})\backslash H$ tends to fill up the fundamental domain as $y \to 0+$.

One can ask analogous questions about geodesics. Before asking these questions, the reader should consider Figs. 3.31–3.34, which give plots of points on the images of various geodesic circles $\overline{C(-x, +x)}$, using the notation of Exercise 20, after the Cayley transform $z \mapsto i(z-i)/(z+i)$, which sends the fundamental domain into the unit disc. Figure 3.31 shows the geodesic $\overline{C(-5,+5)}$. This geodesic has only a finite number of segments and is clearly

3.7. Harmonic Analysis on the Fundamental Domain

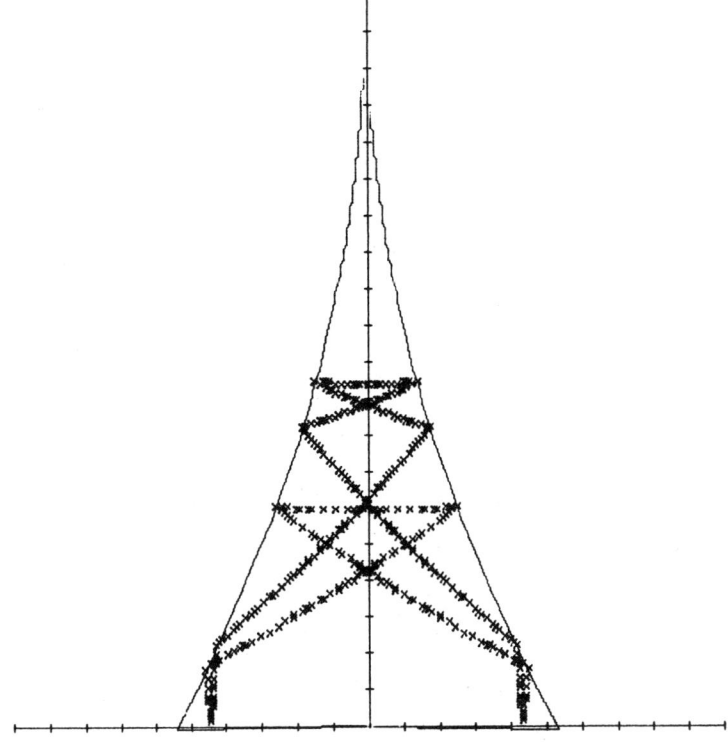

Figure 3.32. Images of points on geodesic circle of center 0 and radius $(15/2)^{1/2}$ after mapping by $SL(2, \mathbb{Z})$ into the fundamental domain and then Cayley transform into the unit disc.

not dense, but it is also not closed. Figure 3.32 shows a genuine closed geodesic $\overline{C(-x, +x)}$ for $x = \sqrt{15/2}$, which is fixed (as is $-x$) by the hyperbolic matrix

$$\begin{pmatrix} 11 & 30 \\ 4 & 11 \end{pmatrix}.$$

Figures 3.33 and 3.34 show points on approximations to $\overline{C(-x, +x)}$ for $x = \sqrt{163}$ and $e = 3.1415...$, respectively. The plots were made with a TRS-80, Model 100 computer.

Questions.

(1) As the length of a closed geodesic approaches infinity, does the geodesic tend to fill up the fundamental domain, in an analogous way to that for the horocycle of Exercise 16 of §3.6?
(2) Can one give a criterion on z, w which will insure that a geodesic $C(z, w)$ is dense in the fundamental domain for $SL(2, \mathbb{Z})$?

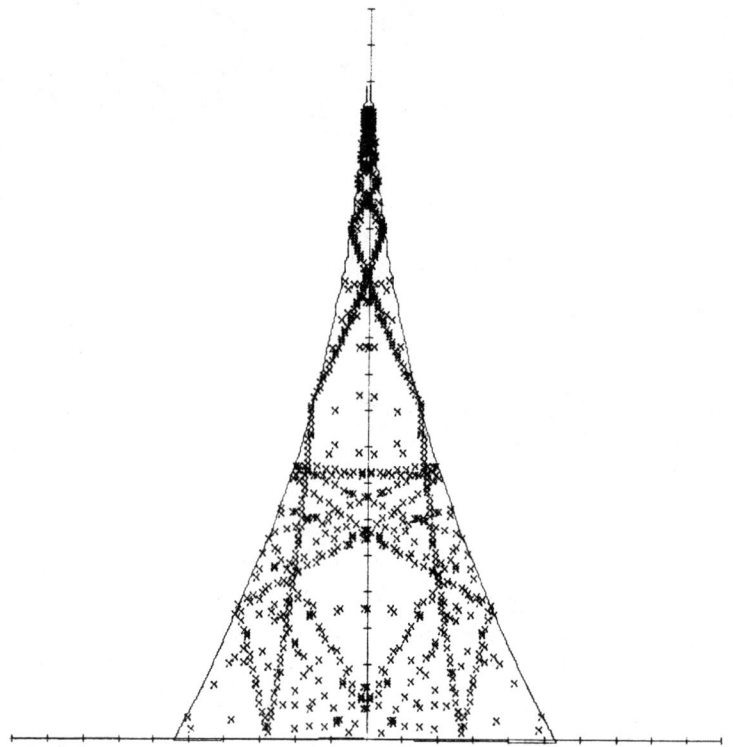

Figure 3.33. Images of points on geodesic circle of center 0 and radius $(163)^{1/2}$ after mapping by $SL(2, \mathbb{Z})$ into the fundamental domain and then Cayley transform into the unit disc.

These questions are related to ergodic theory (as well as continued fraction expansions of real numbers). Let $T: M \to M$ be a measure-preserving transformation of a Riemannian manifold. One says that T is *ergodic* if for every Lebesgue integrable function f on M, one has

$$\lim_{n\to\infty} \frac{1}{n} \sum_{k=0}^{n-1} f(T^k x) = \int_M f(y)\,d\mu(y) \qquad \text{for almost all } x \in M.$$

The average on the left may be considered a time average, while that on the right may be considered a space average. Actually, M is often replaced by its unit tangent bundle in most works on the subject. References for ergodic theory, geodesic flows, etc., are Artin [1], Auslander et al. [1], Bowen [1], Gelfand and Fomin [1], Mautner [2], and Moeckel [1].

If Γ had no parabolic elements, then we could write the trace as follows, using formulas (3.121), (3.122), (3.126)–(3.128):

3.7. Harmonic Analysis on the Fundamental Domain

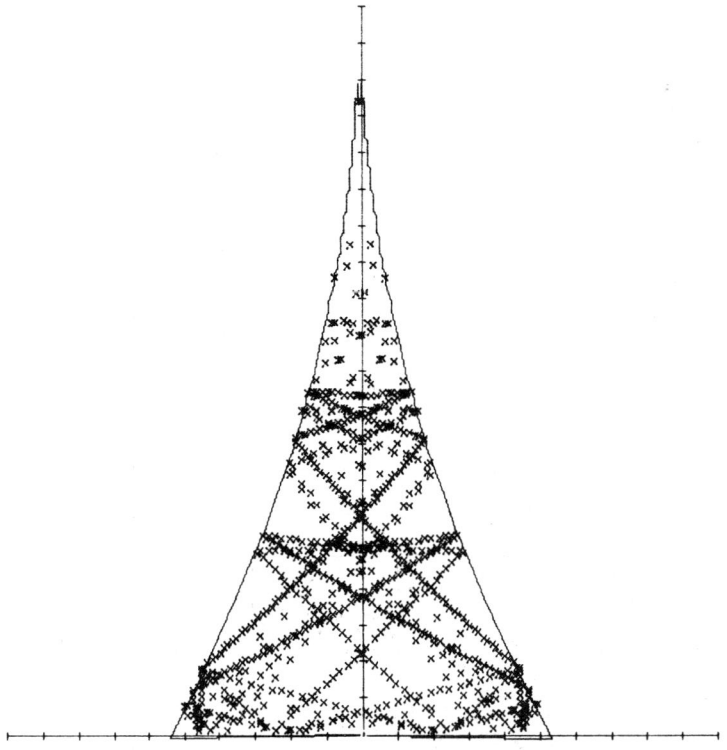

Figure 3.34. Images of points on geodesic circle of center 0 and radius e after mapping by $SL(2, \mathbb{Z})$ into the fundamental domain and then Cayley transform into the unit disc.

$$\sum_{n \geq 0} \hat{g}(s_n) = \int_{\Gamma \backslash H} K_g(b, b) \, db$$

$$= \sum_{\substack{\{\gamma\} \\ \text{distinct in } \bar{\Gamma}}} \sum_{\sigma \in \bar{\Gamma}_\gamma \backslash \bar{\Gamma}} \int_{\Gamma \backslash H} g(a^{-1}\sigma^{-1}\gamma\sigma a) \, da \qquad (3.129)$$

$$= \sum_{\substack{\{\gamma\} \\ \text{distinct in } \bar{\Gamma}}} \int_{\Gamma_\gamma \backslash H} g(x^{-1}\gamma x) \, dx.$$

Exercise 21. Use the relation between K_g and k_g given in formulas (3.121) and (3.122) as well definitions (3.127) and (3.128) to prove formula (3.129) for groups Γ with compact fundamental domain.

However, in the case $\Gamma = SL(2, \mathbb{Z})$, there are parabolic elements and thus we must cancel their contribution to the trace formula against that of the continuous spectrum. This allows us to write the trace formula in the following form:

$$\sum_{n\geq 0} \hat{g}(s_n) = \sum_{\substack{\{\gamma\} \\ \text{distinct nonparabolic} \\ \text{conjugacy classes in } \bar{\Gamma}}} c_g(\gamma) + c_g(\infty), \tag{3.130}$$

where the *orbital integrals* $c_g(\gamma)$ are defined by

$$c_g(\gamma) = \int_{\Gamma_\gamma \backslash G/K} g(a^{-1}\gamma a)\, da \qquad \text{for } \gamma \text{ not parabolic}, \tag{3.131}$$

and

$$\begin{aligned} c_g(\infty) = \lim_{A\to\infty} \Bigg\{ & \int_{y=0}^{A} \int_{|x|\leq 1/2} \sum_{\substack{I \neq \gamma = \left(\begin{smallmatrix}1 & n \\ 0 & 1\end{smallmatrix}\right) \\ n \in \mathbb{Z}}} k_g(\gamma z, z) y^{-2}\, dx\, dy \\ & - \int_{\substack{z\in\Gamma\backslash H \\ y\leq A}} E_g(z,z) y^{-2}\, dx\, dy \Bigg\}. \end{aligned} \tag{3.132}$$

Here E_g is defined by formula (3.123) and $\Gamma\backslash H$ means the standard fundamental domain for $SL(2,\mathbb{Z}) = \Gamma$ (as in Fig. 3.11).

In order to put the trace formula in final form Selberg showed how to evaluate the various $c(\gamma)$ above in terms of $\hat{g}$ or a closely related integral—the *Harish transform* (or Abel transform) defined by

$$Tg(y) = y^{-1/2} \int_{x\in\mathbb{R}} g(x+iy)\, dx = G(\log y). \tag{3.133}$$

We know from Exercise 21 of §3.2 that

$$\hat{g}(s) = MTg(\bar{s} - \tfrac{1}{2}), \tag{3.134}$$

where M denotes the Mellin transform.

Theorem 4 (Selberg's Trace Formula). *Suppose that* $g: H \to \mathbb{C}$ *is in* $C_c^\infty(K\backslash G/K)$; *i.e., g is a compactly supported infinitely differentiable K-invariant function on H. Then, using the notation* (3.124), (3.127), *and* (3.133), *we have*

$$\begin{aligned} \sum_{n\geq 0} \hat{g}(s_n) = {} & \frac{\operatorname{area}(\Gamma\backslash H)}{4\pi} \int_{r\in\mathbb{R}} \hat{g}(\tfrac{1}{2} + ir) r \tanh \pi r\, dr \\ & + \sum_{\substack{\{\gamma_0\} \\ \text{primitive} \\ \text{hyperbolic}}} \sum_{k\geq 1} \frac{\log N\gamma_0}{N\gamma_0^{k/2} - N\gamma_0^{-k/2}} G(k \log N\gamma_0) \\ & + \int_{r\in\mathbb{R}} \left(\frac{1}{4} + \frac{1}{3\sqrt{3}} (e^{\pi r/3} + e^{-\pi r/3}) \right) \frac{\hat{g}(\frac{1}{2} + ir)}{e^{\pi r} + e^{-\pi r}}\, dr \\ & - G(0) \log(2\pi) + \frac{1}{\pi} \int_{r\in\mathbb{R}} \hat{g}(\tfrac{1}{2} + ir) \frac{\zeta'}{\zeta}(-2ir)\, dr. \end{aligned}$$

3.7. Harmonic Analysis on the Fundamental Domain

PROOF. We begin with formulas (3.130)–(3.132). It remains to evaluate the various orbital integrals in terms of the Helgason and Harish transforms of g.

Central Term.

$$c_g(I) = \frac{\text{area}(\Gamma\backslash H)}{4\pi} \int_{r\in\mathbb{R}} \hat{g}(\tfrac{1}{2} + ir) r \tanh \pi r \, dr. \qquad (3.135)$$

Clearly we have

$$c_g(I) = \int_{\Gamma\backslash H} g(a^{-1}Ia) \, da = g(I) \operatorname{area}(\Gamma\backslash H).$$

The inversion formula for the Helgason transform of a K-invariant function (formula (3.29)) completes the proof of formula (3.135).

Hyperbolic Term. Suppose that $\gamma = \gamma_0^k$, where $\gamma_0 =$ a primitive hyperbolic in Γ. Then

$$c_g(\gamma) = \frac{\log N\gamma_0}{N\gamma_0^{k/2} - N\gamma_0^{-k/2}} G(k \log N\gamma_0). \qquad (3.136)$$

Use Exercise 19 to see that if γ is hyperbolic then

$$\Gamma_\gamma = \{\gamma_0^k | k \in \mathbb{Z}\} \quad \text{with } \gamma_0 = \text{primitive hyperbolic}.$$

Then replace γ by $\xi\gamma\xi^{-1}$, $\xi \in G$, with $\xi\gamma\xi^{-1}$ diagonal. Also replace Γ by $\xi\Gamma\xi^{-1}$. Thus, in our computation of $c_g(\gamma)$, we can take

$$\Gamma_\gamma = \{\gamma_0^k | k \in \mathbb{Z}\}, \qquad \gamma_0 = \begin{pmatrix} a & 0 \\ 0 & 1/a \end{pmatrix}, \qquad a^2 = N\gamma_0 \geq 1. \qquad (3.137)$$

We will need the following exercise.

Exercise 22. Show that if Γ_γ is given by (3.137), then we can choose a fundamental domain for $\Gamma_\gamma \backslash H$ to be

$$\{z \in H | 1 \leq y \leq N\gamma_0\}.$$

As usual, we identify z in H with $a_{x,y} \in SL(2,\mathbb{R})$ via

$$z = x + iy \in H \to a_{x,y} = \begin{pmatrix} 1 & x \\ 0 & 1 \end{pmatrix} \begin{pmatrix} y^{1/2} & 0 \\ 0 & y^{-1/2} \end{pmatrix} \in SL(2,\mathbb{R}). \qquad (3.138)$$

Therefore, if $\gamma = \gamma_0^k$, $\gamma_0 = \begin{pmatrix} a & 0 \\ 0 & 1/a \end{pmatrix}$, $a \geq 1$, we have

$$c_g(\gamma) = \int_{\Gamma_\gamma \backslash H} g(a_{x,y}^{-1} \gamma a_{x,y}) y^{-2} \, dx \, dy.$$

Next note that

$$a_{x,y}^{-1}\gamma a_{x,y} = \begin{pmatrix} 1 & -x/y \\ 0 & 1 \end{pmatrix} \begin{pmatrix} a^k & 0 \\ 0 & a^{-k} \end{pmatrix} \begin{pmatrix} 1 & x/y \\ 0 & 1 \end{pmatrix}.$$

So we can make the change of variables $x \to u = x/y$, $du = dx/y$ to obtain

$$c_g(\gamma) = \int_{u \in \mathbb{R}} \int_{y=1}^{a^2 = N\gamma_0} g\left(\begin{pmatrix} 1 & -u \\ 0 & 1 \end{pmatrix} \begin{pmatrix} a^k & 0 \\ 0 & a^{-k} \end{pmatrix} \begin{pmatrix} 1 & u \\ 0 & 1 \end{pmatrix}\right) du\, y^{-1}\, dy$$

$$= \log N\gamma_0 \int_{u \in \mathbb{R}} g(a^{2k}(i+u) - u)\, du.$$

Set $w = (a^{2k} - 1)u$ to find that formula (3.136) is indeed correct.

Elliptic Term. If γ_θ is an elliptic element of Γ with $\mathrm{Tr}\,\gamma_\theta = 2\cos\theta$, then $\theta = \pi/2$ or $\pi/3$. Let m_θ be the order of the centralizer $\bar{\Gamma}\gamma_\theta = \Gamma\gamma_\theta/\pm 1$. Then $m_{\pi/2} = 2$ and $m_{\pi/3} = 3$. And we shall prove that

$$c_g(\gamma_\theta) = \frac{1}{4m_\theta \sin\theta} \int_{r \in \mathbb{R}} \frac{\cosh[(\pi - 2\theta)r]}{\cosh \pi r} \hat{g}(\tfrac{1}{2} + ir)\, dr. \qquad (3.139)$$

Discussions of this term appear in Hejhal [1, Vol. I, pp. 351ff] and Kubota [1, pp. 99ff]. Lang [4, pp. 166–167] gives relations between elliptic and hyperbolic orbital integrals, which may be of use in simplifying the proof.

Upon replacing γ_θ by a conjugate, we can assume that

$$\gamma_\theta = \begin{pmatrix} \cos\theta & \sin\theta \\ -\sin\theta & \cos\theta \end{pmatrix} \in K.$$

Then we use geodesic polar coordinates from §3.1 on the orbital integral, plus the fact that $K = SO(2)$ is abelian, to see that (if $m = m_\theta$)

$$c_g(\gamma_\theta) = \frac{1}{m} \int_{G/K} g(x^{-1}\gamma_\theta x)\, dx$$

$$= \frac{2\pi}{m} \int_{v > 0} g(a_{-v}\gamma_\theta a_v) \sinh v\, dv, \qquad a_v = \begin{pmatrix} e^{-v/2} & 0 \\ 0 & e^{v/2} \end{pmatrix}.$$

Next we want to make use of Fourier inversion on G/K (see formulas (3.27)–(3.29)) which says that

$$g(ke^{-p}i) = \frac{1}{4\pi} \int_{r \in \mathbb{R}} \hat{g}(\tfrac{1}{2} + ir) P_{-1/2+ir}(\cosh p) r \tanh \pi r\, dr.$$

In order to plug this into the elliptic term and evaluate the result, we must remark that $a_{-v}\gamma_\theta a_v = k_1 a_{v(\theta)} k_2$, with $k_i \in K$ and a_v, $a_{v(\theta)}$ as above, then

$$\cosh v(\theta) = (1 - 2\sin^2\theta) + 2\sin^2\theta \cosh^2 v. \qquad (*)$$

It follows that the proof of (3.139) will be completed by the following exercise.

3.7. Harmonic Analysis on the Fundamental Domain

Exercise 23. Show that if $v(\theta)$ is defined by (*), then

$$\int_{v>0} P_{-1/2+ir}(\cosh v(\theta)) \sinh v \, dv = \frac{\cosh[r(\pi-2\theta)]}{2|\sin\theta|r\sinh\pi r}.$$

Hints (see Sneddon [1, pp. 384–387] or Erdélyi et al. [2, Vol. II, p. 330]). Make the change of variables from v to $w = \cosh v(\theta)$. Then you must compute the Mehler-Fock transform

$$\int_{w\geq 1} (w-\cos 2\theta)^{-1/2} P_{-1/2+ir}(w) \, dw \qquad \text{assuming } \cos(2\theta) \geq 0.$$

To compute such a Mehler-Fock transform, you can use the following formulas

(a) $$\int_0^\infty g(t) P_{-1/2+it}(\cosh a) \tanh \pi t \, dt$$
$$= \pi^{-1/2} \int_a^\infty (\cosh t - \cosh a)^{-1/2} Fg(t) \, dt,$$

where

$$Fg(t) = (2/\pi)^{-1/2} \int_0^\infty g(v) \sin(tv) \, dv.$$

(b) When $g(v) = \cosh(bv)/\sinh(\pi v)$, $-\pi < b < \pi$, we have

$$Fg(t) = (2\pi)^{-1/2} \frac{\sinh t}{\cosh t + \cos b}.$$

(c) $$\int_a^\infty \frac{\sinh t \, dt}{(\cosh t + \cos b)(\cosh t - \cosh a)^{1/2}} = \pi (\cosh a + \cos b)^{-1/2}.$$

Parabolic Term.

$$c_g(\infty) = -G(0)\log(2\pi) + \frac{1}{\pi}\int_{r\in\mathbb{R}} \hat{g}(\tfrac{1}{2}+ir)\frac{\zeta'}{\zeta}(-2ir)\,dr. \qquad (3.140)$$

Conjugacy classes of parabolic elements of $PSL(2,\mathbb{Z})$ are represented by $\gamma_n(z) = z+n$, $n\in\mathbb{Z}$, with centralizer

$$\Gamma_{\gamma_n} = \{\gamma_m | m\in\mathbb{Z}\}.$$

Clearly

$$\Gamma_{\gamma_n}\backslash H = \{x+iy \,|\, |x|\leq \tfrac{1}{2}, y>0\}.$$

Then we define, as in (3.132), $c_g(\infty) = \lim_{A\to\infty}\{T_1(A) - T_2(A)\}$, where

$$T_1(A) = \int_{y=0}^{A} \int_{x=-1/2}^{1/2} \sum_{n \neq 0} g\left(a_{x,y}^{-1} \begin{pmatrix} 1 & n \\ 0 & 1 \end{pmatrix} a_{x,y}\right) y^{-2} dx\, dy,$$

$$T_2(A) = \frac{1}{4\pi i} \int_{\substack{\text{Res}=1/2 \\ x^2+y^2 \geq 1, |x| \leq 1/2 \\ y < A}} \hat{g}(s) E_s(z) E_{\bar{s}}(z)\, ds\, y^{-2}\, dx\, dy, \qquad (3.141)$$

$$a_{x,y} = \begin{pmatrix} 1 & x \\ 0 & 1 \end{pmatrix} \begin{pmatrix} y^{1/2} & 0 \\ 0 & y^{-1/2} \end{pmatrix}.$$

We shall follow Selberg's Göttingen lectures [6, p. 79] in evaluating these terms, without using the Poisson summation formula or the Euler-Maclaurin summation formula. This method should be compared with that of Kubota [1, p. 102ff] who uses Euler-Maclaurin. One should also consider the method of Warner [2] involving Poisson's summation formula. See the article of Gelbart and Jacquet in the Corvallis conference proceedings (Borel and Casselman [1, Vol. I, p. 245]) for an expression for a parabolic (or nilpotent) orbital integral as a limit of hyperbolic orbital integrals. Barbasch [1, 2] pursues this matter for general Lie groups.

Let us start with $T_1(A)$. Note that

$$a_{x,y}^{-1} \begin{pmatrix} 1 & n \\ 0 & 1 \end{pmatrix} a_{x,y} = \begin{pmatrix} y^{-1/2} & 0 \\ 0 & y^{1/2} \end{pmatrix} \begin{pmatrix} 1 & n \\ 0 & 1 \end{pmatrix} \begin{pmatrix} y^{1/2} & 0 \\ 0 & y^{-1/2} \end{pmatrix} = \begin{pmatrix} 1 & n/y \\ 0 & 1 \end{pmatrix}.$$

Thus

$$T_1(A) = \int_{y=0}^{A} \sum_{n \neq 0} g(i + n/y) y^{-2}\, dy = \int_{w=1/A}^{\infty} \sum_{n \neq 0} g(i + nw)\, dw,$$

setting $w = 1/y$. It follows that

$$T_1(A) = 2 \int_0^{\infty} g(i + u) \sum_{n=1}^{Au} 1/n\, du. \qquad (3.142)$$

Formula (3.142) is a reasonable one itself, but we want to express everything in terms of G or $\hat{g}$. So we make use of the following exercise.

Exercise 24. Show that

$$\sum_{0 < n \leq x} 1/n = \log x + \gamma + O(1/x) \qquad \text{as } x \to \infty,$$

where γ = Euler's constant = $0.57721566\ldots$.

Hint. This can be proved using the Euler-Maclaurin summation formula (see Edwards [1, p. 285]).

3.7. Harmonic Analysis on the Fundamental Domain

It follows from Exercise 24 and formula (3.142) that

$$T_1(A) = (\log A + \gamma)G(0) + 2\int_0^\infty \log u \, g(i+u)\,du + o(1) \quad \text{as } A \to \infty. \tag{3.143}$$

To complete the evaluation of $T_1(A)$, we need another exercise.

Exercise 25. Show that

$$2\int_0^\infty \log u \, g(i+u)\,du = -\log 2 \, G(0) - \gamma G(0) + \tfrac{1}{4}\hat{g}(\tfrac{1}{2})$$

$$-\frac{1}{2\pi}\int_{r \in \mathbb{R}} \hat{g}(\tfrac{1}{2}+ir)\frac{\Gamma'}{\Gamma}(1-ir)\,dr.$$

Hint. (See Selberg [6].)

Step 1. Let $g(a_{x,y}) = h_g(\text{Tr}({}^t a_{x,y} a_{x,y}) - 2)$, if $a_{x,y}$ is as in formula (3.141); i.e.,

$$g(x+iy) = h_g(y^{-1} + y + x^2/y - 2).$$

Next let

$$H_g(v) = \int_{-\infty}^\infty h_g(v + u^2/2)\,du.$$

Then by Exercise 21 of §3.2, we have

$$h_g(v) = \frac{-1}{2\pi}\int_{w \in \mathbb{R}} H_g'(v + w^2/2)\,dw.$$

We can relate H_g and the Harish transform (3.133) of g as in Exercise 21 of §3.2:

$$H_g(y + y^{-1} - 2)2^{-1/2} = G(\log y) = y^{-1/2}\int_{x \in \mathbb{R}} h_g(y + y^{-1} + x^2 y^{-1} - 2)\,dx.$$

Step 2. So we must deal with

$$2\int_0^\infty \log u \, h_g(u^2)\,du = 2\int_0^\infty \log u \left(\frac{-1}{2\pi}\int_{w \in \mathbb{R}} H_g'(u^2 + w^2/2)\,dw\right)du.$$

Set $u^2 + w^2/2 = v$, and the integral becomes

$$-\frac{2}{\pi\sqrt{2}}\int_0^\infty \log u \int_{v \geq u^2} H_g'(v)(v - u^2)^{-1/2}\,dv\,du.$$

Next reverse the order of integration and obtain

$$-\frac{2}{\pi\sqrt{2}}\int_{v=0}^\infty H_g'(v) \int_{u=0}^{v^{1/2}} \log u \,(v - u^2)^{-1/2}\,du\,dv.$$

Then set $u = v^{1/2}x$ to turn the integral into

$$-\frac{2}{\pi\sqrt{2}} \int_{v=0}^{\infty} H_g'(v) \int_{x=0}^{1} (\log v^{1/2} + \log x)(1 - x^2)^{-1/2}\, dx\, dv$$

$$= -\frac{1}{2\sqrt{2}} \int_{v=0}^{\infty} H_g'(v) \log v\, dv - \log 2\, G(0).$$

For the last equality, you can use Gradshteyn and Ryzhik [1, formula 3.452 on page 335].

Step 3. Note that

$$\int_0^{\infty} \log u\, H_g'(u)\, du = \int_0^{\infty} \log(e^x + e^{-x} - 2)\sqrt{2}G'(x)\, dx$$

$$= \sqrt{2} \int_0^{\infty} xG'(x)\, dx + 2\sqrt{2} \int_0^{\infty} \log(1 - e^{-x})G'(x)\, dx$$

$$= -\sqrt{2} \int_{x=0}^{\infty} G(x)\, dx + 2\sqrt{2} \int_{x=0}^{\infty} \log(1 - e^{-x})G'(x)\, dx$$

$$= -2^{-1/2}\hat{g}(\tfrac{1}{2}) + 2^{3/2} \int_0^{\infty} \log(1 - e^{-x})G'(x)\, dx.$$

Step 4.

$$\int_{x=0}^{\infty} \log(1 - e^{-x})G'(x)\, dx = \gamma G(0) + \frac{1}{2\pi} \int_{r\in\mathbb{R}} \hat{g}(\tfrac{1}{2} + ir)\frac{\Gamma'}{\Gamma}(1 - ir)\, dr.$$

To see this, use

$$G(x) = \frac{1}{2\pi} \int_{r\in\mathbb{R}} e^{ixr}\hat{g}(\tfrac{1}{2} + ir)\, dr$$

and

$$ir \int_{x=0}^{\infty} e^{ixr} \log(1 - e^{-x})\, dx = \frac{\Gamma'}{\Gamma}(1 - ir) + \gamma,$$

from Erdélyi et al. [2, Vol. I, p. 316]. This will complete Exercise 25.

Now we can use formula (3.143) and Exercise 25 to obtain

$$T_1(A) = G(0) \log \frac{A}{2} + \tfrac{1}{4}\hat{g}(\tfrac{1}{2}) - \frac{1}{2\pi} \int_{r\in\mathbb{R}} \hat{g}(\tfrac{1}{2} + ir)\frac{\Gamma'}{\Gamma}(1 - ir)\, dr + o(1)$$

$$\text{as } A \to \infty. \tag{3.144}$$

Now we turn to the second term in $c_g(\infty)$ and use Exercise 12 of §3.5 to show that as $A \to \infty$, we have

3.7. Harmonic Analysis on the Fundamental Domain

$$T_2(A) = \frac{1}{4\pi} \int_{\substack{r \in \mathbb{R} \\ s=1/2+ir}} \hat{g}(s) \left(2 \log A - \frac{\varphi'}{\varphi}(s) + \frac{\varphi(\bar{s})A^{2ir} - \varphi(s)A^{-2ir}}{2ir} \right) dr + o(1),$$

(3.145)

where
$$\varphi(s) = \Lambda(1-s)/\Lambda(s) \quad \text{and} \quad \Lambda(s) = \pi^{-s}\Gamma(s)\zeta(2s).$$

To evaluate this, we need the following exercise.

Exercise 26. Show that

$$\frac{1}{4\pi} \int_{\substack{r \in \mathbb{R} \\ s=1/2+ir}} \hat{g}(s) \left(\frac{\varphi(\bar{s})A^{2ir} - \varphi(s)A^{-2ir}}{2ir} \right) dr = \tfrac{1}{4}\hat{g}(\tfrac{1}{2})\varphi(\tfrac{1}{2}) + o(1), \qquad A \to \infty.$$

Note that $\varphi(\tfrac{1}{2}) = 1$.

Hints. You could use the residue theorem as in Selberg [6], or beginning facts about Fourier integrals (see §1.2 and Kubota [1]).

It follows from (3.145) and Exercise 26 that

$$T_2(A) = \frac{\log A}{2\pi} \int_{r \in \mathbb{R}} \hat{g}(\tfrac{1}{2}+ir) \, dr - \frac{1}{4\pi} \int_{\substack{r \in \mathbb{R} \\ s=1/2+ir}} \hat{g}(s) \frac{\varphi'}{\varphi}(s) \, dr$$
$$+ \tfrac{1}{4}\hat{g}(\tfrac{1}{2}) + o(1), \qquad A \to \infty,$$
$$= \log A \cdot G(0) - \frac{1}{4\pi} \int_{r \in \mathbb{R}} \hat{g}(\tfrac{1}{2}+ir) \frac{\varphi'}{\varphi}(\tfrac{1}{2}+ir) \, dr$$
$$+ \tfrac{1}{4}\hat{g}(\tfrac{1}{2}) + o(1), \qquad A \to \infty.$$

So this combines with (3.144) and (3.132) plus (3.141) to give

$$c_g(\infty) = -G(0)\log 2 - \frac{1}{2\pi} \int_{r \in \mathbb{R}} \hat{g}(\tfrac{1}{2}+ir) \left(\frac{\Gamma'}{\Gamma}(1-ir) - \frac{1}{2}\frac{\varphi'}{\varphi}(\tfrac{1}{2}+ir) \right) dr.$$

(3.146)

Marie-France Vignéras [2] has simplified formula (3.146) as in the following exercise.

Exercise 27. Use the functional equation of Riemann's zeta function to show that

$$c_g(\infty) = -G(0)\log 2\pi + \frac{1}{\pi} \int_{r \in \mathbb{R}} \hat{g}(\tfrac{1}{2}+ir) \frac{\zeta'}{\zeta}(-2ir) \, dr.$$

This completes the proof of formula (3.140) for the parabolic term. Combining formulas (3.135), (3.136), (3.139), and (3.140) gives the formula in Theorem 4 and completes the proof of Selberg's trace formula, once one has done Exercise 29. □

Our discussion of the parabolic term of the trace formula was very indirect. It would be nice to find a simplification. In particular, all the cancellations that occurred during the computation of the parabolic term make one suspect that there is a more direct route to the result given in Exercises 24–27.

Exercise 28. Show that

$$\left|\frac{\varphi'}{\varphi}(s)\right| \leq C(\log(2 + |r|)) \qquad \text{for } s = \tfrac{1}{2} + ir, \qquad r \in \mathbb{R}.$$

Hint. See Titchmarsh [3, pp. 50–53].

Exercise 29. Show that $|\hat{g}(\tfrac{1}{2} + ir)| \leq C(1 + |r|)^{-2-\varepsilon}$, for $\varepsilon > 0$, suffices to make the Selberg trace formula valid.

Selberg's trace formula is surprisingly similar to Weil's exact formulas relating sums over zeros of Riemann's zeta function to sums over the primes (as well as analogues for Hecke L-functions of algebraic number fields). See Lang [3], Cartier and Hejhal [1], and Weil [5, Vol. II, pp. 48–61] for discussions of Weil's result, which was used by Weil to obtain a positivity condition equivalent to the Riemann hypothesis for Hecke L-functions of number fields.

Next we consider various applications of the trace formula. The first is a non-Euclidean analogue of Weyl's result on the asymptotic distribution of the eigenvalues of the Laplace operator (see Theorem 5 of §1.3).

Theorem 5 (Asymptotics of the Eigenvalues of Δ on $L^2(\Gamma \backslash H)$. Set

$$N(x) = \sum_{|\lambda| \leq x} \dim \mathscr{SN}(SL(2, \mathbb{Z}), \lambda),$$

where $\mathscr{SN}(\Gamma, \lambda)$ is the space of Maass waveforms which are cusp forms, as defined in §3.5. Then

$$N(x) \sim \frac{\text{area}(\Gamma \backslash H)}{4\pi} x \qquad \text{as } x \to \infty.$$

PROOF. We use the same sort of argument that worked in the Euclidean case (see Theorem 5 of §1.3). Substitute

$$\hat{g}_t(s) = e^{s(s-1)t}$$

in Selberg's trace formula. Then $g_t(z)$ is the fundamental solution of the non-Euclidean heat equation on H (see formulas (3.32) and (3.33) and note that we have changed the notation slightly).

The left-hand side of the Selberg trace formula in Theorem 4 for $g = g_t$ is

$$\tilde{\theta}_\Gamma(t) = \sum_{n \geq 0} \exp(\lambda_n t),$$

3.7. Harmonic Analysis on the Fundamental Domain

where λ_n runs through the discrete spectrum of Δ on $SL(2,\mathbb{Z})\backslash H$. Like (3.107) the function $\tilde{\theta}_\Gamma$ can be considered to be a non-Euclidean analogue of the theta function, as we mentioned in the introduction to this section (see Molchanov [1] and Exercise 16).

Exercise 30. Show that the Harish transform of the fundamental solution of the heat equation above is

$$G_t(x) = \frac{1}{2\pi}\int_{r\in\mathbb{R}} e^{irx}\hat{g}_t(\tfrac{1}{2}+ir)\,dr = \frac{1}{\sqrt{4\pi t}}\exp\left\{-\left(\frac{t}{4}+\frac{x^2}{4t}\right)\right\}.$$

Next we look as *the various terms of Selberg's trace formula (Theorem 4) with $g = g_t$, as $t \to 0+$.*

Identity Term. Since $\tanh \pi r \sim 1$ as $r \to \infty$, we have

$$\frac{\operatorname{area}(\Gamma\backslash H)}{2\pi}\int_{r=0}^\infty \exp\{-(\tfrac{1}{4}+r^2)t\}r\tanh\pi r\,dr \sim \frac{\operatorname{area}(\Gamma\backslash H)}{4\pi t}, \qquad t\to 0+.$$

Elliptic Term.

$$\int_{r\in\mathbb{R}} (\tfrac{1}{4}+3^{-3/2}(e^{\pi r/3}+e^{-\pi r/3}))\frac{e^{-(1/4+r^2)t}}{e^{\pi r}+e^{-\pi r}}\,dr \le \tfrac{1}{2}e^{-t/4}\int_0^\infty (\tfrac{1}{4}+3^{-3/2})e^{-r^2 t}\,dr$$

$$= O(t^{-1/2}), \qquad t\to 0+.$$

This term is thus negligible compared with the identity term.

Hyperbolic Term.

$$\sum_{\{\gamma\}\,\text{hyperbolic}} \frac{\log N\gamma_0}{N\gamma^{1/2}-N\gamma^{-1/2}}\frac{e^{-t/4}}{\sqrt{4\pi t}}\exp\{-(\log N\gamma)^2/4t\} = o(1), \qquad t\to 0+.$$

Exercise 31. Fill in the details for the stated result on the hyperbolic term.

Parabolic Terms. Note that $G_t(0) = (4\pi t)^{-1/2}e^{-t/4}$ is negligible compared with the identity term. Then use the following exercise to see that the story is the same for the rest of the parabolic term.

Exercise 32. Show that

$$\frac{\zeta'}{\zeta}(-2ir) = O(\log|r|) \qquad \text{as } |r|\to\infty.$$

Hint. Use the functional equation of ζ to change $-2ir$ to $1+2ir$. Then use the Euler-Maclaurin summation formula, for example, to bound $\zeta(1+2ir)$ (see Edwards [1, p. 183]) and Stirling's formula to bound gamma.

It follows from the Selberg trace formula and Exercises 30–32 that

$$\tilde{\theta}_\Gamma(t) \sim \text{area}\,(SL(2,\mathbb{Z})\backslash H)/(4\pi t) \qquad \text{as } t \to 0+.$$

Theorem 5 is a consequence of the Tauberian theorem given as Theorem 5 of §1.2. □

This completes the discussion of Theorem 5. It follows from this theorem that the dimensions of the spaces $\mathcal{SN}(\Gamma, s(s-1)$ cannot grow too rapidly with the imaginary part of s. This answers a question that had worried us in the discussion following Theorem 4 of §3.5. Of course, it still remains open whether the dimensions can be larger than one.

It is also possible to work out error terms for the asymptotic result in Theorem 5 (see Hejhal [1], for example). These applications remind one of the work of analytic number theorists on the distribution of primes (see Davenport [1], for example), using the relationship between primes and zeros of the Riemann zeta function derived from the Euler product for $\zeta(s)$ and the Weierstrass factorization of $\frac{1}{2}s(s-1)\pi^{-s/2}\Gamma(s/2)\zeta(s)$ as a product over the nonreal zeros of $\zeta(s)$. This sort of analogy motivates much of the work in Hejhal [1, 2]. In this trace-formula analogue of analytic number theory, primes are replaced by norms of the primitive hyperbolic conjugacy classes and the zeros of $\zeta(s)$ are replaced with s such that $s(s-1)$ is an eigenvalue of Δ from the discrete spectrum. Next we want to reverse the direction of the flow of information from that in Theorem 5 and use the trace formula to study the $N\gamma$, $\{\gamma\}$ hyperbolic.

Theorem 6 (Sarnak's Theorem on the Asymptotics of Units in Real Quadratic Fields à la Wallace). *Let ε_d, h_d be as in Exercise* 18, *part* (c); *i.e., the ε_d are fundamental units in certain orders $\mathcal{O}$ (not necessarily maximal) in real quadratic fields $\mathbb{Q}(\sqrt{d})$ and h_d is the narrow class number of the order $\mathcal{O}$. Then*

$$\sum_{\varepsilon_d \le x} h_d \sim \frac{x^2}{\log(x^2)} \qquad \textit{as } x \to \infty.$$

Proof. We take our discussion from Wallace [1]. It should be compared with that in Hejhal [1, Vol. II, p. 519], Sarnak [1, 3], and Subia [1]—each of whom chooses a different function to plug into Selberg's trace formula. We choose instead the function of formula (3.108):

$$f_a(z) = \exp[-a\cosh d(i,z)] = \exp[-\tfrac{1}{2}a\,\mathrm{Tr}\,(W_z)],$$

for $z \in H$, and W_z the corresponding positive matrix in $\mathcal{SP}_2$. Then we have (by formula (3.108) the Helgason transform

$$\hat{f}_a(s) = 2(2\pi/a)^{1/2}K_{\bar{s}-1/2}(a).$$

And the Harish transform is

$$F_a(\log y) = y^{-1/2} \int_{x \in \mathbb{R}} \exp\left[-\frac{a}{2}\left(\frac{1}{y} + \frac{x^2 + y^2}{y}\right)\right] dx$$
$$= \left(\frac{2\pi}{a}\right)^{1/2} \exp\left[-\frac{a}{2}\left(y + \frac{1}{y}\right)\right].$$

Now plug this into the Selberg trace formula and use (3.110) to obtain

$$2\left(\frac{2\pi}{a}\right)^{1/2} \sum_{n \geq 0} K_{\bar{s}_n - 1/2}(a)$$
$$\sim \left(\frac{2\pi}{a}\right)^{1/2} \sum_{\substack{\{\gamma\} \\ \text{hyperbolic}}} \frac{\log N\gamma_0}{N\gamma^{1/2} - N\gamma^{-1/2}} \exp[-a(N\gamma + N\gamma^{-1})], \qquad a \to 0+. \tag{3.147}$$

Then, we have

$$\left(\frac{\pi}{a}\right)^{1/2} \sim \sum_{\substack{\{\gamma\} \\ \text{hyperbolic}}} \frac{\log N\gamma_0}{N\gamma^{1/2} - N\gamma^{-1/2}} \exp\left[-\frac{a}{2}(N\gamma + N\gamma^{-1})\right], \qquad a \to 0+.$$

It follows from Exercise 18, since the terms from the primitive hyperbolics dominate the rest, that

$$\left(\frac{\pi}{a}\right)^{1/2} \sim \sum_{\varepsilon_d} h_d \frac{\log \varepsilon_d}{\varepsilon_d - \varepsilon_d^{-1}} e^{-a(\varepsilon_d^2 + \varepsilon_d^{-2})}, \qquad a \to 0+, \tag{3.148}$$

where the sum is over the fundamental units of the theorem.

Exercise 33. Prove formulas (3.147) and (3.148).

Hint. Replacing the sum over all hyperbolic conjugacy classes with that over primitive hyperbolic conjugacy classes is a standard argument in analytic number theory (see Lang [3, p. 159], for example).

One must finally make a Tauberian argument to complete the proof. This time the argument is more complicated than that of Theorem 5 of §1.3. This argument in given in Exercise 34. □

Note that by Exercise 20, Theorem 6 can also be viewed as the prime geodesic theorem, giving the asymptotics of the length spectrum of closed geodesics on the Riemann surface $SL(2, \mathbb{Z})\backslash H$.

Exercise 34 (H. Stark). Complete the proof of Theorem 6 using the following hints.

Step 1. The ε_d^{-1} appearing in (3.148) can be thrown away since $\varepsilon_d \to \infty$ as $d \to \infty$.

Step 2. To apply Theorem 5 of §1.2, set

$$a(t) = \sum_{\varepsilon_d^2 \leq t} h_d \frac{\log \varepsilon_d}{\varepsilon_d},$$

and obtain

$$a(t) \sim 2t^{1/2} \quad \text{as } t \text{ approaches infinity.}$$

Step 3. Change variables via $t = v^2$ and obtain

$$\sum_{\varepsilon_d \leq v} h_d \frac{\log \varepsilon_d}{\varepsilon_d} \sim 2v \quad \text{as } v \text{ approaches infinity.}$$

Step 4. Given any $\delta > 0$, only the terms $x^{1-\delta} < \varepsilon_d < x$ matter and thus you obtain

$$A(x) = \sum_{\varepsilon_d \leq x} \frac{h_d}{\varepsilon_d} \sim \frac{2x}{\log x} \quad \text{as } x \text{ approaches infinity.}$$

Step 5. Use integration by parts to see

$$\sum_{\varepsilon_d \leq x} h_d = \sum_{\varepsilon_d \leq x} h_d \frac{\varepsilon_d}{\varepsilon_d} = \int_0^x t \, dA(t) \sim 2\frac{x}{\log x} x - 2\int_e^x \frac{t}{\log t} dt \sim \frac{x^2}{\log x^2}.$$

It is interesting to compare the result of Theorem 6 with a conjecture made by Gauss which was later proved by Siegel [5, Vol. II, pp. 473–491]:

$$\sum_{d \leq x} h_d \log \varepsilon_d \sim \frac{\pi^2}{18\zeta(3)} x^{3/2}, \quad x \to \infty. \tag{3.149}$$

In both formulas the d's run over positive integers which are not perfect squares and which are congruent to 0 or 1 modulo 4. It does not appear to be possible to use either Theorem 6 or formula (3.149) or both to attack the question of whether an infinite number of h_d are equal to 1.

To emphasize the analogies with prime number theory one can define *Selberg's zeta function* for $SL(2, \mathbb{Z})$ as

$$Z(s) = \prod_{k \geq 0} \prod_d (1 - \varepsilon_d^{-2s-2k})^{h_d} \quad \text{if Re } s \geq 1. \tag{3.150}$$

This function has an analytic continuation as a meromorphic function in the whole complex s-plane with nontrivial zeros at s such that $s(s-1)$ is an eigenvalue in the discrete spectrum of Δ on $\Gamma \backslash H$ and at the poles of the function $\varphi(s) = \Lambda(1-s)/\Lambda(s)$, $\Lambda(s) = \pi^{-s/2}\Gamma(s/2)\zeta(s)$ in Re $s > \frac{1}{2}$ (see Gangolli and Warner [1], Hejhal [1], Selberg [1], Venkov [1], and Vignéras [2]). Marie-France Vignéras [2] shows that the functional equation of Selberg's zeta function involves the Barnes double gamma function, which was used recently by Shintani [3] in the evaluation of zeta functions of totally real number fields at negative integer values.

3.7. Harmonic Analysis on the Fundamental Domain

Selberg's zeta function is shown by Elstrodt [1, p. 68] to arise through the formation of the trace of $R_\lambda R_\mu$, where $R_\lambda = (\Delta - \lambda)^{-1}$ = the resolvent of the Laplacian. Thus the Selberg zeta function actually contains as much information as the trace formula itself.

Ruelle [1] discusses various zeta functions from number theory and geometry. He notes that Selberg's zeta function is a special case of a zeta function considered by Smale for differentiable flows on compact differentiable manifolds, assuming that the fundamental domain $\Gamma\backslash H$ is compact, of course.

There are many other applications of Selberg's trace formula. For example, one can return to the question discussed in the introduction to this section: Can one hear the shape of a drum? That is, can one study questions about the geometry of Riemannian manifolds, such as, Is the set of lengths of closed geodesics determined by the spectrum of the Laplacian? You see the lengths and you hear the eigenvalues. For example, Gangolli [1] shows using the Selberg trace formula that when $\Gamma\backslash H$ is compact the spectrum of the Laplacian determines the lengths of closed geodesics. One can also ask if there are Riemann surfaces whose Laplacians have the same spectrum but which are not isometric surfaces. Marie-France Vignéras [3] shows that there are nonisometric compact Riemann surfaces whose Laplacians have the same spectrum.

Selberg also used the trace formula to compute dimensions of spaces of holomorphic cusp forms (see Selberg [1] and Hejhal [1]). This method has been used with success for higher-rank groups (see later chapters for references). It is closely connected with Hirzebruch's generalization of the Riemann-Roch theorem as well as the Atiyah-Singer index theorem. However, there does not exist a formula for the dimensions of spaces of nonholomorphic cusp forms of the type we have considered.

Selberg [1] and Eichler [3] use the trace formula to obtain a formula involving class numbers of positive definite binary quadratic forms for the traces of Hecke operators (see Hejhal [1] and Lang [6], the chapter by Zagier on the trace formula, with correction in Zagier [4]). This version of the trace formula is called the *Eichler-Selberg trace formula*.

The trace formula is intrinsic to much of the latest work on the Artin conjecture on the holomorphy of Artin L-functions. References are Saito [1] and the article of Gérardin and Labesse in the Corvallis conference proceedings (see Borel and Casselman [1, Vol. II, pp. 115–134]), as well as the article of Shintani in the same conference proceedings ([loc. cit., pp. 97–110]), and finally the notes of Langlands [2] in which a proof of a new case of Artin's conjecture is given. See Tunnell [1], [2] for an extension of Langlands' results.

Tables 3.11 and 3.12 plus Fig. 3.35 summarize the main results of this chapter—at least those which are easily summarized.

Table 3.11. Comparison of Euclidean and Non-Euclidean Harmonic Analysis in Two Dimensions

General	Euclidean	Non-Euclidean
Symmetric space $x \in X \cong G/K$	$x = \begin{pmatrix} x_1 \\ x_2 \end{pmatrix} \in \mathbb{R}^2 \cong \mathbb{R}^2/\{0\}$ Euclidean plane	$z = x + iy \in H, x \in \mathbb{R}, y > 0$ Poincaré upper half-plane
Arc length ds	$ds^2 = dx_1^2 + dx_2^2$	$ds^2 = y^{-2}(dx^2 + dy^2)$
Laplacian Δ	$\Delta = \left(\dfrac{\partial^2}{\partial x_1^2} + \dfrac{\partial^2}{\partial x_2^2}\right)$	$\Delta = y^2\left(\dfrac{\partial^2}{\partial x^2} + \dfrac{\partial^2}{\partial y^2}\right)$
G-invariant area $d\mu$	$d\mu = dx_1\, dx_2$	$d\mu = y^{-2}\, dx\, dy$
G = isometry group	$\mathbb{R}^2$	$g \in G = SL(2, \mathbb{R})$ $g = \begin{pmatrix} a & b \\ c & d \end{pmatrix}, \begin{array}{l} a,b,c,d \in \mathbb{R} \\ ad - bc = 1 \end{array}$
K = subgroup of G of elements fixing the origin	$K = \{0\}$	$k \in K = SO(2), i = \sqrt{-1}$ = origin k orthogonal; i.e., ${}^t\!kk = I$
Action of $g \in G$ on $x \in X$	$x \mapsto x + g$ vector addition	$z \mapsto gz = \dfrac{az + b}{cz + d}$ Fractional linear mapping

3.7. Harmonic Analysis on the Fundamental Domain

Γ = discrete subgroup of G	$\Gamma = \mathbb{Z}^2$	$\gamma \in \Gamma = SL(2,\mathbb{Z})$, $\gamma = \begin{pmatrix} a & b \\ c & d \end{pmatrix}$, $a,b,c,d \in \mathbb{Z}$ $ad - bc = 1$ $\bar{\Gamma} = \Gamma/\pm I$
Elementary eigenfunctions of Δ	$e_y(x) = \exp(2\pi i\, {}^t xy)$	$p_s(z) = (\operatorname{Im} z)^s$, $s \in \mathbb{C}$
Eigenvalues	$\Delta e_y = -4\pi^2 \|y\|^2$	$\Delta p_s = s(s-1) p_s$
Helgason-Fourier Transform	$\hat{f}(y) = \int_{\mathbb{R}} f(x)\overline{e_y(x)}\, dx$	$\mathcal{H}f(s,k) = \int_H f(z)\overline{p_s(kz)}\, d\mu$ $s \in \mathbb{C},\; k \in K$
Fourier inversion or spectral decomposition of Δ (e.g., for $f \in L^2(X)$)	$f(x) = \hat{\hat{f}}(-x)$	$f(z) = \dfrac{1}{4\pi}\int_{t\in\mathbb{R}}\int_{k=K} \mathcal{H}f(s,k) p_s(kz)\, t\tanh\pi t\, dt$
Convolution (defined by convolution on G)	$\widehat{f * g} = \hat{f}\cdot\hat{g}$ (e.g., for $f,g \in L^1(\mathbb{R}^2)$)	$\mathcal{H}(f*g) = (\mathcal{H}f)\cdot(\mathcal{H}g)$ for $f,g \in L^1(H)$ with f or g K-invariant
Differentiation	$\widehat{\Delta f}(y) = -4\pi^2 \|y\|^2 \hat{f}(y)$	$\mathcal{H}(\Delta f)(s,k) = \bar{s}(\bar{s}-1)\mathcal{H}f(s,k)$
Heat equation $\begin{cases} u_t = \Delta u \\ u(x,0) = f(x) \end{cases}$ $u(x,t) = f * g_t$ g_t = fundamental solution	$g_t(x) = (4\pi t)^{-1} \exp(-\|x\|^2/4t)$	$g_t(ke^{-r}i) =$ $\sqrt{2}(4\pi t)^{-3/2} e^{-t/4} \displaystyle\int_r^{\infty} \dfrac{b e^{-b^2/4t}\, db}{\sqrt{\cosh b - \cosh r}}$

Table 3.12. Harmonic Analysis on the Fundamental Domain

General	Euclidean	Non-Euclidean
$\{e_\alpha\}_{\alpha \in A}$ complete orthonormal set of eigenfunctions of Δ on X/Γ	$e_a(x) = \exp(2\pi i\, {}^t a x)$ $a \in \mathbb{Z}^2$ purely discrete spectrum	*Continuous spectrum* of Eisenstein series: $E_s(z) = \sum_{\gamma \in \left(\begin{smallmatrix} * & * \\ 0 & * \end{smallmatrix}\right)\backslash\Gamma} \operatorname{Im}(\gamma z)^s$, $\operatorname{Re} s > 1$, with analytic continuation to other values of s *Discrete spectrum* of cusp forms and constants: $v_0 = (3/\pi)^{1/2}$ $\{v_n\}_{n \geq 1}$, complete orthonormal set of cusp forms (which vanish at ∞ by definition)
$\Delta e_\alpha = \lambda_\alpha e_\alpha$	$\Delta e_a = -4\pi^2 \Vert a \Vert^2 e_a$	$\Delta E_s = s(s-1)E_s$ $\Delta v_0 = 0(0-1)v_0$ $\Delta v_n = s_n(s_n - 1)v_n$
$g(x) = \sum_{\alpha \in A}$ or $\int (g, e_\alpha) e_\alpha(x) \sigma(\alpha)\, d\alpha$ where the integral or sum can be thought of as a Stietjes integral and the spectral measure $\sigma(\alpha)$ comes from the asymptotics and functional equations of the e_α $(g, h) = \int_{X/\Gamma} g(x)\overline{h(x)}\, dx$	$g(x) =$ $\sum_{a \in \mathbb{Z}^2} (g, e_a) e_a(x)$ ordinary Fourier series	$g(z) = \sum_{n \geq 0} (g, v_n) v_n(z)$ $+ \frac{1}{4\pi i} \int_{\operatorname{Re} s = 1/2} (g, E_s) E_s(z)\, ds$ Roelcke-Selberg spectral decomposition of Δ on $L^2(SL(2, \mathbb{Z})\backslash H)$

Poisson summation formula $f \in C_c^\infty(K\backslash X)$ $g(x) = \sum_{\gamma \in \bar{\Gamma}} f(\gamma x)$ $= \sum_{\alpha \in A} \int \hat{f}(\alpha)\overline{e_\alpha(0)}\, e_\alpha(x)$ $\times\, \sigma(\alpha)\, d\alpha.$ (this is also Mercer's theorem)	$\sum_{n \in \mathbb{Z}^2} f(x+n) = \sum_{n \in \mathbb{Z}^2} \hat{f}(n) e_n(x)$ Here $e_n(0) = 1$	$\sum_{\gamma \in \bar{\Gamma}} f(\gamma z) = \sum_{n \geq 0} \hat{f}(s_n)\overline{v_n(i)}\, v_n(z)$ $+ \frac{1}{4\pi i} \int_{\mathrm{Re}\, s = 1/2} \hat{f}(s)\overline{E_s(i)}\, E_s(z)\, ds$ $\hat{f}(s) = \int_H f(z)\overline{p_s(z)}\, d\mu.$
Application to the circle problem	$N_{\mathbb{Z}^2}(x) = \#\{n \in \mathbb{Z}^2 \mid d^2(n,0) \leq x\}$ $\sim \pi x, \qquad x \to \infty$ $d(n,0)^2 = \lVert n \rVert^2 = n_1^2 + n_2^2$	$N_\Gamma(x) =$ $\#\{\gamma \in \bar{\Gamma} \mid \cosh d(\gamma i, i) \leq x) \sim 6x,\ x \to \infty$ $d(z,w) =$ distance z to w $2\cosh d(\gamma i, i) = \mathrm{Tr}({}^t\gamma\gamma)$
Theta function	$\theta_{\mathbb{Z}^2}(a) = \sum_{n \in \mathbb{Z}^2} \exp(-a\,{}^t n n)$ $= \frac{\pi}{a}\theta_{\mathbb{Z}^2}(1/a)$ $\sim \frac{\pi}{a}, \qquad a \to 0+$	$\theta_\Gamma(a) = \sum_{\gamma \in \bar{\Gamma}} \exp\left(-\frac{a}{2}\mathrm{Tr}({}^t\gamma\gamma)\right)$ $f_a(z) = \exp(-a\cosh d(i,z))$ $\hat{f}_a(s) = 2\left(\frac{2\pi}{a}\right)^{1/2} K_{\bar{s}-1/2}(a)$ $K_s = K$-Bessel function $\theta_\Gamma(a) \sim \frac{6}{a}, \qquad a \to 0+$

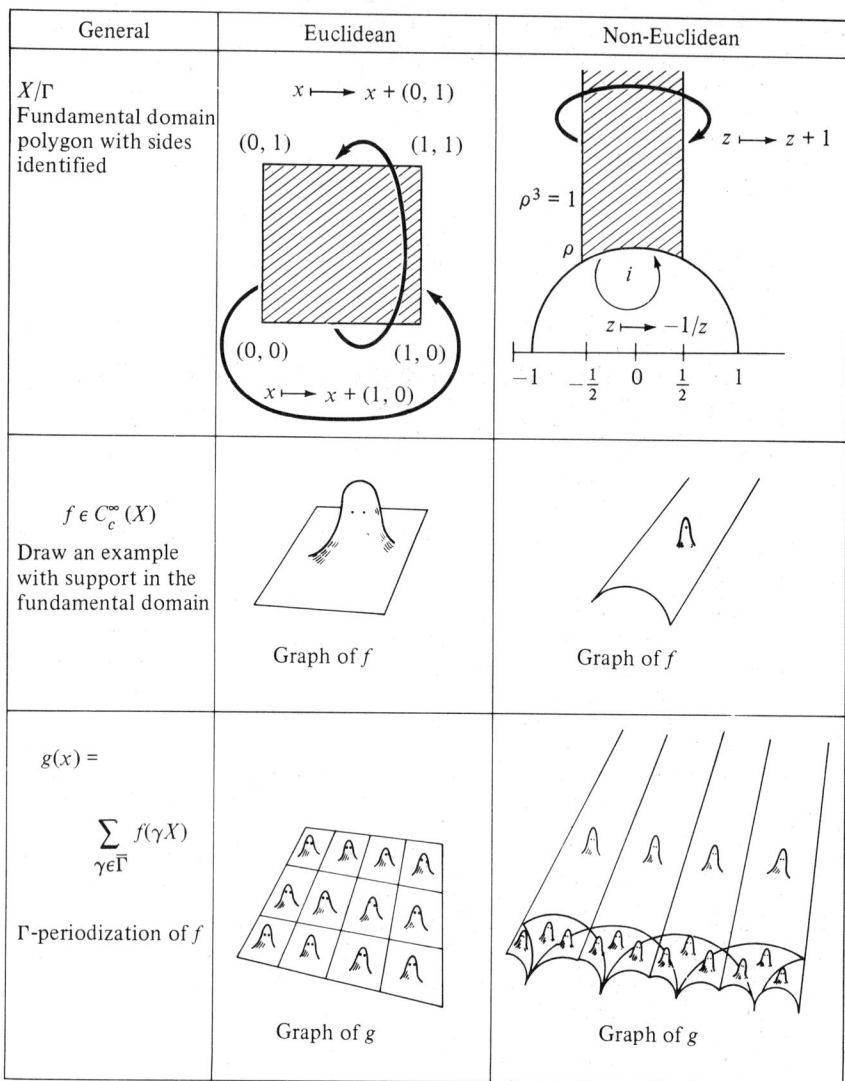

Figure 3.35. Γ-periodizations of compactly supported functions on X.

Bibliography*

L.M. Adleman, C. Pomerance, & R.S. Rumely, On distinguishing prime numbers from composite numbers. *Annals of Math*, (2) *117* (1983), 173–206.

L. Ahlfors, *Complex Analysis*. McCraw-Hill, N.Y., 1979.

———, *Möbius Transformations in Several Dimensions, Lecture Notes*. University of Minnesota, 1981.

J. Altman, *Microwave Circuits*. Van Nostrand, Princeton, N.J., 1964.

R.J. Anderson, On the Mertens conjecture for cusp forms. *Mathematika, 26* (1979), 236–249; *27* (1980), 261.

T.W. Anderson, *An Introduction to Multivariate Statistical Analysis*. Wiley, N.Y., 1958.

A.N. Andrianov, On zeta functions of Rankin type associated with Siegel modular forms, *Lecture Notes in Math. 627*. Springer-Verlag, N.Y., 1977, 325–338.

———.Dirichlet series with Euler product in the theory of Siegel modular forms of genus 2. *Proc. Steklov Inst. Math., 112* (1971), 70–93.

———, Spherical functions for GL(n) over local fields and summation of Hecke series. *Math. U.S.S.R. Sbornik, 12* (1970), 429–452.

———, Euler products corresponding to Siegel modular forms of genus two. *Russian Math. Surveys, 29* (1974), 43–110.

A.N. Andrianov & G.N. Maloletkin, Behavior of theta series of degree N under modular substitutions. *Izv. Akad. Nauk. S.S.S.R., 39* (1975), 243–258.

T. Apostol, *Calculus, Vols. I and II*. Blaisdell, Waltham, Mass., 1967.

———, *Mathematical Analysis*. Addison-Wesley, N.Y., 1974.

———, *Modular Functions and Dirichlet Series in Number Theory*. Springer-Verlag, N.Y., 1976.

T. Arakawa, Dirichlet series corresponding to Siegel's modular forms. *Math. Ann., 238* (1978), 157–174.

———, The dimension of the space of cusp forms on the Siegel upper half plane of degree 2 related to a quaternion unitary group. *J. Math. Soc. Japan, 33* (1981), 125–145.

G. Arfken, *Mathematical Methods for Physicists*. Academic, N.Y., 1970.

J. Arthur, The trace formula in invariant form. *Ann. of Math., 114* (1981), 1–74.

* This bibliography includes many references for Volume II as well as Volume I.

———, Automorphic representations and number theory, *Canadian Math. Soc. Conf. Proc., 1.* A.M.S., Providence, R.I., 1981, 3–54.

E. Artin, *Collected Papers.* Addison-Wesley, Reading, Mass., 1965.

T. Asai, On a certain function analogous to $\log|\eta(z)|$. *Nagoya Math. J.*, *40* (1970), 193–211.

A. Ash, Cohomology of congruence subgroups of $SL(n, \mathbb{Z})$. *Math. Ann.*, *249* (1980), 55–73.

A. Ash & L. Rudolph, The modular symbol and continued fractions in higher dimensions. *Inv. Math.*, *55* (1979), 241–250.

A. Ash, D. Mumford, M. Rapoport, & Y. Tai, *Smooth Compactification of Locally Symmetric Varieties.* Math. Sci. Press, Brookline, Mass., 1975.

J.M. Ash (Ed.), *Studies in Harmonic Analysis, Vol. 13, M.A.A. Studies in Math.* Math. Assoc. of America, Wash., D.C., 1976.

A.O. Atkin & J. Lehner, Hecke operators on $\Gamma_0(N)$. *Math. Ann.*, *185* (1970), 134–160.

L. Auslander, *Differential Geometry.* Harper & Row, N.Y., 1967.

L. Auslander et al., *Flows on Homogeneous Spaces.* Princeton University Press, Princeton, 1963.

A.J. Baden Fuller, *Microwaves.* Pergamon, Oxford, 1979.

W.L. Baily, *Introductory Lectures on Automorphic Forms.* Princeton University Press, Princeton, N.J., 1973.

W.L. Baily & A. Borel, Compactification of arithmetic quotients of bounded symmetric domains. *Ann. of Math.*, *84* (1966), 442–528.

A. Baker, *Transcendental Number Theory.* Cambridge University Press, Cambridge, 1975.

D. Barbasch, Fourier inversion for unipotent invariant integrals. *T.A.M.S.*, *249* (1979), 51–83.

———, Fourier transforms of some invariant distributions on a semi-simple Lie group, *Lecture Notes in Math. 728.* Springer-Verlag, N.Y., 1979, 1–7.

V. Bargmann, Irreducible unitary representations of the Lorentz group. *Ann. of Math.*, *48* (1947), 568–640.

E.S. Barnes, The complete enumeration of extreme senary forms. *Phil. Trans. Royal Soc. London*, *249* (1957), 461–506.

P. Barrucand, H. Williams, & L. Baniuk, A computational technique for determining the class number of a pure cubic field. *Math. Comp.*, *30* (1976), 312–323.

H.J. Bartels, Nichteuklidische Gitterpunktprobleme und Gleichverteilung in linear algebraischen Gruppen. *Comment. Math. Helvetici*, *57* (1982), 158–172.

A.O. Barut & R. Rączka, *Theory of Group Representations and Applications.* Polish Scientific Pub., Warsaw, 1977.

H. Bass, The congruence subgroup problem, in T.A. Springer (Ed.), *Proc. of Conf. in Local Fields.* Springer-Verlag, N.Y., 1967, 16–22.

———, *Algebraic K-Theory, Proc. Internatl. Cong. Math., Vol. I.* Vancouver, 1974, 277–283.

P.T. Bateman & E. Grosswald, On Epstein's zeta function. *Acta Arith.*, *9* (1964), 365–373.

J.G.F. Belinfante & B. Kolman, *A Survey of Lie Groups and Lie Algebras and Computational Methods.* SIAM, Philadelphia, Pa., 1972.

R. Bellman, *A Brief Introduction to Theta Functions.* Holt, Rinehart & Winston, N.Y., 1961.

T. Bengtson, *Matrix K-Bessel Functions.* Ph.D. Thesis, U.C.S.D., 1981.

———, Bessel Functions on $\mathscr{P}_n$. *Pacific J. Math.*, *108* (1983), 19–30.

F.A. Berezin, Laplace operators on semisimple Lie groups. *A.M.S. Transl. (2), Vol. 21* (1962), 239–339.

F.A. Berezin & I.M. Gelfand, Some remarks on the theory of spherical functions on symmetric Riemannian manifolds. *A.M.S. Transl. (2), Vol. 21* (1962), 193–238.

M. Berger, Geometry of the spectrum, I. *Proc. Symp. Pure Math., Vol. 27, Pt. II.* Amer. Math. Soc., Providence, 1975, 129–152.

A. Berman & R.J. Plemmons, *Nonnegative Matrices in the Mathematical Sciences.* Academic, N.Y., 1979.

T.S. Bhanu-Murthy, Plancherel's measure for the factor space $SL(n, \mathbb{R})/SO(n)$. *Dokl. Akad. Nauk. S.S.S.R., 133* (1960), 503–506.

A.T. Bharucha-Reid, *Probabilistic Methods in Applied Math. Vol. 1.* Academic, N.Y., 1968.

L. Bianchi, Geometrische Darstellung der Gruppen Linearer Substitutionen mit ganzen complexen Coefficienten nebst Anwendungen auf die Zahlentheorie. *Math. Ann., 38* (1891), 313–333.

L.C. Biedenharn & J.D. Louck, *Angular Momentum in Quantum Physics, Theory and Application.* Addison-Wesley, Reading, Mass., 1981.

————, *The Racah-Wigner Algebra in Quantum Theory.* Addison-Wesley, Reading, Mass., 1981.

J.L. Birman, Theory of crystal space groups and infra-red Raman lattice processes of insulating crystals, in S. Flugge, (Ed.), *Encyl. of Physics, Vol. XXV/2b.* Springer-Verlag, N.Y., 1974.

G.L. Blankenship, Perturbation theory for stochastic ordinary differential equations with applications to optical waveguide analysis, in *Applications of Lie Group Theory to Nonlinear Network Problems.* Western Periodicals Co., N. Hollywood Calif., 1974, 51–77.

P. Bloomfield, *Fourier Analysis of Time Series.* Wiley, N.Y., 1976.

O. Blumenthal, Über Modulfunktionen von mehreren Veränderlichen. *Math. Ann., 56* (1903), 509–548; *58* (1904), 497–527.

S. Bochner, *Lectures on Fourier Integrals.* Princeton University Press, Princeton, N.J., 1959.

————, Review of L. Schwartz, Théorie des distributions. *Bull. Amer. Math. Soc., 58* (1952), 78–85.

————, Group invariance of Cauchy's formula in several variables. *Ann. of Math., 45* (1944), 686–722.

————, Bessel functions and modular relations of higher type and hyperbolic differential equations. *Comm. Sém. Math. U. Lund, Tome Suppl.,* (1952), 12–20.

S. Bochner & W.T. Martin, *Several Complex Variables.* Princeton University Press, Princeton, N.J., 1948.

H. Boerner, *Group Representations.* Springer-Verlag, N.Y., 1955.

J. Border, *Nonlinear Hardy Spaces and Electrical Power Transfer.* Ph.D. Thesis, U.C.S.D., 1979.

A. Borel, *Introduction aux Groupes Arithmétiques.* Hermann, Paris, 1969.

————, Arithmetic properties of algebraic groups. *Proc. Internatl. Cong. Math.,* Stockholm, 1962.

————, Les fonctions automorphes de plusieurs variables complexes. *Bull. Soc. Math. France, 80* (1952), 167–182.

————, Les espaces hermitiens symétriques. *Séminaire Bourbaki,* Paris, 1952.

A. Borel & W. Casselman, *Automorphic Forms, Representations, and L-Functions, Proc. Symp. Pure Math. 33.* A.M.S., Providence, R.I., 1979.

A. Borel, S. Chowla, et al., *Seminar on Complex Multiplication, Lecture Notes in Math. 21.* Springer-Verlag, N.Y., 1966.

A. Borel & G. Mostow, *Algebraic Groups and Discontinuous Subgroups, Proc. Symp. Pure Math. 9.* A.M.S., Providence, R.I., 1966.

A. Borel & J.-P. Serre, Corners and arithmetic groups, *Comm. Math. Helv., 48* (1973), 436–491.

A. Borel & N. Wallach, *Continuous Cohomology, Discrete Subgroups and Representations of Reductive Groups.* Princeton University Press, Princeton, N.J., 1980.

Z.I. Borevitch & I.R. Shafarevitch, *Number Theory*. Academic, N.Y., 1966.
M. Born & K. Huang, *Dynamical Theory of Crystal Lattices*. Oxford University Press, London, 1956.
P. Bougerol, Comportement asymptotique des puissances de convolution d'une probabilité sur un espace symétrique. *Astérisque, 74* (1980), 29–45.
——————, *Un Mini-cours sur les Couples de Gelfand*. Publications du Laboratoire de Statistiques et Probabilités de l'Université Paul Sabatier, N°01-83, 1983.
P. Bousquet, *Spectroscopy and its Instrumentation*. Hilger, London, 1971.
R. Bowen, The equidistribution of closed geodesics. *Amer. J. Math., 94* (1972), 413–423.
R. Bracewell, *The Fourier Transform and its Applications*. McGraw-Hill, N.Y., 1965.
H. Braun, Konvergenz verallgemeinerter Eisensteinscher Reihen. *Math. Zeitschr., 44* (1939), 387–397.
——————, Der Basissatz für Hermitsche Modulformen. *Abh. aus dem Math. Sem. d. U. Hamburg, 19* (1955), 134–148.
R.P. Brent, On the zeros of the Riemann zeta function. *Math. Comp., 33* (1979), 1361–1372.
E.O. Brigham, *The Fast Fourier Transform*. Prentice-Hall, Englewood Cliffs, N.J., 1974.
R.W. Bruggeman, *Fourier Coefficients of Automorphic Forms, Lecture Notes in Math. 865*. Springer-Verlag, N.Y., 1981.
——————, Fourier coefficients of cusp forms. *Inv. Math., 45* (1978), 1–18.
——————, *Kuznetsov's Proof of the Ramanujan-Petersson conjecture for modular Forms of Weight Zero*. Math. Inst., Rijksuniversiteit, Utrecht (8–1979). (An error has since been found in the proof.)
D. Bump, *L-Series on GL*(3). Preprint, University of Texas, 1983.
——————, *Automorphic forms on GL*(3, $\mathbb{R}$). *Lecture Notes in Math. 1083*. Springer-Verlag, N.Y., 1984.
D. Bump & D. Goldfeld, A Kronecker limit formula for cubic fields, in R.A. Rankin (Ed.), *Modular Forms*. Horwood, Chichester (distrib. Wiley), 1984, 43–49.
H. Burkhardt, *Trigonometrische Reihen und Integrale (bis etwa 1850)*, Enzycl. der Math. Wissen. Teubner, Leipzig, 1916.
R. Burridge & G. Papanicolaou, The geometry of coupled mode propagation in one-dimensional random media. *Comm. Pure Appl. Math., 25* (1972), 715–757.
C.J. Bushnell & I. Reiner, L-functions of arithmetic orders and asymptotic distribution of ideals. *J. fur die reine und angew. Math., 327* (1981), 156–183.
L. Carleson, On the convergence and growth of partial sums of Fourier series. *Acta Math., 116* (1966), 135–157.
G.F. Carrier, M. Krook, & C.E. Pearson, *Functions of a Complex Variable*. McGraw-Hill, N.Y., 1966.
E. Cartan, Sur une classe remarquable d'espaces de Riemann. *Bull. Soc. Math. France, 54* (1926), 214–264.
——————, Sur une classe remarquable d'espaces de Riemann. *Bull. Soc. Math. France, 55* (1927), 114–134.
——————, Sur la détermination d'un système orthogonal complet dans un espace de Riemann symétrique clos. *Rend. Circ. Mat. Palermo, 53* (1929). 217–252.
R.W. Carter, *Simple Groups of Lie Type*. Wiley, N.Y., 1972.
P. Cartier, Some numerical computations relating to automorphic functions, in A.O. Atkin and B. Birch (eds.), *Computers in Number Theory*. Academic, N.Y., 1971.
P. Cartier & D. Hejhal, Sur les zéros de la fonction zêta de Selberg. IHES, preprint (1979).
W. Casselman, *GL*(*n*), *Proc. Durham Symp. Alg. Number Fields*. Academic, London, 1977, 663–704.
J.W.S. Cassels, *An Introduction to the Geometry of Numbers*. Springer-Verlag, Berlin, 1959.

J.W.S. Cassels & A. Fröhlich, *Algebraic Number Theory*. Thompson, Washington, D.C., 1967.

G.W. Castellan, *Physical Chemistry*. Addison-Wesley, Reading, Mass., 1971.

S. Chandrasekhar, *Hydrodynamic and Hydromagnetic Stability*. Oxford University Press, London, 1961.

C. Chevalley, *Theory of Lie Groups*. Princeton University Press, Princeton, N.J., 1946.

Y. Choquet-Bruhat, C. Dewitt-Morette, & M. Dillard-Bleick, *Analysis, Manifolds, and Physics*. North-Holland, N.Y., 1977.

S. Chowla & A. Selberg, On Epstein's zeta function. *J. Reine und Angew. Math.*, 227 (1967), 86–110.

U. Christian, *Siegelsche Modulfunktionen*. Lectures, University of Göttingen, 1974–1975.

———, *Selberg's Zeta-, L- and Eisenstein Series, Lecture Notes in Math. 1030*, Springer-Verlag, N.Y., 1983.

———, Berechnung des Ranges der Schar der Spitzenformen zur Modulgruppe zweiten Grades und Stufe $q > 2$. *J. Reine und Angew. Math.*, 277 (1975), 130–154; 296 (1977), 108–118.

R.V. Churchill & J.W. Brown, *Fourier Series and Boundary Value Problems*. McGraw-Hill, N.Y., 1978.

J.L. Clerc & B. Roynette, Un théorème central limite, *Lecture Notes in Math.*, 739. Springer-Verlag, N.Y., 1979, 122–131.

J. Coates, P-adic L-functions and Iwasawa's theory, in A. Frohlich (Ed.), *Algebraic Number Fields*. Academic, London, 1977, 269–354.

J.E. Cohen, Ergodic theorems in demography. *Bull. Amer. Math. Soc. 1*, (1979), 275–295.

H. Cohn, *A Classical Invitation to Algebraic Numbers and Class Fields*. Springer-Verlag, N.Y., 1978.

L. Cohn, The dimension of spaces of automorphic forms on a certain two-dimensional complex domain, Memoirs *Amer. Math. Soc., 158* (1975).

R.R. Coifman & G. Weiss, Representations of compact groups and spherical harmonics. *L'enseignement Math., 14* (1968), 123–173.

R.E. Collin, *Foundations for Microwave Engineering*. McGraw-Hill, N.Y., 1966.

E.U. Condon & H. Odabaşi, *Atomic Structure*. Cambridge University Press, Cambridge, 1980.

E.U. Condon & G.H. Shortley, *Theory of Atomic Spectra*. Cambridge University Press, Cambridge, 1935.

J.H. Conway, Monsters and moonshine. *Math. Intelligencer, 2* (1980), 165–171.

J.H. Conway & S.P. Norton, Monstrous moonshine. *Bull. London Math. Soc., 11* (1979), 308–339.

J.H. Conway, A.M. Odlyzko, & N.J.A. Sloane, Extreme self-dual lattices exist only in dimensions 1 to 8, 12, 14, 15, 23, and 24. *Mathematika, 25*, (1978), 36–43.

E.T. Copson, *Theory of Functions of a Complex Variable*. Oxford University Press, Oxford, 1960.

COSRIMS, *The Mathematical Sciences: A Collections of Essays*. M.I.T. Press, Cambridge, Mass., 1969.

R. Courant & D. Hilbert, *Methods of Mathematical Physics, Vol. I*. Wiley-Interscience, N.Y., 1961.

H.S.M. Coxeter, *Non-Euclidean Geometry*. University of Toronto Press, Toronto, 1965.

———, *Introduction to Geometry*. Wiley, N.Y., 1961.

H. Cramér, *Mathematical Methods of Statistics*. Princeton University Press, Princeton, N.J., 1946.

C.W. Curtis, Representations of finite groups of Lie type. *Bull. Amer. Math. Soc. (N.S.), 1* (1979), 721–757.

Yu. L. Daletskii, Continual integrals related to operator evolution equations. *Russ. Math. Surveys*, *17* (1962), 3–115.

H. Davenport, *Multiplicative Number Theory*. Springer-Verlag, N.Y., 1981.

———, *Selected Topics in the Geometry of Numbers*. Stanford University Lectures, 1950.

H. Davenport & H. Heilbronn, On the zeros of certain Dirichlet series I, II. *J. London Math. Soc.*, *11* (1936), 181–185, 307–312.

P. De La Torre & L.J. Goldstein, On the transformation of Log $\eta(\tau)$, *Duke Math. J.*, (1973), 291–297.

E.M. De Jager, *Applications of Distributions in Mathematical Physics*. Amsterdam, Math. Centre Tract 10, 1964.

———, Theory of distributions, in E. Roubine (Ed.), *Mathematics Applied to Physics*. Springer-Verlag, N.Y., 1970.

P. Deligne, Formes modulaires et représentations l-adiques. *Sém. Bourbaki*, *21* (1969), exp. 335.

———, La conjecture de Weil, Inst. des Hautes Études Sci. *Pub. Math.*, *53* (1974), 273–307.

B.N. Delone & D.K. Faddeev, *The Theory of Irrationalities of the Third Degree*, Trans. Math. Monographs, 10. Amer. Math. Soc., Providence, R.I., 1964.

B.N. Delone & S.S. Ryskov, Extremal problems in the theory of positive quadratic forms. *Proc. Steklov Inst. Math.*, *112* (1971), 211–231.

J. Delsarte, *Oeuvres*. Éditions du Centre National de la Recherche Scientifique, Paris, 1971.

J.-M. Deshouillers & H. Iwaniec, Kloosterman sums and Fourier coefficients of cusp forms, *Inv. Math*, *70* (1982/83), 219–288.

C.M. Dewitt, L'intégrale fonctionnelle de Feynman. Une introduction. *Ann. Inst. H. Poincaré*, *XI* (1969), 153–206.

B. Diehl, Die analytische Fortsetzung der Eisensteinreihe zur Siegelschen Modulgruppe, *J. für die Reine und Angew. Math.*, *317* (1980), 40–73.

J. Dieudonné, *Treatise on Analysis, Vols. I–VI*. Academic, N.Y., 1969–1978.

P.A.M. Dirac, The physical interpretation of quantum dynamics. *Proc. Royal Soc., London, Sect. A*, *113* (1926–1927), 621–641.

G. Doetsch, *Introduction to the Theory and Application of the Laplace Transform*. Springer-Verlag, N.Y., 1974.

W.F. Donoghue, *Distributions and Fourier Transforms*. Academic, N.Y., 1969.

J.S. Dowker, Quantum mechanics on group space and Huygen's principle. *Ann. of Physics*, *62* (1971), 361–382.

———, Quantum mechanics and field theory on multiply connected and on homogeneous spaces. *J. Phys. A.*, *5* (1972), 936–943.

B.A. Dubrovin, V.B. Matveev, & S.P. Novikov, Non-linear equations of Korteweg-deVries type, finite-zone linear operators, and abelian varieties. *Russ. Math. Surveys*, *31* (1976), 59–146.

R.M. Dudley, Lorentz-invariant Markov processes in relativistic phase space. *Arkiv for Mat.*, *6* (1965), 241–268.

———, Recession of some relativistic Markov processes. *Rocky Mt. J. of Math.*, *4* (1974), 401–406.

J.J. Duistermaat, J.A.C. Kolk, & V.S. Varadarajan, Spectra of compact locally symmetric manifolds of negative curvature. *Inv. Math.*, *52* (1979), 27–93.

N. Dunford & J.T. Schwartz, *Linear Operators, Vols. I, II, III*. Wiley-Interscience, N.Y., 1958, 1963, 1971.

P. Du Val, *Elliptic Functions and Elliptic Curves*. Cambridge University Press, Cambridge, 1973.

H. Dym & H.P. McKean, *Fourier Series and Integrals*. Academic, N.Y., 1972.

F. Dyson, *Symmetry Groups in Nuclear and Particle Physics*. Benjamin, N.Y., 1966.

H. Edwards, *Riemann's Zeta Function*. Academic, N.Y., 1974.

L. Ehrenpreis & F. Mautner, Some properties of the Fourier-Transform on semisimple Lie groups, I, II, III. *Ann. Math.*, *61* (1955), 406–439; *Trans. Amer. Math. Soc.*, *84*, *90* (1957, 1959), 1–55, 431–483.

M. Eichler, *Introduction to the Theory of Algebraic Numbers and Functions*. Academic, N.Y., 1966.

———, The basis problem for modular forms and the traces of the Hecke operators. *Lecture Notes in Math. 320*. Springer-Verlag, N.Y., 1973, 75–151.

———, Modular correspondences and their representations. *J. Indian Math. Soc.*, *20* (1956), 161–206.

———, Eine Verallgemeinerung der Abelschen Integrale. *Math. Z.*, *67* (1957), 267–298.

———, On theta functions of real algebraic number fields. *Acta Arith.*, *33* (1977), 269–292.

———, Über die Anzahl der linear unabhängigen Siegelschen Modulformen von gegebenem Gewicht. *Math. Ann.*, *213* (1975), 281–291.

———, Zur Begründung der Theorie der automorphen Funktionen in mehreren Variablen, *Aeq. Math.*, *3* (1969), 93–111.

J. Elstrodt, Die Resolvente zum Eigenwertproblem der automorphen Formen in der hyperbolischen Ebene I, II, III. *Math. Ann.*, *203* (1973), 295–330; *Math. Z.*, *132* (1973), 99–134; *Math. Ann.*, *208* (1974), 99–132.

———, Die Selbergsche Spurformel für Kompakte Riemannsche Flächen. *Jber d. Dt. Math.-Verein*, *83* (1981), 45–77.

J. Elstrodt, F. Grunewald, & J. Mennicke, On the group $PSL(\mathbb{Z}[i])$, J.V. Armitage (Ed.), *Journées Arithmétiques 1980 Exeter, LMS Lecture Notes*. Cambridge University Press, Cambridge, 1982.

———, Discontinuous groups on 3-dimensional hyperbolic space: Analytical theory and arithmetic applications, *Russian Math. Surveys*, *38* (1983), 137–168.

———, $PSL(2)$ over imaginary quadratic integers, *Astérisque*, *94* (1983), 43–60.

O. Emersleben, Die elektrostatische Gitterenergie endlicher Stücke heteropolarer Kristalle. *Z. Physik Chem.*, *199* (1952), 170–190.

———, Über das Restglied der Gitterenergieentwicklung neutraler Ionengitter. *Math. Nach.*, *9* (1953), 221–234.

P. Epstein, Zur Theorie allgemeiner Zetafunktionen, I, II. *Math. Ann.*, *56*, *63* (1903, 1907), 614–644, 205–216.

A. Erdélyi et al., *Higher Transcendental Functions*, Vols. I, II, III. McGraw-Hill, N.Y., 1953–1955.

———, *Integral Transforms*, Vols. I, II. McGraw-Hill, N.Y., 1954.

B. Ernst, *The Magic Mirror of M.C. Escher*. Ballantine, N.Y., 1976.

P.P. Ewald, Die Berechnung optischer und elektrostatischer Gitterpotentiale. *Ann. d. Phys.*, *64* (1921), 253–287.

H. Eyring, J. Walter, G. & Kimball, *Quantum Chemistry*. Wiley, N.Y., 1944.

L.D. Faddeev, Expansion in eigenfunctions of the Laplace operator on the fundamental domain of a discrete group on the Lobacevskii plane. *Trans. Moscow Math. Soc.*, *17* (1967), 357–386.

J. Faraut, Dispersion d'une mesure de probabilité sur $SL(2, \mathbb{R})$ biinvariante par $SO(1, \mathbb{R})$ et théorème de la limite centrale. Preprint, 1975.

R.H. Farrell, *Techniques of Multivariate Calculus, Lecture Notes in Math.*, *520*. Springer-Verlag, N.Y., 1976.

J.D. Fay, Fourier coefficients of the resolvent for a Fuchsian group. *J. fur die Reine und Angew. Math.*, *293* (1977), 143–203.

L. Fejes Tóth, *Regular Figures*. MacMillan, N.Y., 1964.

W. Feller, *An Introduction to Probability Theory and its Applications*, Vols. I, II. Wiley, N.Y., 1950, 1966.

R. Feynman, *Quantum Mechanics and Path Integrals.* McGraw-Hill, N.Y., 1965.
K.L. Fields, Locally minimal Epstein zeta functions. *Mathematika, 27* (1980), 17–24.
R.A. Fisher, The sampling distribution of some statistics obtained from nonlinear equations. *Ann. Eugenics, 9* (1939), 238–249.
H. Flanders, *Differential Forms.* Academic, N.Y., 1963.
M. Flensted-Jensen, Spherical functions on a real semisimple Lie group. A method of reduction to the complex case. *J. Functional Analysis, 30* (1978), 106–146.
Y.Z. Flicker, *The Trace Formula and Base Change for GL(3), Lecture Notes in Math., 927.* Springer-Verlag, N.Y., 1982.
V.A. Fock, On the representation of an arbitrary function by an integral involving Legendre's functions with a complex index. *Comptes Rendus Acad. Sci. U.R.S.S., Doklady, N.S., 39* (1943), 253–256.
L. Ford, *Automorphic Functions.* Chelsea, N.Y., 1951.
H. Frasch, Die Erzeugenden der Hauptkongruenzgruppen für Primzahlstufen. *Math. Ann., 108* (1933), 229–252.
E. Freitag, *Siegelsche Modulfunktionen.* Springer-Verlag, N.Y., 1982.
———, Zur Theorie der Modulformen zweiten Grades. *Nachr. Akad. Wiss. Göttingen, II, Math.-Phys. Kl.* (1965), 151–157.
R. Fricke & F. Klein, *Vorlesungen über die Theorie der automorphen Functionen, I.* Teubner, Leipzig, 1897.
S. Friedberg, *Theta Functions, Liftings, and Generalized Hilbert Modular Forms.* Ph.D. Thesis, University of Chicago, 1982.
F.G. Friedlander, *Introduction to the Theory of Distributions.* Cambridge University Press, Cambridge, 1982.
———, *The Wave Equation on a Curved Space-Time.* Cambridge University Press, Cambridge, 1975.
R. Fueter, Über automorphe Funktionen der Picard'schen Gruppe I. *Comm. Math. Helve., 3* (1931), 42–68.
H. Funk, Beiträge zur Theorie der Kugelfunktionen. *Math. Ann., 77* (1916), 136–152.
H. Furstenberg, Noncommuting random products. *Trans. Amer. Math. Soc., 108* (1963), 377–428.
———, A Poisson formula for semi-simple Lie groups. *Ann. of Math., 77* (1963), 335–386.
H. Furstenberg & H. Kesten, Products of random matrices. *Ann. Math. Stat., 31* (1960), 459–469.
K. Gangolli, Spectra of discrete uniform subgroups, in W. Boothby and G. Weiss (Eds.), *Geometry and Analysis on Symmetric Spaces.* Dekker, N.Y., 1972, 93–117.
———, Spherical functions on semisimple Lie groups, in W. Boothby and G. Weiss (Eds.), *Geometry and Analysis on Symmetric Spaces.* Dekker, N.Y., 1972, 41–92.
———, Isotropic infinitely divisible measures on symmetric spaces. *Acta Math., 111* (1964), 213–246.
———, On the Plancherel formula and the Paley-Wiener theorem for spherical functions on semisimple Lie groups. *Ann. of Math., 93* (1971), 150–165.
R. Gangolli & G. Warner, On Selberg's trace formula. *J. Math. Soc. Japan, 27* (1975), 328–343.
———, Zeta functions of Selberg's type for some noncompact quotients of symmetric spaces of rank one. *Nagoya Math. J., 78* (1980), 1–44.
P.R. Garabedian, *Partial Differential Equations.* Wiley, N.Y., 1964.
C.F. Gauss, *Werke.* Königlichen Gesellshaft der Wissenshaften, Göttingen, 1870–1927.
S. Gelbart, *Automorphic Forms on Adele Groups.* Princeton University Press, Princeton, N.J., 1975.
———, Bessel functions, representation theory, and automorphic functions, in *Proc. Symp. Pure Math., 26.* A.M.S., Providence, R.I., 1973, 343–345.

———, Automorphic forms and Artin's conjecture, *Lecture Notes in Math., Vol. 627*. Springer-Verlag, N.Y., 1977, 241–276.
———, An elementary introduction to the Langlands program. *Bull. Amer. Math. Soc., 10* (1984), 177–220.
S. Gelbart & H. Jacquet, A relation between automorphic representations of $GL(2)$ and $GL(3)$. *Ann. Sci. Ecole Norm. Sup.*, (*4*), 11 (1978), 471–552.
I.M. Gelfand, Spherical functions on symmetric spaces. *Doklady Akad. Nauk. S.S.S.R., 70* (1950), 5–8.
I.M. Gelfand & S.V. Fomin, Geodesic flows on manifolds. *A.M.S. Translations*, (*2*), *Vol. 1*, (1955), 49–66.
I.M. Gelfand & M.I. Graev, Analogue of the Plancherel formula for the classical groups. *Amer. Math. Soc. Transl., Ser. 2, Vol. 9* (1958), 123–154.
I.M. Gelfand, M.I. Graev, & I.I. Piatetskii-Shapiro, *Representation Theory and Automorphic Functions*. Saunders, Philadelphia, Pa., 1966.
I.M. Gelfand, M.I. Graev, & N. Ya. Vilenkin, *Generalized Functions, Vol. 5*. Academic, N.Y., 1966.
I.M. Gelfand, R.A. Minlos, Z. Ya. Shapiro, *Representations of the Rotation and Lorentz Groups and Their Applications*. MacMillan, N.Y., 1963.
I.M. Gelfand & M.A. Naimark, *Unitary Representations of the Classical Groups* (German translation). Akademie-Verlag, Berlin, 1957.
I.M. Gelfand & G.E. Shilov, *Generalized Functions, Vols. 1–3*. Academic, N.Y., 1964.
I.M. Gelfand & N.Ya. Vilenkin, *Generalized Functions, Vol. 4*. Academic, N.Y., 1964.
P. Gérardin, On harmonic functions on symmetric spaces and buildings. *Canadian Math. Soc. Conf. Proc., 1* (1981), 79—92.
M.E. Gertsenshtein & V.B. Vasil'ev, Waveguides with random inhomogeneities and Brownian motion in the Lobachevsky plane. *Theory of Probability and its Applications, 4* (1959), 391–398.
R.K. Getoor, Infinitely divisible probabilities on the hyperbolic plane. *Pacific J. Math., 11* (1961), 1287–1308.
F. Gilbert, Inverse problems for the earth's normal modes, in W.H. Reid (Ed.), *Mathematical Problems in the Geophysical Sciences, Vol. I*. Amer. Math. Soc., Providence, R.L., 1971, 107–128.
S.G. Gindikin, Analysis in homogeneous domains. *Russian Math. Surveys, 19* (1964), 1–90.
S. Gindikin & F. Karpelevic, Plancherel measures of Riemannian symmetric spaces of non-positive curvature. *Sov. Math. Dokl., 3* (1962), 962–965.
R. Godement, Introduction aux travaux de A. Selberg, *Sém. Bourbaki, exp. 144*. Paris, 1957.
———, La formule des traces de Selberg considerée comme source de problèmes mathématiques. *Sém. Bourbaki, exp. 244*, Paris, 1962.
———, Introduction à la théorie de Langlands, *Sém. Bourbaki, exp. 321* (1966/67). Benjamin, N.Y., 1968.
———, A theory of spherical functions. *Trans. American Math. Soc., 73* (1952), 496–556.
———, Une généralization du théorème de la moyenne pour les fonctions harmoniques, *C.R. Acad. Sci. Paris, 234* (1952), 2137–2139.
———, *Notes on Jacquet-Langlands Theory*. Institute for Advanced Study, Princeton, N.J., 1970.
D. Goldfeld, J. Hoffstein, & S.J. Patterson, Mean values of certain elliptic L-series, in N. Koblitz (Ed.), *Number Theory Related to Fermat's Last Theorem*. Birkhäuser, Boston, 1982.
D. Goldfeld & D. Husemoller, *On* $L^2(SL_2(\mathbb{Z})\backslash H)$ (to appear).
D. Goldfeld & P. Sarnak, Sums of Kloosterman sums. *Inv. Math., 71* (1983), 243–250.
D. Goldfeld & C. Viola, Mean values of L-functions associated to elliptic, Fermat and

other curves at the centre of the critical strip. *J. Number Theory*, *11* (1979), 305–320.
L.J. Goldstein, *Analytic Number Theory*. Prentice-Hall, Englewood Cliffs, N.J., 1971.
———, Dedekind sums for a Fuchsian group I, II. *Nagoya Math. J.*, *50* (1973), 21–47; *53* (1974), 171–187, 235–237.
———, On a conjecture of Hecke concerning elementary class number formulas. *Manuscripta Math. 9* (1973), 245–305.
L.J. Goldstein & M. Razar, A generalization of Dirichlet's class number formula. *Duke Math. J. 43* (1976), 349–358.
R. Goodman, Horospherical functions on symmetric spaces, *Canadian Math. Soc. Conf. Proc.*, *1*. A.M.S., Providence, R.I., 1981, 125–133.
R. Goodman & N. Wallach, Conical vectors and Whittaker vectors. *J. Funct. Anal.*, *39* (1980), 199–279.
F. Götzky, Über eine zahlentheoretische Anwendung von Modulfunktionen zweier Veränderlicher. *Math. Ann.*, *100* (1928), 411–437.
I.S. Gradshteyn & I.M. Ryzhik, *Tables of Integrals, Series and Products*. Academic, N.Y., 1965.
I. Grattan-Guinness & J.R. Ravetz, *Joseph Fourier*. M.I.T. Press, Cambridge, Mass., 1972.
W. Greub, S. Halperin, & R. Vanstone, *Connections, Curvature, and Cohomology*, Vol. II. Academic, N.Y., 1973.
P.R. Griffiths, Fourier transform infrared spectrometry. *Science*, *222* (1983), 297–302.
K.I. Gross, W.J. Holman III, & R.A. Kunze, A new class of Bessel functions and applications in harmonic analysis, *Proc. Symp. Pure Math.*, *35*, Pt. 2. A.M.S., Providence, R. I., 1979, 407–415.
K.I. Gross & R.A. Kunze, Bessel functions and representation theory I, II. *J. Funct. Anal.*, *22* (1976), 73–105; *25* (1977), 1–49.
E. Grosswald, *Topics from the Theory of Numbers*. MacMillan, N.Y., 1966.
———, Relations between values at integer arguments of Dirichlet series satisfying functional equations, *Proc. Symp. Pure Math.*, *24*. Amer. Math. Soc., Providence, R.I., 1973.
F.A. Grünbaum, L. Longhi, & M. Perlstadt, Differential operators commuting with finite convolution integral operators: Some nonabelian examples. Preprint, Center for Pure & Applied Math., U.C. Berkeley, PAM 11.
H. Guggenheimer, *Differential Geometry*. Dover, N.Y., 1977.
K.-B. Gundlach, Die Bestimmung der Funktionen zu einigen Hilbertschen Modulgruppen. *J. fur die Reine und Angew. Math.*, *220* (1965), 109–153.
R.C. Gunning, *Lectures on Modular Forms*. Princeton University Press, Princeton, N.J., 1962.
R. Gupta, *Fields of division points of elliptic curves related to Coates-Wiles*. Ph.D. Thesis, M.I.T., 1983.
M. Gutzwiller, Stochastic behavior in quantum scattering. *Physica D*, *7* (1983), 1–3.
———, The quantization of a classically ergodic system. *Physica D*, *5* (1982), 183–207.
———, Classical quantization of a Hamiltonian with ergodic behavior. *Physical Review Letters*, *45* (1980), 150–153.
H. Haas, *Numerische Berechnung der Eigenwerte der Differentialgleichung $y^2 \Delta u + \lambda u = 0$ für ein unendliches Gebiet im $\mathbb{R}^2$*. Universität Heidelberg, Diplomarbeit, 1977.
M. Hamermesh, *Group Theory and its Applications to Physical Problems*. Addison-Wesley, Reading, Mass., 1962.
R.W. Hamming, *Digital Filters*. Prentice-Hall, Englewood Cliffs, N.J., 1977.
———, *Coding and Information Theory*. Prentice-Hall, Englewood Cliffs, N.J., 1980.

W. Hammond, The modular groups of Hilbert and Siegel. *Amer. J. Math.*, *88* (1966), 497–516.

H. Hancock, *Development of the Minkowski Geometry of Numbers*, Vols. I, II. Dover, N.Y., 1939.

L. Hantsch & W. von Waldenfels, A random walk on the general linear group related to a problem of atomic physics, *Lecture Notes in Math.*, *706*, Springer-Verlag, N.Y., 1979, 131–143.

C. Harder, A Gauss-Bonnet formula for discrete arithmetically defined groups. *Ann. Sci. Éc. Norm. Sup.*, *4* (1971), 409–455.

G.H. Hardy, *Ramanujan*. Chelsea, N.Y., 1940.

———, *Collected Papers, Vols. 1–7*. Oxford Univeristy Press, Oxford, 1966–1979.

G. H. Hardy & E.M. Wright, *An Introduction to the Theory of Numbers*. Oxford University Press, London, 1956.

Harish-Chandra, *Automorphic Forms on Semi-Simple Lie Groups, Lecture Notes in Math. 62*. Springer-Verlag, N.Y., 1968.

———, Spherical functions on a semisimple Lie group I, II. *Amer. J. Math.*, *80* (1958), 241–310, 553–613.

———, Representations of semisimple Lie groups I, II, III. *Trans. Amer. Math. Soc.*, *75* (1953), 185–243; *76* (1954), 26–65, 234–253.

———, Some results on differential equations and their applications. *Proc. Natl. Acad. Sci. U.S.A.*, *45*(1959), 1763–1764.

———, *Collected Papers, I–IV*. Springer-Verlag, N.Y., 1984.

D.C. Harris & M.D. Bertolucci, *Symmetry and Spectroscopy*. Oxford University Press, N.Y., 1978.

C.B. Haselgrove & J.C.P. Miller, *Tables of the Riemann Zeta Function, Royal Society Math. Tables 6*. Cambridge University Press, Cambridge, 1963.

D.R. Heath-Brown & S.J. Patterson, The distribution of Kummer sums at prime arguments. *J. Reine Angew. Math.*, *310* (1979), 111–130.

E. Hecke, *Mathematische Werke*. Vandenhoeck und Ruprecht, Göttingen, 1970.

———, *Lectures on Dirichlet Series, Modular Functions and Quadratic Forms*, given at the Institute for Advanced Study in 1938, B. Schoeneberg & W. Maak (Eds.). Vandenhoeck und Ruprecht, Göttingen, 1983.

———, *Lectures on the Theory of Algebraic Numbers*. Springer-Verlag, N.Y., 1981.

D. Hejhal, *The Selberg Trace Formula for $PSL(2,\mathbb{R})$, Vols. I, II, Lecture Notes in Math.*, *548, 1001*. Springer-Verlag, N.Y., 1976, 1983.

———, The Selberg trace formula and the Riemann zeta function. *Duke Math. J.*, *43* (1976), 441–482.

———, Some observations concerning eigenvalues of the Laplacian and Dirichlet L-Series, in H. Halberstam & C. Hooley (Eds.), *Recent Progress in Analytic Number Theory*. Academic, London, 1981, 95–110.

———, Some Dirichlet series whose poles are related to cusp forms, Pts. I, II. Technical Reports, 1981-14, 1982-13, Chalmers University of Tech., Sweden, 1981, 1982.

———, A Note on the Voronoi summation formula. *Monatshefte fur Math.*, *87* (1979), 1–14.

———, Sur certaines séries de Dirichlet associées aux géodésiques fermées d'une surface de Riemann compacte. *C.R. Acad. Sci. Paris*, *294* (1982), 273–276.

———, Sur quelques propriétés asymptotiques des périodes hyperboliques et des invariants algebriques d'un sous-groupe discret de $PSL(2,\mathbb{R})$. *C.R. Acad. Sci. Paris*, *294* (1982), 509–512.

———, Quelques examples de séries de Dirichlet dont les pôles ont un rapport étroit avec les valeurs propres de l'opérateur de Laplace-Beltrami hyperbolique. *C.R. Acad. Sci. Paris*, *294* (1982), 637–640.

———, Some Dirichlet series with coefficients related to periods of automorphic eigenforms [parts 1–2]. *Proc. Japan Acad. A.*, *58*, *59* (1982, 1983) 413–417, 335–338.

———, A continuity method for spectral theory on Fuchsian groups, in R.A. Rankin (Ed.), *Modular Forms*. Horwood, Chichester (distrib. Wiley), 1984, 107–140.

S. Helgason, Lie Groups and Symmetric Spaces, in C.M. DeWitt & J.A. Wheeler (Eds.), *Battelle Recontres*. Benjamin, N.Y., 1968, 1–71.

———, *Differential Geometry and Symmetric Spaces*. Academic, N.Y., 1962.

———, *Differential Geometry, Lie Groups, and Symmetric Spaces*. Academic, N.Y., 1978.

———, *The Radon Transform*. Birkhäuser, Boston, 1980.

———, *Topics in Harmonic Analysis on Homogeneous Spaces*. Birkhäuser, Boston. 1981.

———, Functions on symmetric spaces. *Proc. Symp. Pure Math.*, *26*, A.M.S., Providence, 1973, 101–146.

———, *Analysis on Lie Groups and Homogeneous Spaces*. A.M.S. Regional Conference 14. A.M.S., Providence, R.I., 1971 (corrected, 1977).

———, An analogue of the Paley-Wiener theorem for the Fourier transform on certain symmetric spaces. *Math. Ann.*, *165* (1966), 297–308.

———, A duality for symmetric spaces with applications to group representations. III. Tangent space analysis, *Adv. in Math.*, *36* (1980), 297–323.

S. Helgason & K. Johnson, The bounded spherical functions on symmetric spaces. *Adv. in Math.*, *3* (1969), 586–593.

J. Helszajn, *Passive and Active Microwave Circuits*. Wiley, N.Y., 1978.

J.W. Helton, Non-Euclidean functional analysis and electronics, *Bull. Amer. Math. Soc.*, *7* (1982), 1–64.

———, The distance of a function to H^∞ in the Poincaré metric. Electrical power transfer. *Journal of Functional Analysis*, *38* (1980), 273–314.

———, A simple test to determine gain bandwidth limitations. *Proc. I.E.E.E. Internatl. Conf. on Circuits and Systems, 1977*.

P. Henrici, *Applied and Computational Complex Analysis*, Vols. *I*, *II*. Wiley, N.Y., 1974, 1977.

G.T. Herman (Ed.), *Image Reconstruction from Projections*. Springer-Verlag, N.Y., 1979.

O. Hermann, Über Hilbertsche Modulfunktionen und die Dirichletschen Reihen mit Eulerschen Produktentwicklung. *Math. Ann.*, *127* (1954), 357–400.

R. Hermann, *Lie Groups for Physicists*. Benjamin, N.Y., 1966.

———, *Fourier Analysis on Groups and Partial Wave Analysis*. Benjamin, N.Y., 1969.

C. Hermite, *Oeuvres*, Vols. *I–IV*. Gauthier-Villars, Paris, 1905–1917.

C. Herz, Bessel functions of matrix argument. *Ann. of Math.*, *61* (1955), 474–523.

Hewlett-Packard Co., *Microwave Theory and Measurements*. Prentice-Hall, Englewood Cliffs, N.J., 1963.

K. Hey, *Analytische Zahlentheorie in Systemen hyperkomplexer Zahlen*. Inaug.-Diss., Hamburg, 1929.

H. Heyer, An application of the method of moments to the central limit theorem, *Lecture Notes in Math. 861*. Springer-Verlag, N.Y., 1981, 65–73.

H. Hijikata, A. Pizer, & T. Shemanske, The basis problem for modular forms on $\Gamma_0(N)$. *Proc. Japan Acad.*, *56* (1980), 280–284.

E. Hilb & M. Riesz, Neuere Untersuchungen über Trigonometrische Reihen. *Enzyklop. d. Math. Wissen.*, Teubner, Leipzig, 1916.

D. Hilbert & S. Cohn-Vossen, *Geometry and the Imagination*. Chelsea, N.Y., 1952.

F. Hirzebruch, Hilbert modular surfaces. *L'Enseignement Math. 21* (1973), Université Genève.

———, *Topological Methods in Algebraic Geometry*. Springer-Verlag, N.Y., 1966.

———, The ring of Hilbert modular forms for real quadratic fields of small discriminant, *Lecture Notes in Math. 627.* Springer-Verlag, N.Y., 1977, 288–323.

J. Hoffstein, Real zeros of Eisenstein series, *Math. Z. 181* (1982), 179–190.

K. Højendahl, Studies in the properties of crystals III. On an elementary method of calculating Madelung's constant. *Danske Vid. Selske, Math.-fys. Medd., 16* (1938), 133–154.

R.T. Hoobler & H.L. Resnikoff, Normal connections for automorphic embeddings. Preprint.

C. Hooley, On the distribution of the roots of polynomial congruences. *Mathematika, 11* (1964), 39–49.

L. Hörmander, On the division of distributions by polynomials. *Arkiv Matem., 3* (1958), 555–568.

———, *An Introduction to Complex Analysis in Several Variables.* Van Nostrand, Princeton, N.Y., 1966.

J. Horvath, An introduction to distributions. *Amer. Math. Monthly, 77* (1970), 227–240.

P.L. Hsu, On the distribution of the roots of certain determinantal equations. *Ann. Eugenics, 9* (1939), 250–258.

L.K. Hua, *Harmonic Analysis of Functions of Several Complex Variables in the Classical Domains, Transl. of Math. Monographs, 6.* A.M.S., Providence, R.I., 1963.

H. Huber, Über eine neue Klasse automorphe Funktionen und eine Gitterpunktproblem in der hyperbolischen Ebene. *Comment. Math. Helvet., 30* (1965), 20–62.

———, Zur analytischen Theorie hyperbolischer Raumformen und Bewegungsgruppen. *Math. Ann., 138* (1959), 1–26.

G. Humbert, Théorie de la réduction des formes quadratiques définis positives dans un corps algébrique K fini. *Comm. Math. Helv., 12* (1939/40), 263–306.

———, Sur la mesure des classes d'Hermite de discriminant donné dans un corps quadratique imaginaire, et sur certains volumes non euclidiens. *Comptes Rendus* (Paris), *169* (1919), 448–454.

J. Humphreys, *Arithmetic Groups, Lecture Notes in Math. 789.* Springer-Verlag, N.Y., 1980.

J. Hunter, *Harmonic Analysis over Imaginary Quadratic Number Fields.* Ph.D. Thesis, U.C.S.D., 1982.

N. Hurt, *Geometric Quantization in Action.* D. Reidel, Amsterdam, 1983.

N. Hurt, Propagators in quantum mechanics on multiply connected spaces, *Lecture Notes in Physics, 50.* Springer-Verlag, N.Y., 1976, 182–192.

N. Hurt & R. Hermann, *Quantum Statistical Mechanics and Lie Group Harmonic Analysis, Part A.* Math. Sci. Press, Brookline, MA. 1982.

A. Hurwitz, Über algebraische Korrespondenzen und das allgemeine Korrespondenzprinzip. *Math. Werke, I.* Eidgenössischen Technischen Hochschule, Basel, 1932, 163–188.

A. Hurwitz & R. Courant, *Funktionentheorie.* Springer-Verlag, N.Y., 1964.

J.-I. Igusa, *Theta Functions.* Springer-Verlag, N.Y., 1974.

———, On Siegel modular forms of genus 2, I, II. *Amer. J. Math., 84* (1962), 175–200; *86* (1964), 392–412.

Y. Ihara, Hecke polynomials as congruence ζ-functions in elliptic modular case. *Ann. of Math.,* (2)*85* (1967), 267–295.

K. Imai, Generalization of Hecke's correspondence to Siegel modular forms. *Amer. J. Math., 102* (1980), 903–936.

K. Imai & A. Terras, Fourier expansions of Eisenstein series for $GL(3, \mathbb{Z})$. *Trans. Amer. Math. Soc., 273* (1982), 679–694.

L. Infeld & T.E. Hull, *Rev. Mod. Phys., 23* (1951), 21.

A.E. Ingham, An integral which occurs in statistics. *Proc. Cambridge Phil. Soc., 29* (1933), 271–276.

K. Iwasawa, *Lectures on p-adic L-Functions*. Princeton University Press, Princeton, N.J., 1972.

———, On some types of topological groups. *Ann. of Math.*, *50* (1949), 507–558.

C.G.J. Jacobi, *Gesammelte Werke, Vol. I*. Chelsea, N.Y., 1969.

H. Jacquet, Les fonctions de Whittaker associées aux groupes de Chevalley. *Bull. Soc. Math. France*, *95* (1967), 243–309.

H. Jacquet & R.P. Langlands, Automorphic Forms on $GL(2)$, *Lecture Notes in Math. 114*. Springer-Verlag, N.Y., 1970.

H. Jacquet, I.I. Piatetskii-Shapiro & J. Shalika, Automorphic forms on $GL(3)$, I, II. *Ann. of Math.*, *109* (1979), 169–212, 213–258.

H. Jacquet & J. Shalika, A non-vanishing theorem for zeta functions of $GL(2)$. *Inv. Math.*, *38* (1976), 1–16.

———, On Euler products and the classification of automorphic representations I, II. *Amer. J. of Math.*, *103* (1981), 499–558, 777–815.

A.J. James, Special functions of matrix and single argument in statistics, in R. Askey (Ed.), *Theory and Applications of Special Functions*. Academic, N.Y., 1975, 497–520.

T. Janssen, *Crystallographic Groups*. North Holland-Elsevier, N.Y., 1973.

J.M. Jauch, Projective representations of the Poincaré group, in E.M. Loebl (Ed.), *Group Theory and Its Applications*. Academic, N.Y., 1968, 131–182.

H. & B. Jeffreys, *Methods of Mathematical Physics*. Cambridge University Press, Cambridge, 1972.

G.M. Jenkins & D.G. Watts, *Spectral Analysis and its Applications*. Holden-Day, San Francisco, 1968.

N. Jochnowitz, The index of the Hecke ring, T_k, in the ring of integers of $T_k \otimes \mathbb{Q}$. *Duke Math. J.*, *46* (1979), 861–869.

F. John, *Plane Waves and Spherical Means Applied to Partial Differential Equations*. Wiley-Interscience, N.Y., 1955.

G.A. Kabatiansky & V.I. Levenshtein, Bounds for packings on the sphere and in space. *Problems of Information Transmission*, *14* (1978), 1–17.

M. Kac, *Selected Papers*. M.I.T. Press, Cambridge, Mass., 1979, 474–496.

———, The search for the meaning of independence, in J. Gani, (Ed.), *The Making of Statisticians*. Springer-Verlag, N.Y., 1982, 62–72.

V.G. Kac, An elucidation of "infinite-dimensional algebras ... and the very strange formula." $E_8^{(1)}$ and the cube root of the modular invariant j. *Advances in Math.*, *35* (1980), 264–273.

———, Infinite-dimensional algebras, Dedekind's η-function, classical Möbius function and the very strange formula. *Advances in Math.*, *30* (1978), 85–136.

E.R. Kanasewich, *Time Sequence Analysis in Geophysics*. University of Alberta Press, Alberta, Canada, 1975.

M. Karel, Eisenstein series and fields of definition. *Compositio Math.*, *32* (1976), 225–291.

F.I. Karpelevich, V.N. Tutubalin, & M.G. Shur, Limit theorems for the compositions of distributions in the Lobachevsky plane and space. *Theory of Probability and its Applications*, *4* (1959), 399–402.

M. Kashiwara, A. Kowata, K. Minemura, K. Okamoto, T. Oshima, & M. Tanaka, Eigenfunctions of invariant differential operators on a symmetric space. *Ann. Math.*, *107* (1978), 1–39.

K. Katayama, Kronecker's limit formulas and their applications. *J. Fac. Sci. U. Tokyo*, *13* (1966), 1–44.

S. Kato, *Dynamics of the Upper Atmosphere*. Reidel, Amsterdam, 1980.

S. Katok, *Modular Forms Associated to Closed Geodesics and Arithmetic Applications*, Ph.D. thesis, University of Maryland, 1983.

N. Katz, An overview of Deligne's proof of the Riemann hypothesis for varieties over function fields (Hilbert's Problem 8). *Proc. Symp. in Pure Math.*, *Vol. 28*.

American Math. Soc., Providence, R.I., 1976, 286–288.
———, Sommes exponentielles. *Astérisque*, 79 (1980), available from Amer. Math. Soc. in U.S.
Y. Katznelson, *An Introduction to Harmonic Analysis*. Wiley, N.Y., 1968.
G. Kaufhold, Dirichletsche Reihen mit Funktionalgleichung in der Theorie der Modulfunktion 2. Grades. *Math. Ann.*, 137 (1959), 454–476.
O.-H. Keller, Geometrie der Zahlen. *Enzyklop. der Math. Wissenschaften*, *I.2.2. Aufl. Heft. 11, III*.
J.B. Keller & G.C. Papanicolaou, Stochastic differential equations with applications to random harmonic oscillators and wave propagation in random media. *SIAM J. Appl. Math.*, 21 (1971), 287–305.
J.F.C. Kingman, Random walks with spherical symmetry. *Acta Math.*, 109 (1963), 11–53.
A.B. Kirillov, *Elements of the Theory of Representations*. Springer-Verlag, N.Y., 1976.
———, Unitary representations of nilpotent Lie groups. *Russ. Math. Surveys*, 17 (1962), 53–104.
F. Klein & R. Fricke, *Vorlesungen über die Theorie der elliptischen Modulfunktionen, Vol. I*. Johnson Reprint Corp., N.Y., 1966.
H. Klingen, Volumbestimmung des Fundamentalbereichs der Hilbertschen Modulgruppe n-ten Grades. *J. für die Reine und Angew. Math.*, 206 (1961), 9–19.
———, Eisensteinreihen zur Hilbertschen Modulgruppe n-ten Grades. *Nachr. Akad. Wiss. Göttingen*, (1960), 87–104.
———, Uber die Werte der Dedekindschen Zetafunktion. *Math. Ann.*, 145 (1962), 265–272.
———, Zum Darstellungssatz für Siegelsche Modulformen, *Math. Z.*, 102 (1967), 30–43; *Correction*, 105 (1968), 399–400.
H.D. Kloosterman, Thetareihen in total reellen algebraischen Zahlkörpern. *Math. Ann.*, 103 (1930), 279–299.
M. Knopp, *Modular Functions in Analytic Number Theory*. Markham, Chicago, 1970.
D. Knuth, *The Art of Computer Programming, Vol. 2*. Addison-Wesley, Reading, Mass., 1981.
S. Kobayashi, *Hyperbolic Manifolds and Holomorphic Mappings*. Dekker, N.Y., 1970.
M. Koecher, Über Dirichlet-Reihen mit Funktionalgleichung. *J. Reine Angew. Math.*, 192 (1953), 1–23.
———, Über Thetareihen indefiniter quadratischer Formen. *Math. Nachr.*, 9 (1953), 51–85.
———, Gruppen und Lie Algebren rationaler Funktionen, to appear.
A.N. Kolmogorov, Une série de Fourier-Lebesgue divergente partout. *C.R. Acad. Sci. Paris, Ser. A.-B.*, 183 (1926), 1327–1328.
———, *Foundations of the Theory of Probability*. Chelsea, N.Y., 1950.
A.N. Kolmogorov & S.V. Fomin, *Introductory Real Analysis*. Dover, N.Y., 1975.
M.J. Kontorovich & N.N. Lebedev, *J. Exper. Theor. Phys. U.S.S.R.* 8 (1938), 1192–1206.
———, *Acad. Sci. U.S.S.R., J. Physics*, 1 (1939), 229–241.
———, *J. Exper. Theor. Phys. U.S.S.R.*, 9 (1939), 729–741.
A. Koranyi, A survey of harmonic functions on symmetric spaces. *Proc. Symp. Pure Math.*, 35. Amer. Math. Soc., Providence, R.I., 1979, 323–344.
J. Korevaar, *Mathematical Methods*. Academic, N.Y., 1968.
A. Korkine & G. Zolotareff, Sur les formes quadratiques positives. *Math. Ann.*, 11 (1877), 242–292.
B. Kostant, On Whittaker vectors and representation theory, *Inv. Math.*, 48 (1978), 101–184.
S. Kovalevsky, Sur le problème de la rotation d'un corps solide autour d'un point fixe. *Acta Math.*, 12 (1889), 177–232.
L. Kronecker, *Werke*. Leipzig, 1929.

T. Kubota, *Elementary Theory of Eisenstein Series*. Wiley, N.Y., 1973.

———, Some results concerning reciprocity law and real analytic automorphic functions. *Proc. Symp. Pure Math.*, *Vol. 20*. American Math. Soc., Providence, R.I., 1971, 382–395.

———, *On Automorphic Functions and the Reciprocity Law in a Number Field*, Lectures in Math. Kyoto University Kinokuniya Book Store Co. Ltd., Tokyo, Japan, 1969.

———, Über diskontinuierliche Gruppen Picardschen Typus und zugehörige Eisensteinsche Reihen. *Nagoya Math. J.*, *32* (1968), 259–271.

———, On a special kind of Dirichlet series. *J. Math. Soc. Japan*, 20 (1968), 193–207.

S. Kudla, Relations between automorphic forms produced by theta functions. *Lecture Notes in Math. 627*. Springer-Verlag, N.Y., 1977, 277–285.

———, Theta functions and Hilbert modular forms. *Nagoya Math. J.* 69 (1978), 97–106.

S. Kudla & J. Millson, Harmonic differentials and closed geodesics on Riemann surfaces. *Inv. Math.*, 54 (1979), 193–211.

N.V. Kuznetsov, Petersson's conjecture for cusp forms of weight zero and Linnik's conjecture. Sums of Kloosterman sums. *Math. U.S.S.R. Sbornik*, *39* (1981), 299–342.

G. Lachaud, Analyse spectrale des formes automorphes et séries d'Eisenstein, *Inv. Math.*, 46 (1978), 39–79.

J. Lagarias & A.M. Odlyzko, On computing Artin L-functions in the critical strip. *Math. Comp.*, 33 (1979), 1081–1095.

———, Effective versions of the Chebotarev density theorem, in A. Fröhlich (Ed.), *Algebraic Number Fields*. Academic, London, 1977, 377–407.

I. Lakatos, *Proofs and Refutations*. Cambridge University Press, Cambridge, 1976.

J.-L. Lagrange, *Oeuvres*, *Vols. I–XIV*. Gauthier-Villars, Paris, MDCCCXCII–MDCCCLVII.

A.L. Lance, *Introduction to Microwave Theory and Measurements*. McGraw-Hill, N.Y., 1964.

S. Lang, *Analysis I*. Addison-Wesley, Reading, Mass., 1968.

———, *Real Analysis*. Addison-Wesley, Reading, Mass., 1983.

———, *Algebraic Number Theory*. Addison-Wesley, Reading, Mass., 1968.

———, $SL(2, \mathbb{R})$. Addison-Wesley, Reading, Mass., 1975.

———, *Elliptic Functions*. Addison-Wesley, Reading, Mass., 1973.

———, *Introduction to Modular Forms*. Springer-Verlag, N.Y., 1976.

R.P. Langlands, *Eisenstein Series*. *Lecture Notes in Math. 544*. Springer-Verlag, N.Y., 1976.

———, *Base Change for GL(2)*. Princeton University Press, Princeton, N.J., 1980.

———, Problems in the theory of automorphic forms. *Lecture Notes in Maths.*, *Vol. 170*. Springer-Verlag, N.Y., 1970, 18–61.

———, The dimension of spaces of holomorphic forms. *Amer. J. Math.*, 85 (1963), 99–125.

P.S. De Laplace, Théorie des attractions des sphéroides et de la figure des planètes, *Mém de l'Acad.* (1782), Paris, 1785.

A.F. Lavrik, The principle of the theory of nonstandard functional equations for the Dirichlet functions. *Proc. Steklov Inst. Math.*, *132* (1973), 77–85.

———, On functional equations of Dirichlet functions. *Math. U.S.S.R. Izvestija*, *31* (1967), 421–432.

P. Lax & R. Phillips, *Scattering Theory for Automorphic Functions*. Princeton University Press, Princeton, N.J., 1976.

N.N. Lebedev, *Special Functions and their Applications*. Dover, N.Y., 1972.

———, Sur une formule d'inversion. *Dokl. Akad. Nauk. S.S.S.R.*, *52* (1946), 655–658.

P. Le Corbeiller, *Crystals and the Future of Physics. The World of Mathematics, Vol. II* (J.R. Newman, Ed.). Simon & Schuster, N.Y., 1956, 876.

A.M. Legendre, Recherches sur l'attraction des sphéroides homogènes. *Mém. math. phys. prés. à l'Acad. Aci. par. divers savantes, 10* (1785), 411–434.

———, Recherches sur la figure des planètes. *Mém. math. phys., Reg. de l'Acad. Sci. 1784, 1787,* 370–389.

D.H. Lehmer, Note on the distribution of Ramanujan's tau function. *Math. Comp.,* 24 (1970), 741–743.

———, Ramanujan's function $\tau(n)$. *Duke Math. J.,* 10 (1943), 483–492.

———, Some functions of Ramanujan. *Math. Student,* 27 (1959), 105–116.

———, Note on the distribution of Ramanujan's tau function. *Math. Comp.,* 24 (1970), 741–743.

J. Lehner, Discontinuous groups and automorphic functions. *Math. Survey 8.* American Math. Soc., Providence, R.I., 1964.

———, The Picard theorems, *Math. Monthly,* 76 (1969), 1005–1012.

E. Le Page, Théorèmes limites pour les produits de matrices aléatoires. *Lecture Notes in Math., 928.* Springer-Verlag, N.Y., 1982, 258–303.

G. Letac, Problèmes classiques de probabilité sur un couple de Gelfand. *Lecture Notes in Math. 861.* Springer-Verlag, N.Y., 1981, 93–120.

W. Leveque, *Reviews in Number Theory.* American Math. Soc., Providence, R.I., 1974.

B.M. Levitan & I.S. Sargsjan, *Introduction to Spectral Theory. American Math. Soc. Transl. of Math. Monographs, 39.* Amer. Math. Soc., Providence, R.I., 1975.

W.-C. W. Li, Hecke-Weil-Jacquet-Langlands theorem revisited, *Lecture Notes in Math. 751.* Springer-Verlag, N.Y., 206–220.

———, On a theorem of Hecke-Weil-Jacquet-Langlands, in H. Halberstam & C. Hooley (Eds.), *Recent Progress in Analytic Number Theory.* Academic, London, 1981, 119–152.

———, Newforms and functional equations. *Math. Ann., 212* (1975), 285–315.

S. Lichtenbaum, Values of zeta functions, etale cohomology, and algebraic K-theory. Preprint.

M.J. Lighthill, *Introduction to Fourier Analysis and Generalized Functions.* Cambridge University Press, Cambridge, 1978.

G. Lion & M. Vergne, *The Weil Representation, Maslov Index and Theta Series.* Birkhäuser, Boston, 1980.

A.L. Loeb, A systematic survey of cubic crystal structures. *J. Solid State Chem., 1* (1970), 237–267.

M. Loeve, *Probability Theory.* Van Nostrand, Princeton, N.J., 1963.

S. Łojasiewicz, Sur le problème de la division. *Stud. Math., 18* (1959), 87–136.

J.S. Lomont *Applications of Finite Groups.* Academic, N.Y., 1959.

K. Lonngren & A. Scott (Eds.), *Solitons in Action.* Academic, N.Y., 1978.

L.H. Loomis & S. Sternberg, *Advanced Calculus.* Addison-Wesley, Reading, Mass., 1968.

O. Loos, *Symmetric Spaces.* Benjamin, N.Y., 1969, Vols. I, II.

J.D. Louck & N. Metropolis, Hidden symmetry and the number-theoretic structure of the energy levels of a perturbed harmonic oscillator. *Adv. in Appl. Math., 1* (1980), 182–220.

A.K. Louis, Orthogonal function series expansions and the null space of the Radon transform. *S.I.A.M. J. Math. Analysis,* May, 1984.

———, Optimal sampling in nuclear magnetic resonance (NMR) tomography. *J. Computer Assisted Tomography, 6* (1982), 334–340.

D. Ludwig, The Radon transform on Euclidean space. *Comm. on Pure and Appl. Math., 29* (1966), 49–81.

H. Maass, *Lectures on Modular Functions of One Complex Variable.* Tata Institute of Fundamental Research, Bombay, 1964.

———, *Siegel's Modular Forms and Dirichlet Series. Lecture Notes in Math. 216.*

Springer-Verlag, N.Y., 1971.

———, Über eine neue Art von nichtanalytischen automorphen Funktionen und die Bestimmung Dirichletscher Reihen durch Funktionalgleichung. *Math. Ann.*, *121* (1949), 141–183.

———, Modulformen zweiten Grades und Dirichletreihen. *Math. Ann.*, *122* (1950), 90–108.

———, Die Primzahlen in der Theorie der Siegelschen Modulfunktionen. *Math. Ann.*, *124* (1951), 87–122.

———, Zur Theorie der automorphen Funktionen von n Veränderlichen. *Math. Ann.*, *117* (1940), 538–578.

———, Über Gruppen von hyperabelschen Transformationen. *Sitz.-Ber. der Heidelberg Akad. Wiss., Math.-Nat. Kl. 2 Abh.* (1940).

———, Modulformen zu indefiniten quadratischen Formen. *Math. Scand.*, *17* (1965), 41–55.

———, Automorphe Funktionen und indefinite quadratische Formen. *Sitz. Heidelberg Akad. Wiss., Math.-Nat. Kl. 1 Abh.* (1949).

———, Über die räumliche Verteilung der Punkte in Gittern mit indefiniter Metrik. *Math. Ann.*, *138* (1959), 287–315.

I.G. Macdonald, *Symmetric Functions and Hall Polynomials*. Clarendon Press, Oxford, 1979.

———, Affine root systems and Dedekind's eta function. *Inv. Math.*, *15* (1972), 91–143.

G. Mackey, *Unitary Group Representations in Physics, Probability, and Number Theory*. Benjamin/Cummings, Reading, Mass., 1978.

———, *Harmonic Analysis as the Exploitation of Symmetry—A Historical Survey*. Rice University Studies #64, Houston, Texas, 1978.

———, *The Theory of Group Representations*. University of Chicago Press, Chicago, 1976.

F.J. Macwilliams & N.J.A. Sloane, *The Theory of Error-Correcting Codes*. North-Holland, Amsterdam, 1978.

W. Magnus, *Noneuclidean Tesselations and Their Groups*. Academic, N.Y., 1974.

P. Magnusson, *Transmission Lines and Wave Propagation*. Allyn & Bacon, Boston, 1965.

K. Mahler, On Minkowski's theory of reduction of positive definite forms. *Quart J. Math.*, *9* (1938), 259–262.

J.I. Manin, Cyclotomic fields and modular curves, *Russ. Math. Surveys*, *26* (1971), 7–78.

———, Parabolic points and zeta functions of modular curves. *Izv. Akad. Nauk. S.S.S.R.*, *6* (1972), 19–64.

———, Correspondences, motives and monoidal transformations. *Mat. Sb.*, *6* (1968), 439–470.

———, Modular forms and number theory. *Proc. Internatl. Cong. Math., 1978, Helsinki*, Vol. I, 177–186.

J. Masley & H. Montgomery, Cyclotomic fields with unique factorization. *J. Reine Angew. Math.*, *286* (1976), 248–256.

A. Matsumoto (Ed.), *Microwave Filters and Circuits*. Academic, N.Y., 1970.

K. Maurin, *Analysis, Vols. I, II*. Reidel, Dordrecht, Holland, 1980.

———, *General Eigenfunction Expansions and Unitary Representations of Topological Groups*. Polish Scientific Publ., Warsaw, 1968.

F. Mautner, Spherical functions and Hecke operators, in *Lie groups and their Representations, Proc. Summer School of the Bolya Janos Math. Soc., Budapest, 1971*. Halsted, N.Y., 1975, 555–576.

———, Geodesic flows on symmetric Riemann spaces, *Ann. Math.*, *65* (1957), 416–431.

H.P. McKean, Selberg's trace formula as applied to a compact Riemann surface. *Comm. Pure & Appl. Math.*, 25 (1972), 225–246; 27 (1974), 134.

H.P. McKean, & E. Trubowitz, Hill's operator and hyperelliptic function theory in the presence of infinitely many branch points, *Comm. Pure Appl. Math.*, 29 (1976), 143–226.

F.G. Mehler, Über eine mit den Kugel und Cylinderfunctionen verwandte Funktion und ihre Anwendung in der Theorie der Elektricitätsvertheilung, *Math. Ann.*, 18 (1881), 161–194.

M.L. Mehta, *Random Matrices and the Statistical Theory of Energy Levels*. Academic, N.Y., 1967.

J. Mennicke, *Vorträge über Selbergs Spurformel I*. University of Bielefeld, W. Germany.

———, Lectures on discontinuous groups on 3-dimensional hyperbolic space, Modular Form Conf., Durham, 1983.

A. Messiah, *Quantum Mechanics, Vols. I, II*, Wiley, N.Y., 1961, 1962.

C. Meyer, *Die Berechnung der Klassenzahl abelscher Körper über quadratischen Zahlkörpern*, Akademie-Verlag, Berlin, 1957.

J. Milnor, Hilbert's problem 18, *Proc. Symp. Pure Math., Vol. 28.* American Math. Soc., Providence, R.I., 1976, 491–506.

———, Hyperbolic geometry: the first 150 years, *Bull. American Math. Soc. N.S.*, 6 (1982), 9–24.

J. Milnor & D. Husemoller, *Symmetric Bilinear Forms*. Springer-Verlag, N.Y., 1973.

S. Minakshisundaram & A. Pleijel, Some properties of the eigenvalues of the Laplacian. *J. Diffl. Geom.*, 1 (1967), 43–69.

H. Minkowski, *Gesammelte Abhandlungen*. Chelsea, N.Y., 1911 (reprinted 1967).

———, *Geometrie der Zahlen*. Leipzig & Berlin, 1896.

C.W. Misner, K.S. Thorne, & J.A. Wheeler, *Gravitation*. W.H. Freeman, San Francisco, 1973.

R. Moeckel, Geodesics on modular surfaces and continued fractions. *Ergodic Theory and Dynamical Systems*, 2 (1982), 69–83.

S.A. Molchanov, Diffusion processes and Riemannian geometry. *Russ. Math. Surveys*, 30 (1975), 1–63.

M.I. Monastyrsky & A.M. Perelomov, Coherent states and bounded homogeneous domains. *Reports on Math. Phys.* 6 (1974), 1–14.

H. Montgomery, *Adam's Prize Essay*. Preprint.

———, The pair correlation of zeros of the zeta function, *Proc. Symp. Pure Math. 24.* A.M.S., Providence, R.I., 1973, 181–193.

C.C. Moore, Representations of solvable and nilpotent groups and harmonic analysis on nil and solvmanifolds. *Proc. Symp. Pure Math. 26.* A.M.S., Providence, R.I., 1973, 3–44.

———, Compactifications of symmetric spaces. *Amer. J. Math.*, 86 (1964), 201–218.

L.J. Mordell, On Mr. Ramanujan's empirical expansions of modular functions. *Proc. Cambridge Phil. Soc.*, 19 (1917), 117–124.

———, On Hecke's modular functions, zeta functions, and some other analytic functions in the theory of numbers. *Proc. London Math. Soc.* (2)32 (1931), 501–556.

C. Moreno, Explicit formulas in the theory of automorphic forms. *Lecture Notes in Math. 626.* Springer-Verlag, N.Y., 1977, 73–216.

C. Moreno & F. Shahidi, The fourth moment of the Ramanujan tau function. *Math. Ann.*, 266 (1983), 233–239.

Y. Morita, An explicit formula for the dimension of spaces of Siegel modular forms of degree 2, *J. Fac. Sci. U. Tokyo*, 21 (1974), 167–248.

D.F. Morrison, *Multivariate Statistical Methods*. McGraw-Hill, N.Y., 1976.

P. Morse & H. Feshbach, *Methods of Theoretical Physics*, Vols. *I*, *II*. McGraw-Hill, N.Y., 1953.
G. Mostow, Some new decomposition theorems for semisimple Lie groups. *Memoirs Amer. Math. Soc.*, *14* (1955), 31–54.
G.D. Mostow & T. Tamagawa, On the compactness of arithmetically defined homogeneous spaces. *Ann. Math.*, *76* (1962), 440–463.
R.J. Muirhead, Systems of partial differential equations for hypergeometric functions of matrix argument. *Ann. Math. Stat.*, *41* (1970), 991–1001.
———, *Aspects of Multivariate Statistical Theory*. Wiley, N.Y., 1978.
C. Müller, *Spherical Harmonics, Lecture Notes in Math. 17*. Springer-Verlag, N.Y., 1966.
D. Mumford, *Curves and Their Jacobians*. University of Michigan Press, Ann Arbor, Mich., 1976.
———, *Geometric Invariant Theory*. Springer-Verlag, N.Y., 1965.
———, On the equations defining Abelian Varieties, I, II, III, *Inv. Math.*, *1* (1966), 287–354; *3* (1967), 75–135, 215–244.
———, *Tata Lectures on Theta*, *I*, *II*. Birkhäuser, Boston, 1982, 1984.
M. Ram Murty, Oscillations of Fourier coefficients of modular forms. *Math. Ann.*, *262* (1983), 431–446.
———, On the Sato-Tate conjecture, in N. Koblitz (Ed.), *Number Theory Related to Fermat's Last Theorem*. Birkhäuser, Boston, 1982, 195–205.
M.A. Naimark, *Linear Differential Operators*, Vols. *I*, *II*. Ungar, N.Y., 1968.
Y. Namikawa, *Toroidal Compactification of Siegel Spaces, Lecture Notes in Math., 812*. Springer-Verlag, N.Y., 1980.
W. Narkiewicz, *Algebraic Numbers*. Polish Sci. Pub., Warsaw, 1973.
Z. Nehari, *Conformal Mapping*. McGraw-Hill, N.Y., 1952.
E. Nelson, *Dynamical Theories of Brownian Motion*. Princeton University Press, Princeton, N.Y., 1967.
H. Neuenhöffer, Über die analytische Fortsetzung von Poincaréreihen. *Sitz. Heidelberg Akad. Wiss.* (1973).
S.P. Novikov, A method for solving the periodic problem for the KdV equation and its generalizations. *Rocky Mt. J. of Math.*, *8* (1978), 83–93.
M.E. Novodvorsky & I.I. Piatetskii-Shapiro, Rankin-Selberg method in the theory of automorphic forms. *Proc. Symp. Pure Math. 30*. American Math. Soc., Providence, R.I., 1976, 297–301.
A. Nussbaum, *Applied Group Theory for Chemists, Physicists, and Engineers*. Prentice-Hall, Englewood Cliffs, N.J., 1971.
F. Oberhettinger, *Tables of Mellin Transforms*. Springer-Verlag, N.Y., 1974.
———, *Tables of Bessel Transforms*. Springer-Verlag, N.Y., 1972.
———, *Fourier Transforms of Distributions and Their Inverses: A Collection of Tables*. Academic, N.Y., 1973.
———, *Fourier Expansions: A Collection of Formulas*. Academic, N.Y., 1973.
F. Oberhettinger & L. Badii, *Tables of Laplace Transforms*. Springer-Verlag, N.Y., 1973.
A. Ogg, *Modular Forms and Dirichlet Series*. Benjamin, N.Y., 1969.
———, Survey of modular functions of one variable. *Lecture Notes in Math. 320*. Springer-Verlag, N.Y., 1973, 1–35.
———, A remark on the Sato-Tate conjecture, *Inv. Math.*, *9* (1970), 198–200.
F.W.J. Olver, *Asymptotics and Special Functions*. Academic, N.Y., 1974.
M.S. Osborne & G. Warner, *The Theory of Eisenstein Systems*. Academic, N.Y., 1981.
G.C. Papanicolaou, Wave propagation in a one-dimensional random medium. *SIAM J. Appl. Math.*, *21* (1971), 13–18.
S.J. Patterson, A lattice-point problem in hyperbolic space, *Mathematika*, *22* (1975), 81–88; *23* (1976), 227.

H. Petersson, Über eine Metrisierung der ganzen Modulformen. *Jber. Deutsche Math. Verein.*, *49* (1939), 49–75.

———, Über die Entwicklungskoeffizienten der Automorphen Formen. *Acta Math.*, *58* (1932), 170–215.

W. Pfetzer, Die Wirkung der Modulsubstitutionen auf mehrfache Thetareihen zu ungerader Variablenzahl, *Arch. Math.* 6 (1953), 448–454.

R. Phillips & P. Sarnak, *On cusp forms for co-finite subgroups of PSL*(2, ℝ). Preprint.

I.I. Piatetskii-Shapiro, Euler subgroups, Lie groups and their representations, in I.M. Gelfand (Ed.), *Summer School of the Bolyai Janos Math. Soc.* Halsted, N.Y., 1975, 597–620.

———, *Automorphic Functions and the Geometry of the Classical Domains*. Gordon and Breach, N.Y., 1969.

———, Cuspidal automorphic representations associated to parabolic subgroups and Ramanujan conjecture, in N. Koblitz (Ed.), *Number Theory related to Fermat's Last Theorem*. Birkhäuser, Boston, 1982.

E. Picard, Sur un groupe de transformations des points de l'espace situés du meme coté d'un plan. *Bull. Soc. Math. de France*, *12* (1844), 43–47.

H. Poincaré, *Oeuvres*, *Vols. I–XI*. Gauthier-Villars, Paris, 1951–53.

L. Pontryagin, *Topological Groups*. Moscow, 1957.

S.J. Press, *Applied Multivariate Analysis*. Holt, Rinehart & Winston, N.Y., 1972.

N.V. Proskurin, Expansions of automorphic functions. *Proc. Steklov Inst. Math.*, *116* (1982), 119–141.

N.J. Pullman, *Matrix Theory and its Applications*. Dekker, N.Y., 1976.

G. Purdy, R. & A. Terras, & H. Williams, Graphing L-Functions of Kronecker Symbols in the Real Part of the Critical Strip. *Math. Student*. (Journal issue dated 1979 but in press as of Nov., 1984.)

H. Rademacher, *Topics in Analytic Number Theory*. Springer-Verlag, N.Y., 1973.

———, Eine Bemerkung über die Heckeschen Operatoren $T(n)$. *Abh. Math. Sem. U. Hamburg*, *31* (1967), 149–151.

J. Radon, Über die Bestimmung von Funktionen durch ihre Integralwerte längs gewisser Mannigfaltigkeiten. *Berichte Sächsische Akad. Wissen. zu Leipzig*, *69* (1917), 262–277.

M.S. Raghunathan *Discrete Subgroups of Lie Groups*. Springer-Verlag, N.Y., 1972.

K. Ramachandra, Some applications of Kronecker's limit formula, *Ann. of Math*, *80* (1964), 104–148.

K.G. Ramanathan, Quadratic forms over involutorial division algebras II. *Math. Ann.*, *143* (1961), 293–332.

———, Zeta functions of quadratic forms. *Acta Arith.*, *7* (1961), 39–69.

B. Randol, Small eigenvalues of the Laplace operator on compact Riemann surfaces, *Bull. American Math. Soc*, *80* (1974), 996–1000.

———, A remark on the multiplicity of the discrete spectrum of congruence subgroups. *Proc. Amer. Math. Soc.*, *81* (1981), 339–340.

R.A. Rankin, *Modular Forms and Functions*. Cambridge University Press, Cambridge, 1977.

———, On horocyclic groups, *Proc. London Math. Soc.*, (3)4 (1954), 219–234.

———, Contributions to the theory of Ramanujan's function $\tau(n)$ and similar arithmetical functions, I, II, III. *Proc. Cambridge Phil. Soc.*, *35* (1939), 351–356, 357–372; *36* (1940), 150–151.

R. Rankin & J.M. Rushforth, The coefficients of certain integral modular forms, *Proc. Cambridge Phil. Soc.*, *50* (1954), 305–308.

H. Rauch & A. Lebowitz, *Elliptic Functions, Theta Functions, and Riemann Surfaces*. Williams & Wilkins, Baltimore, Maryland, 1973.

J.W.S. Rayleigh, On the character of the complete radiation at a given temperature, *Phil. Mag.*, *27* (1889) and *Scientific Papers*. Cambridge University Press, Cambridge, 1902, and Dover, N.Y., 1964, Vol. 3, 273.

M.J. Razar, Values of Dirichlet series at integers in the critical strip. *Lecture Notes in Math.*, *627*. Springer-Verlag, N.Y., 1977, 1–10.

———, Modular forms for $\Gamma_0(N)$ and Dirichlet series. *Trans. Amer. Math. Soc. 231* (1977), 489–495.

M. Reed & B. Simon, *Methods of Modern Mathematical Physics, Vol. I, Functional Analysis.* Academic, N.Y., 1972.

A. Regev, letter of 8/30/79, from Math. Dept., Weizmann Inst. Science, Rehovot, Israel.

R. Remak, Über die Minkowskische Reduktion der definiten quadratischen Formen. *Compositio Math.*, *5* (1938), 368–391.

H.L. Resnikoff, On the graded ring of Hilbert modular forms associated with $\mathbb{Q}(\sqrt{5})$. *Math. Ann.*, *203* (1974), 161–170.

———, Automorphic forms of singular weight are singular forms. *Math. Ann.*, *215* (1975), 175–193.

———, Theta functions for Jordan algebras. *Inv. Math.*, *31* (1975), 87–104.

———, Differential geometry and color perception. *J. Math. Biology*, 1 (1974), 97–131.

H.L. Resnikoff & Y.-S. Tai, On the structure of a graded ring of automorphic forms on the 2-dimensional complex ball, I, II. *Math. Ann.*, *238* (1978), 97–117; 258 (1982), 367–382.

B. Riemann, *Collected Works.* Dover, N.Y., 1953.

L. Robin, *Functions Sphériques de Legendre et Functions Sphéroidales, Vol. 3.* Gauthier-Villars, Paris, 1959.

W. Roelcke, Das Eigenwertproblem der automorphen Formen in der hyperbolischen Ebene I, II, *Math. Ann.*, *167* (1966), 292–337; *168* (1967), 261–324.

———, Über die Wellengleichung bei Grenzkreisgruppen erster Art. *Sitzber. Akad. Heidelberg, Math.-naturwiss. Kl. 1953/55.*

———, Über den Laplace-Operator auf Riemannschen Mannigfaltigkeiten mit diskontinuierlichen Gruppen. *Math. Nachr.*, *21* (1960), 131–149.

C.A. Rogers, *Packing and Covering.* Cambridge Univerisity Press, Cambridge, 1964.

J. Rosenberg, A quick proof of Harish-Chandra's Plancherel theorem for spherical functions on a semisimple Lie group. *Proc. Amer. Math. Soc.*, *63* (1977), 143–149.

S.N. Roy, P-statistics or some generalizations in analysis of variance appropriate to multivariate problems. *Sankyha*, *4* (1939), 381–396.

D. Ruelle, Zeta functions and statistical mechanics. *Astérisque*, *40* (1976), 167–176.

W. Rühl, *The Lorentz Group and Harmonic Analysis.* Benjamin, N.Y., 1970.

———, An elementary proof of the Plancherel theorem for the classical groups. *Comm. Math. Phys.*, *11* (1969), 297–302.

S.S. Ryskov, The geometry of positive quadratic forms. *A.M.S. Transl. (2) 109.* A.M.S., Providence, R.I., 1977, 27–32.

S.S. Ryskov & E.P. Baranovskii, Classical methods in the theory of lattice packings, *Russ. Math. Surveys*, *34* (1979), 1–68.

A.A. Sagle & R.E. Walde, *Introduction to Lie Groups and Lie Algebras.* Academic, N.Y., 1973.

H. Saito, *Automorphic Forms and Algebraic Extensions of Number Fields. Lectures in Math. 8.* Kinokuniya Book Store, Tokyo, Japan, 1975.

Y. Sakamoto, Calculation of Madelung's coefficient of NaCl, *J. Sci. Hiroshima U., A, 16* (1953), 569.

P. Samuel, *Algebraic Theory of Numbers.* Houghton-Mifflin, Boston, 1970.

P. Sarnak, *Prime Geodesic Theorems.* Thesis, Stanford University, 1980.

———, Asymptotic behavior of periodic orbits of the horocycle flow and Eisenstein series. *Comm. Pure Appl. Math.*, *34* (1981), 719–739.

———, Class numbers of indefinite binary quadratic forms. *J. Number Theory, 15*

(1982), 229–247.
———, The arithmetic and geometry of some hyperbolic three-manifolds. *Acta Math.* *5* (1984), 253–295.
I. Satake, Theory of spherical functions on reductive algebraic groups over p-adic fields. *Publ. Math.* *18*. Inst. des Hautes Études Scient., Paris, 1963.
———, Review of Ash, Mumford, Rapoport and Tai in *Math. Reviews*, *56*, #15642.
———, On the compactification of the Siegel space. *J. Indian Math. Soc.*, *20* (1956), 259–281.
G. Schiffman, Intégrales d'entrelacement et fonctions de Whittaker. *Bull. Soc. Math. France*, *99* (1971), 3–72.
B. Schoeneberg, *Elliptic Modular Functions*. Springer-Verlag, N.Y., 1974.
———, Das Verhalten von mehrfachen Thetareihen bei Modulsubstitutionen. *Math. Ann.*, *116* (1939), 511–523.
A. Schuster, On the investigation of hidden periodicities with application to a supposed 26 day period of meteorological phenomena, *Terr. Magn.*, *3* (1898), 13–41.
E. Schwandt, Non-degenerate wavefunctions with wave parameter $\rho = 1$, *J. London Math. Soc.*, *3* (1971), 183–186.
H.L. Schwartz, *Mathematics for the Physical Sciences*. Hermann, Paris, 1966.
———, *Théorie des Distributions*. Hermann, Paris, 1959.
———, Some applications of the theory of distributions, in T.L. Saaty (Ed.), *Lectures on Modern Mathematics. Wiley, N.Y., 1963, 23–58.*
R.L.E. Schwarzenberger, *N-dimensional Crystallography*. Pitman, San Francisco, 1980.
J. Schwermer, Eisensteinreihen und die Kohomologie von Kongruenzuntergruppen von $SL(n, \mathbb{Z})$, *Bonner Math. Schriften*, *99* (1977).
L.A. Seeber, *Untersuchungen über die Eigenschaften der positive ternaren quadratischen Formen*. Freiburg, 1831.
A. Selberg, Harmonic analysis and discontinuous groups in weakly symmetric Riemannian spaces with applications to Dirichlet series. *J. Indian Math. Soc.*, *20* (1956), 47–87.
———, On the estimation of Fourier coefficients of modular forms. *Proc. Symp. Pure Math.* *8*. American Math. Soc., Providence, R.I., 1965, 1–15.
———, Discontinuous groups and harmonic analysis. *Proc. Internatl. Cong. of Math.*, Stockholm, 1962, 177–189.
———, Remarks on a multiple integral. *Norsk Mat. Tidsskr.*, *26* (1944), 71–78 (in Norwegian).
———, Bemerkungen über eine Dirichletsche Reihe, die mit der Theorie der Modulformen nahe verbunden ist. *Arch. Math. Naturvid.*, *43* (1940), 47–50.
———, *Lectures on the Trace Formula*. University of Göttingen, 1954.
———, A new type of zeta function connection with quadratic forms. Report of the Institute in The Theory of Numbers, Univ. of Colorado, Boulder, Colorado, 1959, 207–210.
Séminaire Henri Cartan, 1957/1958, *Fonctions Automorphes*. Benjamin, N.Y., 1967.
J.-P. Serre, *A Course in Arithmetic*. Springer-Verlag, N.Y., 1973.
———, Modular forms of weight one and Galois representations, in A. Fröhlich (Ed.), *Algebraic Number Fields*. Academic, N.Y., 1977, 193–268.
———, Cohomologie des groupes discrets. *Prospects in Math.*, Princeton University Press, Princeton, N.J., 1971.
———, Lectures on coefficients of modular forms, Modular Forms Conf., Durham, 1983.
———, *Abelian l-adic Representations and Elliptic Curves*. Benjamin, N.Y., 1968.
J.-P. Serre & H.M. Stark, Modular forms of weight $\frac{1}{2}$. *Lecture Notes in Math. 627*. Springer-Verlag, N.Y., 1977, 28–67.

I.R. Shafarevitch, *Basic Algebraic Geometry*. Springer-Verlag, N.Y., 1974.

F. Shahidi, On certain L-series. *Amer. J. Math.*, *103*. (1981), 297–355.

D. Shale, Linear symmetries of free boson fields. *Trans. Amer. Math. Soc.*, *103* (1962), 149–167.

J.A. Shalika, The multiplicity one theorem for GL(n), *Ann. of Math.*, *100* (1974), 171–193.

D. Shanks, Five number-theoretic algorithms, in *Proc. 2nd Manitoba Conf. Numerical Math.*, *Winnipeg, Manitoba, 1972*, 51–70.

C.E. Shannon, Communication in the presence of noise. *Proc. I.R.E.*, *37* (1949), 10–21.

L.A. Shepp & J.B. Kruskal, Computerized tomography: the new medical x-ray technology. *Amer. Math. Monthly*, *85* (1978), 420–429.

H. Shimizu, On discontinuous groups operating on the product of the upper half planes. *Ann. of Math.*, *77* (1963), 33–71.

G. Shimura, *Introduction to the Arithmetic Theory of Automorphic Functions*. Princeton University Press, Princeton N.J., 1971.

―――――, On modular forms of half integral weight. *Ann. of Math.*, *97* (1973), 440–481.

―――――, On the holomorphy of certain Dirichlet series. *Proc. London Math. Soc.*, *31* (1975), 79–98.

―――――, Sur les intégrales attachées aux formes automorphes. *J. Math. Soc. Japan*, *11* (1959), 291–311.

T. Shintani, On evaluation of zeta functions of totally real algebraic number fields at non-positive integers. *J. Fac. Sci. U. Tokyo*, *23* (1976), 393–417.

―――――, A proof of the classical Kronecker limit formula. *Tokyo J. Math.*, *3* (1980), 191–199.

―――――, On special values of zeta functions of totally real algebraic number fields. *Proc. Internatl. Cong. of Math.*, *Helsinki, 1978*, 591–597.

―――――, On zeta-functions associated with the vector space of quadratic forms. *J. Fac. Sci. U. Tokyo*, *22* (1975), 25–65.

―――――, On "liftings" of holomorphic automorphic forms (a representation theoretic interpretation of the recent work of Saito). *U.S.-Japan Sem.*, Ann Arbor, Michigan, 1975.

―――――, On an explicit formula for class 1 "Whittaker functions" on GL_n over $\wp$-adic fields. *Proc. Japan Acad.*, *52* (1976), 180–182.

C.L. Siegel, *Topics in Complex Function Theory*. Wiley-Interscience, N.Y., 1969–73.

―――――, *Lectures on Quadratic Forms*. Tata Institute of Fundamental Research, Bombay, 1957.

―――――, *Analytische Zahlentheorie, Lecture Notes*. University of Göttingen, 1963–64.

―――――, *Lectures on Advanced Analytic Number Theory*. Tata Institute of Fundamental Research, Bombay, 1963.

―――――, *Gesammelte Abhandlungen, Vols. I–IV*. Springer-Verlag, N.Y., 1966, 1979.

―――――, *Geometry of Numbers*. New York University Lectures, 1945–46.

―――――, On the history of the Frankfurt mathematics seminar. *Math. Intelligencer*, *1* (1979), 223–232.

G.J. Simmons, Cryptology: The mathematics of secure communication. *Math. Intelligencer*, *1* (1979), 233–246.

Ya. G. Sinai, *Introduction to Ergodic Theory*. Princeton University Press, Princeton, 1977.

I.M. Singer, Eigenvalues of the Laplacian and invariants of manifolds. *Proc. Internatl. Cong. of Math.*, *Vancouver, 1974, Vol. I*, 187–200.

I.M. Singer & J.A. Thorpe, *Lectures on Elementary Topology and Geometry*. Scott, Foresman & Co., Glenview, Illinois, 1967.

D. Slepian & H.O. Pollak, Prolate spheroidal wave functions, Fourier analysis &

uncertainty, I, II. *Bell System Tech. J. 1* (1961), 43–84.

N.J.A. Sloane, Binary codes, lattices, and sphere-packings. *Combinatorial Surveys, Proc. 6th British Comb. Conf.* Academic, London, 1977, 117–164.

————, Error-correcting codes and invariant theory: New applications of a 19th century technique. *Math. Monthly, 84* (1977), 82–107.

————, The packing of spheres, *Scientific American, 250* (1984), 116–125.

R.A. Smith, The L^2-norm of Maass wave functions, *Proc. Amer. Math. Soc., 82* (1981), 179–182.

I.H. Sneddon, *The Use of Integral Transforms.* McGraw-Hill, N.Y., 1972.

L. Solomon, Partially ordered sets with colors. *Proc. Symp. Pure Math. 34,* A.M.S., Providence, R.I., 1979, 309–329.

A. Sommerfeld, *Partial Differential Equations in Physics.* Academic, N.Y., 1949.

C. Soulé, The cohomology of $SL(3, \mathbb{Z})$. *Topology, 17* (1978), 1–22.

R. Spira, Calculation of Dirichiet L-functions, *Math. Comp. 23* (1969), 489–497.

M. Spivak, *Calculus on Manifolds.* Benjamin, N.Y., 1965.

————, *A Comprehensive Introduction to Differential Geometry, Vol. II.* Publish or Perish, Berkeley, Ca., 1979.

F.D. Stacey, *Physics of the Earth.* Wiley, N.Y., 1977.

I. Stakgold, *Green's Functions and Boundary Value Problems.* Wiley N.Y., 1979.

————, *Boundary Value Problems of Mathematical Physics, Vols. I, II.* MacMillan, N.Y., 1967.

J.A. Staniforth, *Microwave Transmission.* The English Universities Press, London, 1972.

R.J. Stanton & R.A. Tomas, Expansions for spherical functions on noncompact symmetric spaces. *Acta Math.* , *140* (1978), 251–276.

H. Stark, The analytic theory of algebraic numbers. *Bull. American Math. Soc., 81* (1975), 961–972.

————, On the problem of unique factorization in complex quadratic fields. *Proc. Symp. Pure Math.* , *7.* American Math. Soc., Providence, 1969, 41–56.

————, Values of L-functions at s = 1, I–IV. *Advances in Math 7* (1971), 301–343; *17* (1975), 60–92; *22* (1976), 64–84; *35* (1980), 197–235.

————, Class fields and modular forms of weight one. *Lecture Notes in Math. 601.* Springer-Verlag, N.Y., 1977, 277–287.

————, M.I.T. & U.C.S.D. number theory course lecture notes (unpublished).

————, On the zeros of Epstein's zeta function. *Mathematika, 14* (1967), 47–55.

————, On the transformation formula for the symplectic theta function and applications. *J. Fac. Sci. U. Tokyo, 29* (1982), 1–12.

————, On modular forms from L-functions in number theory, I, II, to appear.

————, The Coates-Wiles theorem revisited, in N. Koblitz (Ed.), *Number Theory Related to Fermat's Last Theorem.* Birkhäuser, Boston, 1982, 349–362.

————, Fourier coefficients of Maass wave forms, in R.A. Rankin (Ed.), *Modular Forms.* Horwood, Chichester (distrib. Wiley), 1984, 263–269.

E. Stein & G. Weiss, *Fourier Analysis on Euclidean Spaces.* Princeton University Press, Princeton, N.J., 1971.

T.J. Stieltjes, Recherches sur les fractions continues. *Ann. Fac. Sci. Toulouse, 8* (1894), 1–22.

G. Strang, *Linear Algebra and its Applications.* Academic, N.Y., 1976.

G. Strang & G.J. Fix, *An Analysis of the Finite Element Method.* Prentice-Hall, Englewood Cliffs, N.J., 1973.

H. Strassberg, L-functions for $GL(n)$. *Math. Ann., 245* (1979), 23–36.

R. Styer, *Hecke Theory over Complex Quadratic Fields.* Ph.D. Thesis, M.I.T., 1981.

N. Subia, Formule de Selberg et formes d'espaces hyperboliques compactes. *Lecture Notes in Math. 497.* Springer-Verlag, N.Y., 674–700.

M. Sugiura, *Unitary Representations and Harmonic Analysis.* Wiley, N.Y., 1975.

P. Swinnerton-Dyer, Conjectures of Birch and Swinnerton-Dyer and of Tate, in T.A. Springer (Ed.), *Proc. Conf. Local Fields.* Springer-Verlag, N.Y., 1967, 132–157.

G. Szegö, Über einige asymptotische Entwicklungen der Legendreschen Funktionen. *Proc. London Math. Soc*, *36* (1932), 427–450.

L.A. Takhtadzhyan & A.I. Vinogradov, Theory of Eisenstein series for the group $SL(3, \mathbb{R})$ and its application to a binary problem. *J. Soviet Math.*, *18* (1982), 293–324.

———, The Gauss-Hasse hypothesis on real quadratic fields with class number one. *J. Reine Angew. Math.*, *335* (1982), 40–86.

J.D. Talman, *Special Functions.* Benjamin, N.Y., 1968.

T. Tamagawa, On Selberg's trace formula. *J. Fac. Sci. U. Tokyo, Sec. I*, *8* (1960), 363–386.

———, On the zeta-functions of a division algebra. *Ann. of Math.*, (2)*77* (1963), 387–405.

———, On some extensions of Epstein's Z-series. *Proc. Internatl. Symp. on Alg. No. Theory.* Tokyo-Nikko, 1955, 259–261.

Y. Tanigawa, Selberg trace formula for Picard groups. *Proc. Intl. Symp. Number Theory.* Japan Society for the Promotion of Science, Tokyo, 1977, 229–242.

J. Tate, The work of David Mumford. *Proc. Internatl. Cong. of Math.*, *Vancouver, 1974, Vol. I.* 11–15.

———, The general reciprocity law. *Proc. Symp. Pure Math. 28.* American Math. Soc., Providence, R.I., 1976, 311–322.

O. Taussky, On a theorem of Latimer and MacDuffee. *Canadian Math. J.*, *1* (1949), 300–302.

———, On matrix classes corresponding to an ideal and its inverse. *Illinois J. Math.*, *1* (1957), 108–113.

A. Terras, Applications of special functions for the general linear group to number theory. *Sém. Delange-Pisot-Poitou, 1976–1977*, exp. 23.

———, Integral formulas and integral tests for series of positive matrices. *Pacific J. of Math.*, *89* (1980), 471–490.

———, The minima of quadratic forms and the behavior of Epstein and Dedekind zeta functions. *J. Number Theory*, *12* (1980), 258–272.

———, Bessel series expansions of the Epstein zeta function and the functional equation. *Trans. American Math. Soc.*, *183* (1973), 477–486.

———, Some formulas for the Riemann zeta function at odd integer argument resulting from Fourier expansions of the Epstein zeta function. *Acta Math.*, *29* (1976), 181–189.

———, A generalization of Epstein's zeta function. *Nagoya Math. J.*, *42* (1971), 173–188.

———, Functional equations of generalized Epstein zeta functions in several variables. *Nagoya Math. J.*, *44* (1971), 89–95.

———, On automorphic forms for the general linear group. *Rocky Mt. J. of Math.*, *12* (1982), 123–143.

———, Fourier-coefficients of Eisenstein series of one complex variable for the special linear group. *Trans. Amer. Math. Soc.*, *205* (1975), 97–114.

———, The Fourier expansion of Epstein's zeta function for totally real algebraic number fields and some consequences for Dedekind's zeta function. *Acta Arith.*, *30* (1976), 187–197.

———, The Fourier expansion of Epstein's zeta function over an algebraic number field and its consequences for algebraic number theory. *Acta Arith.*, *32* (1977), 37–53.

———, A relation between $\zeta(s)$ and $\zeta(s - 1)$ for any algebraic number field, in A. Frohlich (Ed.), *Algebraic Number Fields.* Academic, N.Y., 1977, 475–483.

———, Special functions for the symmetric space of positive matrices. *SIAM J.*

Math. Analysis, in press.

R. Terras, On the convergence of the continued fraction for the incomplete gamma function and an algorithm for the Riemann zeta function. Preprint.

―――, The determination of incomplete gamma functions through analytic integration. *J. Comp. Phys.*, *31*, (1979), 146–151.

―――, Generalized exponential operators in the continuation of the confluent hypergeometric functions. *J. Comp. Physics*, *44* (1981), 156–166.

E.C. Titchmarsh, *Introduction to the Theory of Fourier Integrals*. Oxford University Press, Oxford, 1937.

―――, *Eigenfunction Expansions Associated with Second Order Differential Equations*, Vols. I, II. Oxford University Press, Oxford, 1946.

―――, *The Theory of Riemann's Zeta Function*. Oxford University Press, Oxford, 1951.

J.S. Trefil, *Introduction to the Physics of Fluids and Solids*. Pergamon, N.Y., 1975.

F. Treves, Applications of distributions to PDE theory. *Amer. Math. Monthly*, *77* (1970), 241–248.

K. Trimèche, Probabilitiés indéfiniment divisibles et théorème de la limite centrale pour une convolution généralisée sur la demi-droite, *C.R. Acad. Sci. Paris*, *286* (1978), 63–66.

L.-C. Tsao, The rationality of the Fourier coefficients of certain Eisenstein series on tube domains. *Comp. Math.*, *32* (1976), 225–291.

S. Tsuyumine, Construction of modular forms by means of transformation formulas for theta series. *Tsukuba J. Math.*, *3* (1979), 59–80.

J. Tunnell, Artin's conjecture for representations of octahedral type. *Bull. Amer. Math. Soc.*, *5* (1981), 173–175.

―――, On the local Langlands conjecture for $GL(2)$. *Inv. Math.*, *46* (1978), 179–200.

V.N. Tutubalin, On limit theorems for the product of random matrices. *Theory of Probability and its Applications*, *15* (1965), 15–27.

A. Van Der Poorten, A proof that Euler missed ... Apery's proof of the irrationality of $\zeta(3)$. *Math. Intelligencer*, *1* (1979), 195–203.

B.L. Van Der Waerden, *Group Theory and Quantum Mechanics*. Springer-Verlag, N.Y., 1974.

―――, *Studien zur Theorie der Quadratischen Formen*. Birkhäuser, Basel, 1968.

―――, Die Reduktionstheorie der positiven quadratischen Formen. *Acta Math.*, *96* (1956), 265–309.

―――, Punktverteilungen auf der Kugel und Informationstheorie. *Die Naturwissenschaften*, *48* (1961), 189–192.

G.A. Vanasse, *Spectrometric Techniques*, *I*. Academic, N.Y., 1977.

V.S. Varadarajan, *Lie Groups, Lie Algebras and their Representations*. Prentice-Hall, Englewood Cliffs, N.J., 1974.

―――, *Harmonic Analysis on Real Reductive Groups, Lecture Notes in Math.*, *576*. Springer-Verlag, N.Y., 1977.

―――, Eigenfunction expansions on semisimple Lie groups, in A.F. Talamanca (Ed.), *Harmonic Analysis and Group Representations*, C.I.M.E. (*International Math. Summer Center*). Liguori editore, Napoli, Italy, 1982, 351–422.

A.B. Venkov, Spectral theory of automorphic functions, the Selberg zeta function, and some problems of analytic number theory. *Russian Math. Surveys*, *34* (1979), 79–153.

―――, On the trace formula for $SL(3,\mathbb{Z})$, *J. Sov. Math.*, *12* (1979), 384–424.

―――, *Spectral Theory of Automorphic Functions*, Proc. Steklov Inst. Math. *153* (*1981*) Amer. Math. Soc. translation, 1982.

A.B. Venkov, V.L. Kalinin, & L.D. Faddeev, A nonarithmetic derivation of the Selberg trace formula. *J. Soviet Math.*, *8* (1977), 171–199.

M. Vergne, Representations of Lie groups and the orbit method, in B. Srinivasan & J.

Sally (Eds.), *Emmy Noether in Bryn Mawr*. Springer-Verlag, N.Y., 1983, 59–101.

M.-F. Vignéras, Séries théta des formes quadratiques indéfinies, *Lecture Notes in Math. 627*. Springer-Verlag, N.Y., 1977, 227–240.

———, L'équation fonctionelle de la fonction zêta de Selberg de la groupe modulaire $PSL(2, \mathbb{Z})$, *Astérisque*, *61* (1979), 235–249.

———, Variétés riemanniennes isospectrales et non isométriques. *Ann. of Math.*, (2)*112* (1980), 21–32.

———, *Arithmétique des Algèbres de Quaternions*, Lecture Notes in Math. 800, Springer-Verlag, N.Y., 1980.

———, Quelques remarques sur la conjecture $\lambda_1 \geq \frac{1}{4}$. *Sém Théorie des Nombres*, Paris 1981–82, M.-J. Bertin (Ed.), Birkhäuser, Boston, 1983, 321–343.

N.J. Vilenkin, *Special Functions and the Theory of Group Representations*, Translations of Math. Monographs, *22*. American Math. Soc., Providence, R.I., 1968.

A.D. Virtser, Central limit theorem for semisimple Lie groups. *Theory of Probability and its Applics.*, *15* (1970), 667–687.

V.S. Vladimirov, *Equations of Mathematical Physics*. Dekker, N.Y., 1971.

G.F. Voronoï, Propriétés des formes quadratiques positives parfaites. *J. Reine Angew. Math.*, *133* (1908), 97–178.

———, Nouvelles applications des paramètres continus à la theorie des formes quadratiques. *J. Reine Angew. Math.*, *134* (1908), 198–287; *136* (1909), 67–178.

S.M. Voronin, On analytic continuation of certain Dirichlet series. *Proc. Steklov Inst. Math.*, *157* (1983), 25–30.

D. Wallace, *Selberg's trace formula and units in higher degree number fields*. Ph.D. Thesis, U.C.S.D., 1982.

———, Conjugacy classes of hyperbolic matrices in $SL(n, \mathbb{Z})$ and ideal classes in an order, *Trans. Amer. Math. Soc. 283* (1984), 177–184.

N. Wallach, Lecture Notes, 1981 NSF-CBMS Regional Conf. on Representations of Semi-Simple Lie Groups and Applics. to Analysis, Geometry & Number Theory (unpublished).

———, *Harmonic Analysis on Homogeneous Spaces*. Dekker, N.Y., 1973.

———, *Symplectic Geometry and Fourier Analysis*. Math. Sci. Press, Brookline, Mass., 1977.

A. Wangerin, Theorie der Kugelfunktionen und der verwandten Funktionen, insbesondere der Laméschen und Besselschen (Theorie spezieller durch lineare Differential-gleichungen definierter Funktionen). *Enzykl. der Mathematischen Wissenschaften*, Vol. 2, Pt. 1, No. 2, Teubner, Leipzig, 1904–1916, 699–762.

G. Warner, *Harmonic Analysis on Semi-Simple Lie Groups*, Vols. I, II. Springer-Verlag, N.Y., 1972.

———, Selberg's trace formula for non-uniform lattices: the R-rank one case. *Studies in Algebra & Number Theory*, Advances in Math., Supp. Studies, *6*. Academic, N.Y., 1979, 1–142.

G.N. Watson, *A Treatise on the Theory of Bessel Functions*. Cambridge University Press, London, 1962.

S. Watson, *Statistics on Spheres*. Wiley, N.Y., 1983.

N. Wax (Ed.), *Selected Papers on Noise and Stochastic Processes*. Dover, N.Y., 1954.

H. Weber, *Lehrbuch der Algebra*, Vols. I–III. Chelsea, N.Y., 1961.

A. Weil, *L'intégration dans les Groupes Topologiques*. Hermann, Paris, 1965.

———, *Basic Number Theory*. Springer-Verlag, N.Y., 1974.

———, *Elliptic Functions According to Eisenstein and Kronecker*. Springer-Verlag, N.Y., 1976.

———, *Dirichlet Series and Automorphic Forms*, Lecture Notes in Math., *189*. Springer-Verlag, N.Y., 1971.

———, *Oeuvres Scientifiques, Collected Papers (1926–1978)*, Vols. I–III. Springer-Verlag, N.Y., 1979.

———, *Discontinuous Subgroups of Classical Groups*. University of Chicago, 1958.
H. Weyl, *Gesammelte Abhandlungen*. Springer-Verlag, N.Y., 1968.
———, *The Theory of Groups and Quantum Mechanics*. Dover, N.Y., 1928.
———, *The Classical Groups, Their Invariants and Representations*. Princeton University Press, Princeton, N.J., 1946.
H. E. White, *Introduction to Atomic Spectra*. McGraw-Hill, N.Y., 1934.
E.T. Whittaker & G.N. Watson, *A Course of Modern Analysis*. Cambridge University Press, Cambridge, 1927.
D.V. Widder, *The Laplace Transform*. Princeton University Press, Princeton, N.J., 1941.
N. Wiener, *The Fourier Integral and Certain of its Applications*. Cambridge University Press, London, 1933.
———, *Selected Papers*. M.I.T. Press, Cambridge, Mass., 1964.
E.P. Wigner, *Group Theory and its Application to the Quantum Mechanics of Atomic Spectra*. Academic, N.Y., 1959.
———, On a generalization of Euler's angles, in E.M. Loebl (Ed.), *Group Theory and its Applications*. Academic, N.Y., 1968, 119–129.
H. Williams & J. Broere, A computational technique for evaluating $L(1,\chi)$ and the class number of a real quadratic field. *Math. Comp.*, 30 (1976), 887–893.
J. Wishart, The generalized product moment distribution in samples from a normal multivariate population. *Biometrika*, 20A (1928), 32–43.
K. Wohlfahrt, Über Operatoren Heckescher Art bei Modulformen reeller Dimension. *Math. Nachr.* 16 (1957), 233–256.
B.G. Wybourne, *Classical Groups for Physicists*. Wiley, N.Y., 1974.
T. Yamazaki, On Siegel modular forms of degree 2. *Amer. J. of Math.*, 98 (1973), 39–53.
K. Yosida, *Functional Analysis*. Springer-Verlag, N.Y., 1968.
D. Zagier, Eisenstein series and the Riemann zeta function, Eisenstein series and the Selberg trace formula I. *Automorphic Forms, Representation Theory and Arithmetic, Bombay Colloquium, 1979*. Springer-Verlag, N.Y., 1981, 275–302, 303–355.
———, A Kronecker limit formula for real quadratic fields. *Math. Ann.*, 231 (1975), 153–184.
———, Modular forms whose coefficients involve zeta functions of quadratic fields. *Lecture Notes in Math.* 627. Springer-Verlag, N.Y., 1977, 105–169.
———, Correction to "The Eichler-Selberg trace formula on $SL(2,\mathbb{Z})$." *Lecture Notes in Math.* Springer-Verlag, N.Y., 1977, 171–173.
D.P. Zhelobenko, The classical groups. Spectral analysis of their finite dimensional representations. *Russian Math. Surveys*, 17 (1962), 1–94.
J.M. Ziman, *Principles of the Theory of Solids*. Cambridge University Press, Cambridge, 1972.
H. Zirin, *The Solar Atmosphere*. Blaisdell, Waltham, Mass., 1966.
I.J. Zucker, Madelung constants and lattice sums for invariant cubic lattice complexes and certain tetragonal structures. *J. Phys. A:Math. Gen.*, 8 (1975), 1734–1745.
A. Zygmund, *Trigonometric Series*. Warsaw, 1935.

Index

Abelian extension (*see* Class field)
Abelian integral, 196
Abelian theorem, 22
Abelian variety, 192, 197, 236
Abel's integral equation, 23
Addition formula, 104–105, 108
Adelic theory, 182, 227, 236
Adjoint of
  constant coefficient operator, 255
  differential operator, 110
Algebraic integer, 66
Algebraic number field, 65–76, 194, 203, 210, 217, 233, 236, 239–240, 276, 292–293
ALGOL program, 62
Aliasing, 44–45
Almost periodic function, 210
Analytic
  continuation, 58–59, 189, 207, 216, 229, 232–234
  function, 167, 191, 233
  functional, 146
Arc length on a symmetric space, 86–87, 121, 296
Area element on a symmetric space, 86–87, 124, 296
Area of fundamental domain, 171
Area of unit sphere in $\mathbb{R}^m$, 5
Artin conjecture, 236, 295

Artin L-function, 68, 236, 250, 295
Artin reciprocity law, 76, 191, 210
Arzelà-Ascoli theorem, 263
Associated Legendre function, P_s^a, 89–90, 141
Asymptotics of eigenvalues of the Laplacian, 40, 228, 290
Asymptotics of special functions, 60, 108, 112, 137, 139, 142, 158–159, 254, 268, 270
Asymptotics/functional equations principle, 113, 138–139, 144, 208, 254, 298
Atiyah-Singer index theorem, 295
Atomic physics, 93–96, 135, 156, 162
Autocovariance or autocorrelation, 44
Automorphic form, 167, 182–183, 204–205
Automorphic function, 165, 182, 190–191

Balmer series of spectral lines, 94, 96
Band limited function, 34, 51
Barnes double gamma function, 294
Base change, 236
Basis problem for modular forms, 187, 194–195, 201, 240
Bernoulli number, B_n, 75, 203, 211

Bessel function
   $H_s^{(1)}$, 112–114
   I_s, 114, 190, 208, 268, 272
   J_s, 107–110, 112–114, 139–140, 158–160
   K_s, 16, 39, 57, 114, 136–141, 208–210, 219–220, 223–227, 231–232, 267–268, 272, 292–293, 298
Beta function, 142
Binary code, 199
Bochner-Hecke formula, 108
Body-centered cubic lattice, 77
Borel sets, 100–101
Borel-Weil-Bott theorem, 102
Boson pair creation, 169
Boundary of a symmetric space, 146
Bounded variation, 22
Bound states, 93
Branched Riemann surface, 179
Brauer-Siegel theorem, 68, 74
Brownian motion, 28, 252
Bruhat decomposition, 209

Calculus of variations, 85–86
CAT scanner, 114–119
Cauchy principal value, 115
Cauchy problem, 19
Cauchy residue theorem, 22, 185–186, 211–212, 231, 254, 257
Cauchy-Riemann equations, 122
Cayley transform, 127, 163, 171, 223, 277–281
Celestial mechanics, 88
Central element, 275, 283
Centralizer, 201, 276
Central limit theorem, 26, 92, 161
Cesàro sum, 3, 33
Chaos, 250
Characteristic function, 24, 157
Character of a representation, 102, 191, 201, 235–236
Circle method, 190
Circle problem, 40, 266–268, 299
Class field or abelian extension, 56, 68, 75, 168, 191, 210
Class group or ideal class group, 66–75, 191
Classical mechanics, 251

Class number, 66–75, 168, 171–174, 190, 211, 213, 276, 295
Class one representation, 104
Clebsch-Gordon series, 102
Closed or periodic geodesic, 201, 266, 277–281, 292
Coding theory, 199–200
Coherent states, 169
Cohomology, 248
Compact fundamental domain, 179, 251
Compact group, 101
Compactification of fundamental domain, 163, 179, 201
Compact operator, 48–49, 251, 253, 263
Compact symmetric space, 83–88
Completely reducible representation, 100–101
Computerized tomography, 107, 114–119
Confluent hypergeometric function, 57, 63
Conformal mapping, 122, 168, 190, 197–198
Congruence subgroup, 167, 174, 176–181, 188, 193–195, 201, 203, 205, 225–228, 235–238, 241, 245, 247, 265
Conical function, 142
Conjugacy class, 276
Conjugate of an algebraic number, 66
Constant term in Fourier expansion of Maass wave form, 214–216, 255
Continued fraction, 62, 171, 248
Continuous spectrum, 93, 110–111, 205, 220, 224, 253–254
Contour integral, 185, 212
Convergence of
   sequence of distributions, 6
   trigonometric series of distributions, 37–38
Convolution, 3, 7–12, 14–16, 21, 24, 26, 104, 106, 149, 259–262, 297
Correspondence between spaces of automorphic forms, 197, 228
Courant mini-max principle, 224
Covering transformations, 167
Cryptography, 65
Crystallography, 41, 76–82, 163, 165
CsCl, 77

Cubic equation, 187, 197
Cubic lattice, 77
Cubic number field, 69
Curvature, 124
Cusp form, 184, 187–190, 201–203, 213–214, 217–229, 240, 247, 249, 253–254, 258–259, 262–263, 266, 274, 290, 298
Cusp of fundamental domain, 178, 206, 248
Cyclic group, 76
Cyclotomic field, 65, 67, 69, 75

$\mathscr{D}$, test functions, 2
d'Alembert solution of wave equation, 14
Dedekind eta function, 189–190, 199, 210
Dedekind sum, 189, 236, 294
Dedekind zeta function, 68–75, 217, 235–236
Degeneracy, 94
Degree of an algebraic number field, 65
Degree of a representation, 99, 191
δ, Dirac delta distribution, 2, 15–19, 37, 113–114, 139, 144, 155, 251
Δ, discriminant modular form, 184, 186–191, 197, 205, 234, 240, 249, 296
Δ, Laplace operator (see Laplace operator)
Demography, 135, 156
Densely would line in torus, 278
Densest lattice packing of spheres, 200
Density of a random variable, 24, 157
Derivative of a distribution, 4
Diamond, 77
Different, 74
Differentiable manifold, 84–85
Dihedral group, 76
Dimensions of spaces of
  holomorphic modular forms, 186–187, 295, 214
  Maass wave forms, 227–228, 292
Dirac comb, 37
Dirac delta distribution, 2
Dirac sequence or family, 3, 12, 139, 144, 262
Direct sum of representations, 100
Dirichlet kernel, 4, 32
Dirichlet L-function, 217–218, 224, 233, 272
Dirichlet problem, 140–141, 167, 151–152, 217–219, 228, 233
Dirichlet unit theorem, 67
Discontinuous subgroup, 119, 121, 163–181, 297
Discrete or point spectrum, 93, 110, 113, 205, 217–224, 253–254, 258–259, 263–265
Discrete subgroup, 163–181, 297
Discriminant function Δ (see Δ)
Discriminant of a number field, 66
Dispersion, 159
Distribution
  modulo one, 34, 278
  of random variable, 15, 24
  of solutions of quadratic congruences, 272
Distribution or generalized function, 1–8, 14–18, 37–39
Divisor function, 184, 203–204, 209, 234
Doubly periodic function, 195–196, 198, 241
Drum, 251–252, 295
Dual group, 102
Dual ideal, 73

E_s, nonholomorphic Eisenstein series, 206–210, 214, 216, 253–254, 256–258
E_s^*, nonholomorphic Eisenstein series with gamma and zeta factors, 208, 234
Earthquake, 83
Eichler-Selberg trace formula, 241, 250, 295
Eigenfunctions, 49, 51–53, 90, 93, 99, 104, 109–110, 113, 135–149, 204–229, 233, 239, 242, 297–298
Eigenvalues, 49, 51–53, 82, 90, 93, 99, 104, 110, 136–137, 206, 213, 217–227, 239–240, 242, 250, 262, 290, 297

Index 333

Eisenstein series, 183–184, 186–187, 206–216, 225, 240, 246–248, 253–254, 256–258, 298
Electromagnetic spectrum, 43, 127
Electrostatic field, 121, 152
Elementary divisor theory, 66, 72
Elementary row and column operations, 66, 72–73
Elliptic curve, 192, 210, 213, 236
Elliptic element, 178, 275–276, 284–285
Elliptic function, 187, 195–199, 204
Elliptic geometry, 88, 120, 124
Elliptic integral, 167, 187–188, 195–199
Elliptic modular function or modular function for $SL(2, \mathbb{Z})$, 75, 167, 182, 190–192, 210
Elliptic modular group or modular group, 163
Elliptic partial differential equation, 205
Elliptic point in fundamental domain, 170, 178
Energy level, 93
Epstein zeta function, 31, 39, 53, 58–65, 67, 70, 79–82, 205–207, 217, 225, 229, 258, 271–272
Equations of fifth degree, 168
Equivalent positive matrices, 193
Equivalent representations, 100, 134
Ergodic theorem, 34, 280
Ergodic transformation, 280
Escher, 163
η, Dedekind eta function, 189–190, 199, 210
Euclidean algorithm, 171
Euclidean group, $E(n)$, 109, 163
Euclid's fifth postulate, 88, 123
Euclid's second postulate, 88
Euler angles, 91
Euler constant, γ, 210–211, 286
Euler formula for $\zeta(2n)$, 75, 211
Euler-Lagrange equation, 86
Euler-Maclaurin summation formula, 286, 291
Euler product, 58, 68, 238–246
Even integral positive matrix, 192
Ewald's method of theta functions, 81–82
Exceptional Lie group, 193, 200

Excited states, 94
Expectation or mean, 24, 158–159

Face-centered cubic lattice, 77, 200
Factorization method of Infeld and Hull, 54
Farey fractions, 248–249
Fast Fourier transform, 30, 119
Fejér kernel, 3, 33
Fermat conjecture, 65
Feynman integrals, 252
Finite dimensional representation, 99, 194
Finite element method, 49, 225
Finite Fourier series, 190
Finite simple group, 191, 228
Fischer-Griess monster group, 191
Fourier analysis on
 fundamental domain, 28–55, 204–205, 235, 250–295, 297–298
 symmetric space, 8–28, 91–92, 103–104, 109, 134–156, 296
Fourier-Bessel series, 113
Fourier coefficient, 30, 202–203, 208–209, 220–227, 239–242, 247
Fourier expansions of automorphic forms, 182, 184, 188–189, 202–203, 208–216, 219–227, 229–234, 239, 242, 247
Fourier inversion, 10, 15, 43, 103, 110, 115, 138, 143, 146–149, 162, 297
Fourier series, 21, 30, 36–38, 91–92, 145
Fourier transform, 9, 11, 15, 55, 92, 104, 107–109, 134–156, 296
Fourier transform spectroscopy, 41–44
Fractional ideal, 67
Fractional linear transformation, 121–122, 127, 177–178, 296
Frobenius reciprocity law, 102
Fuchsian group, 175
Functional equation, 59, 73, 108, 113, 137, 141, 147, 207–208, 229–236, 254
Fundamental domain, 163–181
Fundamental function or Hauptmodul, 190
Fundamental group or Poincaré group, 84

Fundamental solution
   heat equation, 13–14, 17–18, 28, 152–155, 272, 290, 296
   Laplace equation, 5, 18, 51–52, 270
   wave equation, 14, 18–19, 252
Fundamental unit, 67, 277
Funk-Hecke theorem, 105

G, semisimple real, noncompact usually, Lie group, 121, 296
G/K, symmetric space, 124
G_k, holomorphic Eisenstein series, 183–184, 240
$GL(n)$ (see General linear group)
Galois group, 168, 191, 236
γ, element of discrete group Γ or Euler's constant, 163–181, 210–211
Γ, discrete subgroup of G, usually $SL(2, \mathbb{Z})$, 163–181, 297
Γ, gamma function, 16, 57, 59, 70–73, 139, 144, 208, 229, 231, 234–236, 258
$\Gamma(N)$, congruence subgroup, 174, 176–177, 179–181, 237–238, 245
$\Gamma_0(N)$, congruence subgroup, 194, 201, 235–237, 241, 266
$\Gamma_1(N)$, congruence subgroup, 237
Gauss-Bonnet formula, 88, 211
Gauss class number conjectures, 67, 73
Gauss hypergeometric function, 142–144, 270
Gaussian curvature, 124
Gauss kernel or heat kernel, 4, 13, 25–26, 28, 152–155, 251–252, 272–273, 290
Gauss sum, 212, 235
General linear group of nonsingular $n \times n$ matrices, $GL(n, K)$, ix, 68, 103, 193, 227, 236, 266
Generators of discrete groups, 169–171, 176
Genus, 175, 179, 181, 266
Geodesic, 87–88, 122–123, 127, 249, 277–281
Geodesic polar coordinates, 125–126, 141–144, 149–150, 284
Geodesic-reversing isometry, 88, 125
Geophysics, 98
Gibb's phenomenon, 32–34, 92

Girard formula, 88
Great circle, 88
Greatest common divisor, 207
Green's function or resolvent kernel, 5, 52–53, 55–56, 92, 105, 111–114, 201, 259, 268–272, 295
Green's theorem, 5, 90, 155, 213, 215, 269
Grenzkreisgruppe, 175
Grossencharacter, 75, 225, 233, 236
Ground state, 94
Group extension, 79
Group representation, 19–20, 41, 99–107, 134, 162, 227–228, 235–236

H, Poincaré upper half-plane, 121
$\mathscr{H}$, Helgason (Fourier, Mellin) transform on a symmetric space, 146–152, 157–158, 260, 262, 265, 267, 275, 297
Haar measure or G-invariant measure, 100–101, 260
Hamming code, 200
Hamming distance, 199
Hankel function, 112
Hankel's inversion formula, 109–110, 112–113, 139–140
Hardy-Littlewood circle method, 190
Harish or Abel transform, 150, 282–283, 287, 291–293
Harmonic analysis (see Fourier analysis)
Harmonic function, 98, 214
Harmonic polynomial, 90
Hasse-Weil zeta function, 236
Hauptkreisgruppe, 175
Hauptmodul, 190
Heat equation or diffusion equation, 13–14, 17, 39, 106, 152–155, 167, 251–252, 272–273, 290–291, 297
Heaviside step function, 5, 17
Hecke correspondence between modular forms and Dirichlet series, 229–250
Hecke L-function with grossencharacter, 75, 225, 233, 236
Hecke operators, 220–225, 238–242, 249, 295
Hecke's relation between Dedekind and Epstein zeta functions, 70

Index

Heisenberg uncertainty principle, 20
Helgason transform (see $\mathcal{H}$)
Hermite function, 54–55
Hidden periodicities, 41, 45
Highest point method, 169
Heighest weight, 102
Hilbert problems 12 and 18, 68, 75, 163, 168
Hilbert-Schmidt operator, 48
Hilbert space or complete inner product space, 99, 111
Hilbert transform, 116
Hirzebruch-Riemann-Roch theorem, 295
Højendal's method for Madelung constant, 80
Homogeneous space, 84, 124–125
Homology, 249
Horocycle, 248–250
Huyghen's principle, 19, 162
Hydrogen, 93–96, 162
Hyperbolic element of $SL(2, \mathbb{R})$, 178, 201–202, 272, 275–280, 292–294
Hyperbolic geometry, 120, 123–124
Hyperbolic space, 120
Hyperfunction, 146
Hypergeometric function, 142–144, 270

I_s (see Bessel function)
Ideal, 66
Ideal class group, 66
Images, method of, 52, 55, 268–269
Impedance, 128, 132
Incomplete gamma function, 57–65, 73–74, 82, 230, 232, 234
Incomplete theta series, 254–258
Independent random variables, 24, 157
Indicator function of a set, 161
Induced representation, 102
Infeld and Hull factorization method, 54
Inner product (truncated) of Eisenstein series, 216
Instrument function, 43
Integral basis or $\mathbb{Z}$-basis of an ideal, 66
Integral test, 59, 271
Interferogram function, 41
Intertwining operator, 100

Invariant differential operator, 61, 87, 99, 124, 146
Invariant integral operator, 259–263
Invariant random variable under K, 156–157
Inversion in a sphere, 52–53, 198
Inversion of transforms
  Fourier, 10, 15, 43, 110, 115
  Hankel, 110
  Helgason, 147, 283–284
  Kontorovich-Lebedev, 114
  Laplace, 21
  Mehler-Fock, 144
  Mellin, 56
  Radon, 114–116
Irreducible representation, 99, 101
Isometric circle, 175
Iwasawa decomposition, 264

J and j, modular invariant, 168, 190–192, 199
J_s (see Bessel function)
Jacobi-Abel functions, 196
Jacobian elliptic function, 197–198
Jacobi derivative formula, 196
Jacobi identity for Δ, 189
Jacobi theta function, 39, 195–196
Jacobi transformation formula for θ, 39, 59, 81, 195
Jacobi triple product formula, 199
Jordan form, 177–178, 275

K, maximal compact subgroup of G, 124, 296
K_s (see Bessel function)
Kernel of integral operator, 3–4, 47, 48–49
Kirchoff's formula for wave equation, 19
Kleinian group, 176
Kloosterman sum, 201, 247
Kodaira-Titchmarsh formula, or Stieltjes, Stone, Kodaira, Titchmarsh formula, 111, 259
Kontorovich-Lebedev transform, 114, 138–141, 145
Korteweg-deVries equation, 204

Kronecker delta, δ_{ij}, 73, 105
Kronecker limit formula, 210–211
Kronecker symbol, 194, 217
K-theory, 75

$\mathscr{L}$, Laplace transform, 20–23, 56, 143, 252
L-function, 75–76, 189, 218, 229–250, 272
$L^p(X, d\mu)$, space of Lebesgue integrable functions, 11, 30–31, 89, 100–101, 111, 213–214, 253–254, 256, 261
$L_0^2(\Gamma \backslash H)$, 253, 256, 258–259, 263–265
Laguerre polynomials, 96
$\Lambda(s) = \pi^{-s}\Gamma(s)\zeta(2s)$, 59–61, 208, 216, 253, 258, 289
Lamb shift, 94
Lanczos smoothing, 33–34
Landau kernel, 3
Langlands philosophy, 236
Laplace equation, 5, 18, 51–52
Laplace operator, Δ, 5, 19, 86–87, 124, 126, 135–155, 204–206, 213–215, 217–225, 233, 242, 250–251, 253–265, 268–272, 290–292, 294–298
Laplace series of spherical harmonics, 91–92
Laplace-Stieltjes transform, 22
Laplace transform, $\mathscr{L}$, 20–23, 56, 143, 252
Lattice, 74, 76–82, 200, 251
Leaky torus, 179–181, 249
Learning theory, 135, 156
Least squares, 45
Lebesgue dominated convergence theorem, 6, 27, 161
Legendre function,
 P_s, 141–149, 152–154, 158–161, 262, 269–270
 Q_s, 269–270
Legendre polynomial, P_n, 89–92, 105–106, 109
Legendre symbol, 194
Lehmer's problem on $\tau(n)$, 188–189
Lemniscate integral, 167
Length spectrum of a Riemannian manifold, 277, 292–293, 295

Length standard, 40
Level of a congruence subgroup, 176, 194, 236
Level of a positive integral quadratic form, 194
Lichtenbaum conjectures, 75, 211
Lie group, 99, 121, 139, 156, 162, 191, 193, 199–200, 286
Limit point and limit circle in ODE'S, 112–113
Limit point of a discrete group, 175
Liouville's theorem on conformal mapping in 3 dimensions, 198
Lobatchevsky upper half-plane (see Poincaré upper half-plane)
Locally Euclidean space, 99
Locally integrable function, 2
Lorentz-type group, 97, 121, 132–134
Lyman series of spectral lines, 94

M, Mellin transform, 56–59, 139, 144, 150, 167, 194, 209, 229–235, 248, 254, 256–257, 282
$\mathcal{M}(\Gamma, k)$, space of holomorphic modular forms of weight k, 182
Maass-Selberg relations, 216
Maass wave form or nonholomorphic modular form, 204–229
Madelung constant, 79–82
Magnetic field, 96–98, 119
Matched load in an electrical network, 128–131
Matrices associated with electrical circuits, 132–133
Matrix entry of a representation, 99, 104
Maximum principle, 214
Maxwell's equations, 97, 121
Mean, 20, 24, 158–159
Mehler-Fock transform, 143–149, 285
Mellin transform (see M)
Mercer's theorem, 50, 265, 275, 299
Mertens conjecture, 240
Method of images, 52, 55, 268, 273
Method of least squares, 46
Method of theta functions (Riemann, Ewald), 59–61, 81–82
Microwave engineering, 127–133, 156, 161–162

Minakshisundaram-Pleijel zeta functions, 82
Mini-max principle, 224
Minimum of a positive quadratic form over the integer lattice, 62–65
Minkowski's fundamental lemma in the geometry of numbers, 68
Modular function associated with Haar measure, 101
Modular group, $SL(2,\mathbb{Z})$, 163, 192
Modular invariant (see J, j)
Modular or automorphic form, 167, 182–229
Modular or automorphic function, 75–76, 167, 190–192
Modular symbol, 248–249
Moduli variety, 192
Modulus, λ, 188, 198
Monster group, 191–192
Moonshine, 192
Multiplication formula for Fourier transform, 10
Multiplier system for a discrete group, 182–183, 189, 194–195

$\mathcal{N}(\Gamma, \lambda)$, space of Maass wave forms, 205, 214
NaCl (see Salt)
Narrow class number, 276
Neumann problem, 217–220, 233
Non-Euclidean central limit theorem, 161
Non-Euclidean distance, 123
Non-Euclidean Eisenstein series or Epstein zeta function, 271
Non-Euclidean Fourier expansion, 254
Non-Euclidean geometry, 88, 120–124
Non-Euclidean lattice point or circle problem, 266–268
Non-Euclidean normal distribution, 158, 273, 290–291
Non-Euclidean Poisson sum formula, 253–273, 275, 299
Non-Euclidean shock wave, 249
Non-Euclidean theta function, 267, 273, 290–291, 298
Normal density, 25–26, 28, 158
Normal fundamental region, 175

Norm of a hyperbolic element of $SL(2,\mathbb{R})$, 276
Norm of an ideal, 66
Nuclear magnetic resonance tomography, 119
Numerical integration, 82

$o(y^p)$, 216
$O(y^p)$, 248
$O(n)$, orthogonal group or rotation group, 84, 119, 163
Octahedral group, 76
Orbital integral, 282–290
Order, 241, 276

$\wp$, Weierstrass elliptic function, 187, 196, 204
$\mathcal{P}_n$, space of positive $n \times n$ real matrices, ix, 58, 60, 79, 125
p_s, power function, 136, 206, 260–262
P_s (see Legendre function)
p-adic number, 56, 75, 238
p-adic symmetric space, 156
Paley-Wiener theorem, 15, 138, 144, 147
Parabolic elements of $SL(2,\mathbb{R})$, 178, 275–276, 281, 285–290
Parseval's equality, 10, 30–31
Partitions, 189–190
Pell's equation, 210
Periodic geodesic (see Closed geodesic)
Periodization of test function or distribution, 35, 265, 300
Periodogram, 48
Perpendicular bisectors method, 174–175
Petersson inner product, 239, 242
Peter-Weyl theorem, 104
Phragmen-Lindelof theorem, 258
Picard's theorem, 168, 191
Plancherel formula, 10, 103, 147
Plancherel theorem, 10–11, 103–104, 147, 155, 161
Plancherel measure, 11, 103–104, 113, 147, 297
Planck's constant, 93
Planck's law, 41
Poincaré generators and relations theorem, 169–171

Poincaré group (*see* Fundamental group)
Poincaré polygon, 163, 170
Poincaré series, 200–202, 229, 247, 254, 266
Poincaré upper half-plane, viii, 121
Point group, 76
Point spectrum (*see* Discrete spectrum)
Poisson integral formula, 111, 146
Poisson summation formula, 35–39, 50–51, 55, 119, 167, 196, 209, 264–273, 275, 286, 299
Polar coordinates, 86–87, 125–126
Positive definite matrix, 58, 60, 79, 124–125, 168, 192–194, 200, 203, 206–207
Positive operator, 49
Potential theory, 146, 197–198
Power function (*see* p_s)
Power spectrum, 44
Prime geodesic theorem, 293
Prime number or ideal theorem, 70, 74, 167, 212, 238, 290, 292–293
Primitive hyperbolic element of $SL(2, \mathbb{Z})$, 277, 292
Principal ideal, 66
Probability density, 24, 95, 157
Probability distribution, 24, 95, 98, 156–157
Product of distributions, 6
Projection-valued measure, 110
Projective linear group, $PSL(2, \mathbb{R})$, 274
Projective plane, 88

Quadratic form, 58, 73, 168, 171, 174, 192–194, 197, 203–204, 213, 253, 277, 295
Quadratic number field, 67, 69, 73, 168, 171–174, 210–211, 217, 225, 235, 241, 250, 276–277, 292
Quadratic reciprocity law, 194–195
Quantum mechanics, 39, 41, 53, 79, 93–95, 98, 102, 121, 162, 252
Quantum numbers, 94
Quantum statistical mechanics, 39, 58, 82, 252
Quaternion algebras, 179, 251
Quiche and salad, 208
Quotient (*see* Homogeneous space)

R, Radon transform, 107, 114–119
Rademacher formula for the partition function, 190
Radon-Nikodym theorem, 24
Radon transform (*see* R)
Ramanujan-Petersson conjecture, 188, 202–203, 220, 224–227, 240–241, 247, 266
Ramanujan sum, 209–210
Ramanujan τ-function, 180, 240, 249
Random variable, 24, 156
Rankin-Selberg method, 211–212, 246–248, 255
Reciprocity law, 191
Regular polyhedra, 119
Regulator of a number field, 67
Relative Poincaré series attached to a hyperbolic element, 201–202
Representation of a group, 19–20, 68, 79, 98–104, 132–134, 162, 168, 194, 199
Representation of an integer by a quadratic form, 197, 203
Residual spectrum, 258
Residues of zeta functions, 70, 73, 207, 258
Resolvent, 51–52, 111, 259, 268–272, 295
Resolvent kernel (*see* Green's function)
Riemann hypothesis, 61, 74, 208, 224–225, 240, 248, 250, 258, 290
Riemannian manifold, 86, 121–124
Riemann-Lebesgue lemma, 12, 31, 115
Riemann mapping theorem, 167
Riemann method of theta functions (*see* Method of theta functions)
Riemann-Roch theorem, 184, 187, 295
Riemann sphere, 190
Riemann surface, 175, 177, 179, 266
Riemann zeta function, 56, 58–61, 69, 167, 183–184, 194, 207–212, 217–218, 224–225, 234, 238, 240–241, 285, 289–292, 294
Right invariant integral, 101
Roelcke-Selberg conjecture, 228, 258
Roelcke-Selberg spectral resolution of Δ on $L^2(\Gamma\backslash H)$, 254–265, 298
Rotation group (*see* $O(n)$ or $SO(N)$)
Rydberg constant, 94

Index   339

$\mathscr{S}$, Schwartz space or inverse of $\mathscr{H}$, 9, 148–149
$SL(n)$, special linear group of $n \times n$ matrices of determinant one,
  $SL(n, \mathbb{R})$, viii, 101, 121, 147, 158
  $SL(n, \mathbb{Z})$, viii, 163, 171, 176, 192, 211, 236, 266, 272, 274
  $SL(n, \mathbb{Z}/N\mathbb{Z})$, 177, 194, 228, 235
  $SL(2, \mathbb{Z}[i])$, 274
$sn(z)$, Jacobian elliptic function, 197–198
$SO(N)$, special orthogonal group, 84, 103–104, 162
$SO(p, q)$, Lorentz-type group, viii, 121, 132–134, 162
$Sp(n, \mathbb{Z})$, symplectic group, Siegel modular group, viii, 273
$SU(n)$, special unitary group, 97
$SU(p, q)$, viii, 121, 127, 162
$\mathscr{SM}(\Gamma, k)$, space of holomorphic cusp forms, 184
$\mathscr{SN}(\Gamma, \lambda)$, space of Maass cusp forms, 213
$\mathscr{SP}_n$, determinant one surface in positive matrix space, 124–125, 206
Salt, 79–82
Sampling formula, 34–35, 43–44
Satellites in spectroscopy, 44
Sato-Tate conjecture, 188
Schrödinger operator, 53–54, 93–95, 102, 252
Schur's lemma, 100
Schwarz-Christoffel transformation, 167, 197–198
Schwartz space, 9
Second moment, 158
Selberg trace formula, 45–51, 105, 201, 205, 216, 228, 241, 250, 252, 274–295
Selberg transform, 146, 265, 275
Selberg zeta function, 294
Self-adjoint operator, 48, 110, 213, 224, 242, 261, 263
Semidirect product of groups, 109
Semisimple Lie group, 101
Separation of variables, 90, 93, 109, 138, 141–142, 152, 208, 269
Serre conjecture on $\tau(n)$, 189
Shah functional, 37

Shannon sampling, 34–35, 43–44
Siegel modular form, 194, 196, 273
Siegel modular group (see $Sp(n, \mathbb{Z})$)
Siegel upper half-plane or space, 194, 196
Siegel zero, 68, 74
$\sigma_s(n)$ (see Divisor function)
Singular differential operator, 53–54, 110–111
Singular eigenvalue problem, 53–54, 111, 253
Singular series, 209
Slash operator, 238
Smith chart, 129
Smoothing operator, 33–34
Solitons, 204
Source spectral density, 41
Space group, 76–79
Special linear group (see $SL(n, \mathbb{R})$)
Special orthogonal group (see $SO(n)$)
Special unitary group (see $SU(n)$)
Spectral lines, 93–96, 102
Spectral measure, 111–114, 138–139, 143–144, 147, 254
Spectral resolution of differential operators, 19, 49–50, 53–54, 57–58, 91–92, 110–114, 138–139, 143–144, 147, 253–254, 297–298
Spectral theorem, 49, 91–92, 110–114, 125, 254, 263
Spectroscopy, 40–44, 93–95
Spectrum, 93–94, 110, 251–252, 271
Sphere, 84
Spherical Bessel function, 108
Spherical function, 83, 90, 106, 141, 238, 262
Spherical harmonic, 83, 88, 90, 92, 105–106, 194
Spherical polar coordinates, 86–87
Spheroidal wave function, 51
Spurious eigenvalues, 217–220
Standard deviation, 24, 158
Star, 45–48, 94
Stark conjectures, 68, 76, 210
Stieltjes integral, 22, 40, 114, 298
Stieltjes, Stone, Kodaira, Titchmarsh formula, 110–111
Stochastic differential equations, 156
Stone-Weierstrass theorem, 31

Sturm-Liouville operator, SLOP, 111–114
Sun's magnetic field, 95–97
Support of a distribution or function, 4
Surface spherical harmonic, 88, 90–91, 108–109
Symmetric power representation, 132–134, 189
Symmetric space, 88, 109, 120–127, 146, 156, 168–169

T, Harish transform, 150, 282
Tables of eigenvalues and Fourier coefficients of Maass wave forms, 218–222
Tables of transforms,
  Fourier, 16
  Helgason, 151
  Kontorovich-Lebedev, 140
  Mehler-Fock, 146
  Mellin, 57
Taniyama-Weil conjecture, 236
Tauberian theorem, 22, 40, 211, 248, 268, 271, 292–293
Tautachrone, 23
Tchebychef polynomial, 241
Tempered distribution, 15
Tensor product of representations, 102
Tessellation, 163–166, 171–173
Test functions, 2
Tetrahedral group, 76
θ, theta function, 39, 59, 61, 167, 192–200, 203, 210, 225, 228, 234, 241, 267, 273, 291, 298
Theta group, 167, 177–178
Time limited function, 51
Time series analysis, 41, 44
Titchmarsh-Kodaira formula (*see* Kodaira-Titchmarsh formula)
Topological group, 99
Torus, $\mathbb{R}^n/\mathbb{Z}^n$, 102, 251, 278
Trace formula (*see* Selberg trace formula)
Trace of operator, 50, 102–103
Transcendental number, 68, 75
Transformation formula for θ, 39, 59, 81, 192, 195
Translate of a distribution, 8

Transmission line, 156, 161–162
Twisted L-function, 235, 237
Twisted trace formula, 236

Uncertainty principle, 20, 51–52
Uniform distribution, 34, 224, 248–250, 272, 279–280
Unimodular group, 101, 260
Unique factorization, 58, 65–67
Unitary representation, 99
Unit disc, 126–127
Units in an algebraic number field, 67–68, 76, 266, 276–277, 292–294
Universal covering surface, 167
Unramified extension, 191

Values of L-series and zeta functions, 75–76, 211, 249, 294
Variance, 20, 24, 158–159
Venus spectrum, 42
Vibrating membrane or drum, 251–252
Voltage reflection coefficient, 128
Volume
  fundamental domain, 171, 207, 211
  Riemannian manifold, 85, 87, 124
Von Neumann spectral theorem, 110

Water waves, 204
Wave equation, 14, 18–19, 119, 162, 251
Wave number, 41
Weierstrass elliptic function, $\wp$, 187, 196, 204
Weight enumerator of a code, 200
Weight of an automorphic form, 182–183
Weights of a representation, 102
Weil conjectures, 188, 202, 241, 247
Weil-Hecke theory, 235–236
Weyl asymptotics of the eigenvalues of Δ, 40, 228, 290
Weyl character formula, 102
Weyl criterion for uniform distribution, 34
Weyl ergodic theorem, 34
Wiener expression for the Fourier transform on $\mathbb{R}$, 55

Wiener integral, 252
Wiener-Khintchine formula, 44

$\mathbb{Z}$-basis or integral basis of an ideal, 66
Zeeman effect, 94
Zeros of
    automorphic forms, 184
    doubly periodic functions, 196, 199
    zeta and L-functions, 61, 68, 74, 208, 218, 240, 272, 290, 292
$\zeta(2n)$, 75
$\zeta(3)$, 211, 241
$\zeta(s)$ (*see* Riemann zeta function)
Zeta function (*see* Dedekind, Epstein, Riemann, and Selberg zeta function)
Zeta function of a flow on a compact manifold, 295
Zeta functions of a variety, 202
ZnS, 77
Zonal spherical function, 90–91, 106